D1159785

Prentice Hall Advanced Reference Series 

Physical and Life Sciences

PRENTICE HALL
Biophysics and Bioengineering Series

Abraham Noordergraaf, Series Editor

AGNEW AND MCCREERY, EDS. *Neural Prostheses: Fundamental Studies*
ALPEN *Radiation Biophysics*
DAWSON *Engineering Design of the Cardiovascular System of Mammals*
GANDHI, ED. *Biological Effects and Medical Applications of Electromagnetic Energy*
LLEBOT AND JOU *Introduction to the Thermodynamics of Biological Processes*
RIDEOUT *Mathematical and Computer Modeling of Physiological Systems*
WOLAVER *Phase-Locked Loop Circuit Design*

FORTHCOMING BOOKS IN THIS SERIES *(tentative titles)*

COLEMAN *Integrative Human Physiology: A Quantitative View of Homeostasis*
FOX *Fundamentals of Medical Imaging*
GRODZINSKY *Fields, Forces, and Flows in Biological Tissues and Membranes*
HUANG *Principles of Biomedical Image Processing*
MAYROVITZ *Analysis of Microcirculation*
SCHERER *Respiratory Fluid Mechanics*
VAIDHYANATHAN *Regulation and Control in Biological Systems*
WAAG *Theory and Measurement of Ultrasound Scattering in Biological Media*

Phase-Locked Loop Circuit Design

Dan H. Wolaver
Worchester Polytechnic Institute

Prentice Hall
Englewood Cliffs, New Jersey 07632

Library of Congress Cataloging-in-Publication Data

Wolaver, Dan H.
Phase-locked loop circuit design / Dan H. Wolaver.
p. cm. -- (Prentice Hall advanced reference series)
Includes bibliographical references and index.
ISBN 0-13-662743-9
1. Phase-locked loops. 2. Electronic circuit design. I. Title.
TK7872.P38W65 1991
621.381'5--dc20 90-23685
CIP

Editorial/production supervison
and interior design: Rick DeLorenzo
Cover design: Wanda Lubelska Design
Manufacturing buyers: Kelly Behr and Susan Brunke
Acquisitions Editor: Karen Gettman

Prentice Hall Advanced Reference Series
Prentice Hall Biophysics and Bioengineering Series

A Division of Simon & Schuster
Englewood Cliffs, New Jersey 07632

Printed in the United States of America
10 9 8 7 6 5 4 3 2 1

ISBN 0-13-662743-9

Prentice-Hall International (UK) Limited, *London*
Prentice-Hall of Australia Pty. Limited, *Sydney*
Prentice-Hall Canada Inc., *Toronto*
Prentice-Hall Hispanoamericana, S.A., *Mexico*
Prentice-Hall of India Private Limited, *New Delhi*
Prentice-Hall of Japan, Inc., *Tokyo*
Simon & Schuster Asia Pte. Ltd., *Singapore*
Editora Prentice-Hall do Brasil, Ltda., *Rio de Janeiro*

Contents

PREFACE

This book provides a practical introduction to phase-locked loops for the practicing electrical engineer. Beginning with basic principles, it covers applications such as clock recovery, FM and PM modulation and demodulation, and frequency synthesis. Each application includes the development of design formulas for the system parameters—bandwidth, noise, acquisition range and speed, dynamic range, stability, and accuracy. While providing the necesssary system theory, the book's main emphasis is the practical realization of phase-locked loop circuits. For example, it addresses stray coupling, current limitations, offset voltages, and bandwidth limitations. Many alternative circuits are described with extensive use of examples and figures.

The experienced specialist in phase-lock loops will find material here that extends his knowledge. Several new digital phase detectors are described. The choice between lock acquisition techniques is clarified. The often confounding problem of injection locking is treated in depth.

To simplify the connection between phase-locked loop theory and design, the text abandons the traditional natural frequency ω_n and damping factor ζ of control theory. The parameter ω_n is often misleading since it has little relation to system behavior in a highly damped system. The parameters used in this text are the bandwidth K and the zero frequency ω_2, which give a better description of system behavior. K is the 3–dB bandwith for all dampings except those near instability. The value of ω_2 in relation to K essentially gives the damping through the expression $\zeta = 0.5\sqrt{K/\omega_2}$, and it is closely tied to the circuit elements. Both K and ω_2 are clearly evident in Bode plots of frequency responses, providing a visual link between design and performance.

This text has been used for a course on phase-locked loop circuit design at the graduate level, where it has served those with immediate applications for phase-locked loops and those who wish to consolidate their facility with circuit design in general. The study of phase-locked loops is an excellent vehicle for putting to use various disciplines of electrical engineering: communication theory, control theory, signal analysis, noise characterization, design with transistors and op amps, digital circuit design, and nonlinear circuit analysis.

The author is grateful to his students at Worcester Polytechnic Institute for their help in refining the contents of this book. The work assignments at Bell Telephone Laboratories and at Tau-tron, Inc. have provided the anvil on which to shape his understanding of phase-locked loops. The author has found the study and design of phase-locked loops to be a rich area for providing challenges to innovation and solutions to practical problems. It is his hope that this text will shorten the path for other design engineers and help them to enjoy the discovery and creativity available in phase-locked loop circuit design.

CHAPTER

1

INTRODUCTION

Phase-locked loops are used primarily in communication applications. For example, they recover clock from digital data signals, recover the carrier from satellite transmission signals, perform frequency and phase modulation and demodulation, and synthesize exact frequencies for receiver tuning. In this chapter we look at the basic principles of phase-locked loop operation in these applications. The approach here is informal and non-numeric in order to provide a quick overview. The intent is to provide heuristic descriptions that will raise questions to be answered in the following chapters.

A phase-locked loop (PLL) is basically an oscillator whose frequency is locked onto some frequency component of an input signal v_i. This is done with a feedback control loop, as shown in Fig. 1–1. The frequency of this component in v_i is ω_i (in rad/s), and its phase is θ_i. The oscillator signal v_o has a frequency ω_o and a phase θ_o. The phase detector (PD) compares θ_o with θ_i, and it develops a voltage v_d proportional to the phase difference. This voltage is applied as a control voltage v_c to the voltage-controlled oscillator (VCO) to adjust the oscillator frequency ω_o. Through negative feedback, the PLL causes $\omega_o = \omega_i$, and the phase error is kept to some (preferably small) value. Thus, both the phase and the frequency of the oscillator are "locked" to the phase and the frequency of the input signal.

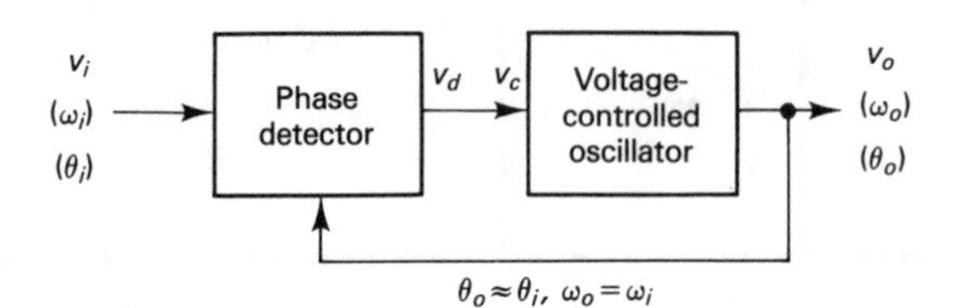

FIGURE 1–1 Basic phase-locked loop

In Chapters 2 and 3, we will add another component—a loop filter—to the simple PLL in Fig. 1–1. This will serve to modify the PLL bandwidth and reduce the phase error. For now, we simplify the PLL by omitting the loop filter in order to better understand the basic operation of the PLL.

Seven applications are discussed briefly in this chapter. They will be more thoroughly covered in Chapters 9 through 11.

1–1 CARRIER RECOVERY

Figure 1–2 shows a received signal v_i consisting of bursts of a sinusoid. This is similar to the ''color bursts'' in a TV signal. The frequency and phase of an oscillator in the TV receiver must be locked to those of the bursts. This oscillator signal v_o is then used to demodulate the color information in the TV signal. When a burst occurs, the PD (phase detector) has a chance to compare the phase of v_o with that of v_i. Any error produces a voltage v_d that is applied to the VCO (voltage-controlled oscillator) to correct the phase. (A brief change in frequency changes the phase.)

A spectrum of the input shows that the input signal v_i has a component at ω_i, the frequency of the sinusoid during the burst (see Fig. 1–2). But there are many other spectral components nearby—some only 10% away from ω_i. One question is whether the PLL will choose the correct frequency to lock onto. How does it acquire lock in the first place? Once it is locked to the proper frequency, will the VCO phase drift too much between the bursts? Every communications signal is corrupted by noise to some extent.

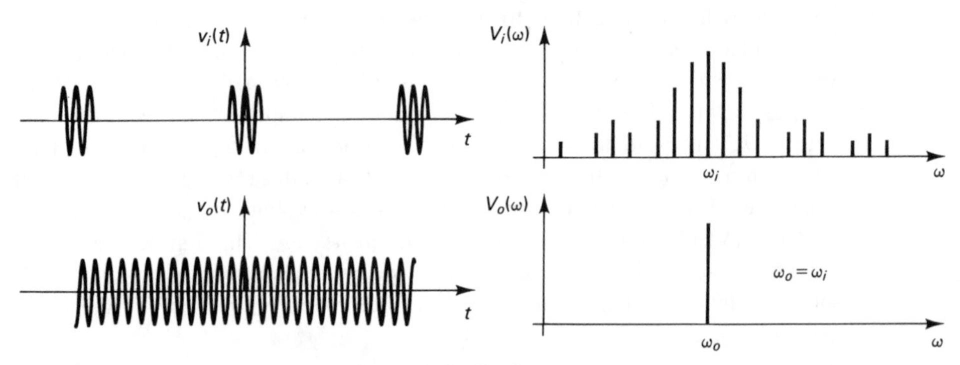

FIGURE 1–2 Carrier recovery

How will the noise affect the purity of the VCO signal v_o? Will the PLL be able to average the phase of v_i over many bursts, thereby reducing the effect of the noise?

1–2 CLOCK RECOVERY

In this application, a clock signal v_o is to be synchronized to a digital data signal v_i. For the example in Fig. 1–3, v_i represents a logic "1" by a pulse and a "0" by the absence of a pulse. The data sequence here is 1,0,1,1,0,1,1,1,0,0,1. An analysis of the spectrum of this data signal shows that there is a component at ω_i, where $2\pi/\omega_i$ is the spacing between logic symbols. There is also a background broad spectral density due to the random gaps representing 0's in the data. A PLL can be used to lock an oscillator frequency ω_o to the ω_i component, producing the clock signal v_o shown.

The clock could have been recovered with a narrow-band filter rather than a PLL. However, the background spectral density in the vicinity of ω_i will also pass through the filter, corrupting the clock. What effect will this background spectral density have on the PLL? Can the effective Q of a PLL match that of a crystal filter with a Q of 10,000? Can clock be recovered if there is no space between the pulses representing adjacent "1's"?

1–3 TRACKING FILTER

One advantage of a PLL over a narrow-band filter is its ability to track an input frequency ω_i that is drifting with time. Figure 1–4 shows an ω_i that is ramping downward with time, perhaps due to doppler shift, as with a satellite passing overhead. The PLL tracks the component at ω_i and continues to recover the clock. If a narrow-band filter rather than a PLL were used to recover the ω_i component, the component would quickly drift out of the narrow passband. In this application, the PLL acts about like a narrow-band filter whose center frequency can move.

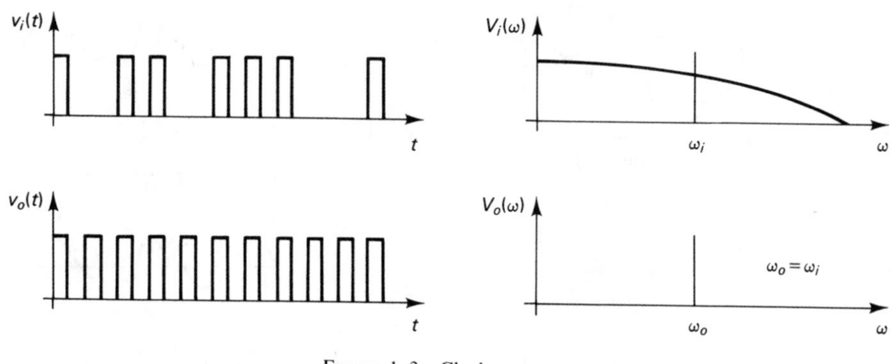

FIGURE 1–3 Clock recovery

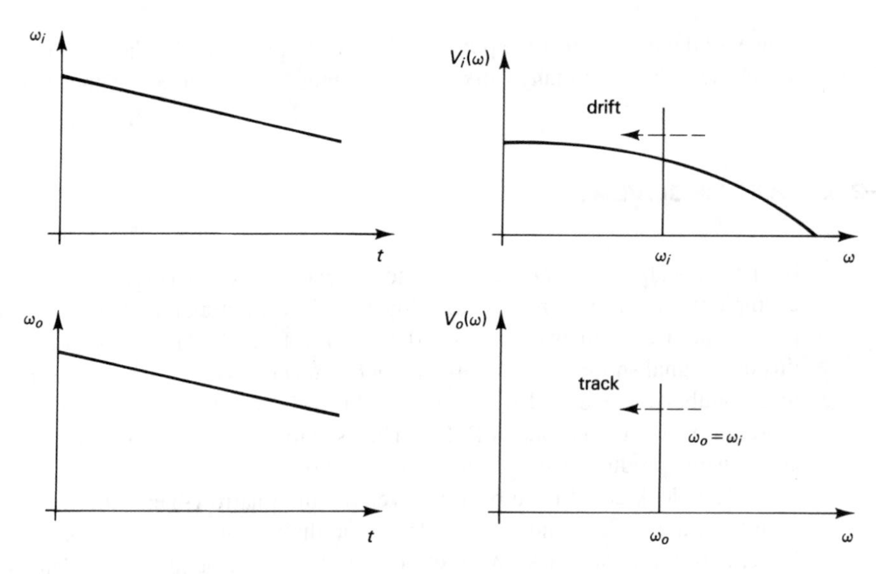

FIGURE 1–4 Tracking filter

How fast can ω_i move without the PLL losing track? What is the relationship between the speed with which the PLL can track ω_i and the effective bandwidth of the PLL? What limits the range over which the PLL can track ω_i?

1–4 FREQUENCY DEMODULATION

Almost all FM receivers today use a PLL for frequency demodulation. In this application, the PLL output frequency ω_o tracks the input frequency ω_i as it varies according to the modulation (see Fig. 1–5). If the VCO control voltage v_c is proportional to ω_o, it is also proportional to ω_i. Therefore, v_c is the demodulated signal.

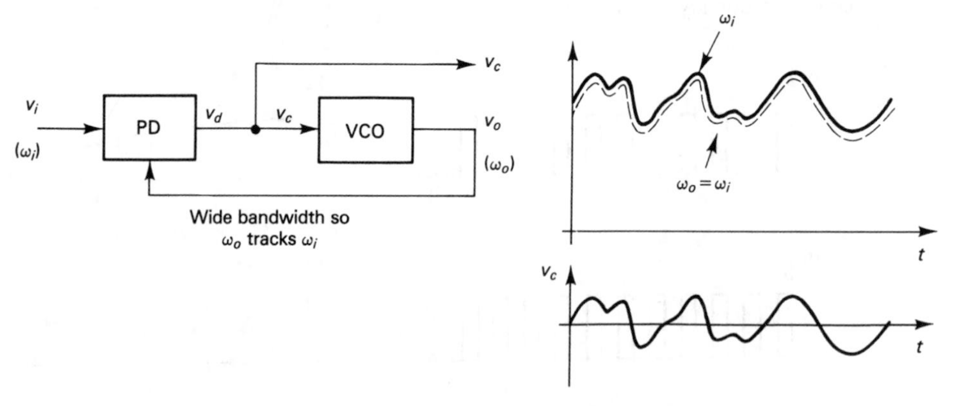

FIGURE 1–5 Frequency demodulation

Note that the bandwidth of the PLL must be wide enough that it has the necessary speed to track the variations in ω_i. How wide must the PLL bandwidth be? What happens if it is too wide? How much noise does it take to cause the PLL to temporarily lose lock at times? (These "cycle slips" are heard as "clicks.")

1–5 PHASE DEMODULATION

In a similar application, a PLL can be used for phase demodulation. Here, the received signal v_i is a carrier whose modulated phase θ_i conveys the information (see Fig. 1–6). In this application, the PLL bandwidth is so small (the PLL is so "sluggish") that θ_o sits at the average of θ_i rather than following it. The output phase θ_o is nearly constant; that is, v_o is the recovered unmodulated carrier. This serves as a reference for the phase detector to demodulate θ_i. If the phase detector has a linear characteristic, its output v_d is proportional to θ_i, and v_d is the demodulated output.

How narrow must the PLL bandwidth be? What is the relationship between the bandwidth and the length of time it takes the PLL to reach lock initially? How does the strength of carrier component of v_i depend on the amplitude of the phase modulation? Can the PLL still recover a carrier if there is no carrier component in v_i?

1–6 PHASE MODULATION

In Fig. 1–7, a PLL is modified by summing a modulation signal v_m into the circuit to modulate the phase θ_o. The voltage v_m tries to change the frequency of the VCO. But if the bandwidth of the PLL is wide enough, it can respond quickly and adjust v_d to cancel the effect of v_m. Thus, $v_d \approx -v_m$, and $v_c = v_d + v_m$ remains essentially constant. If the

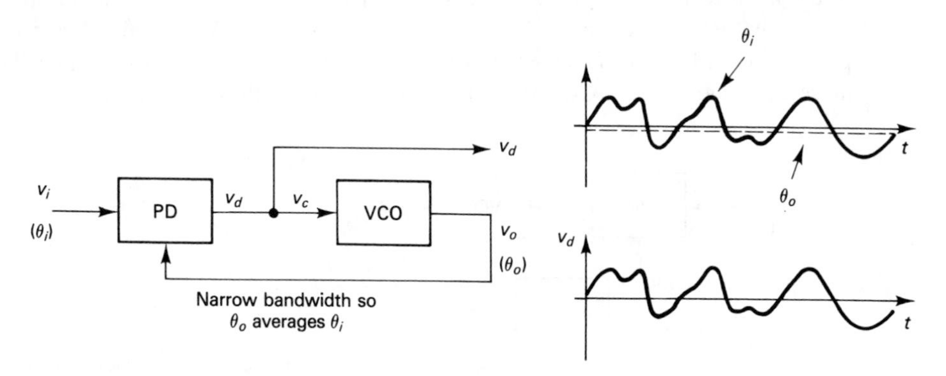

FIGURE 1–6 Phase demodulation

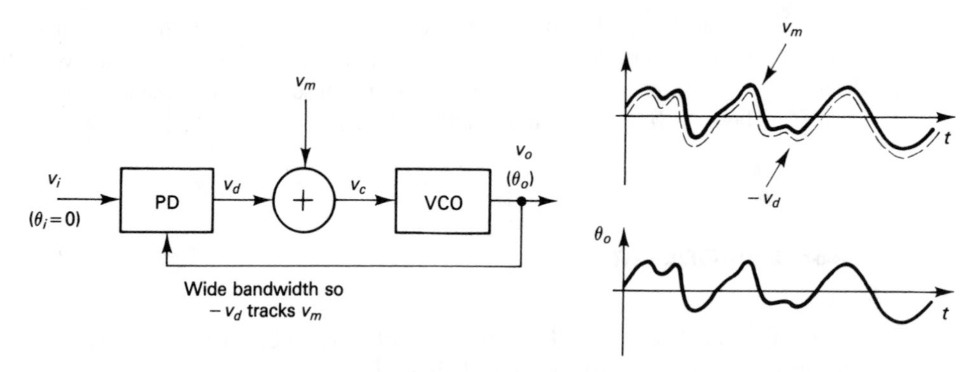

FIGURE 1–7 Phase modulation

input phase θ_i is constant (zero), v_d is proportional to $-\theta_o$, and θ_o is therefore proportional to v_m. Thus, the signal v_m modulates the phase θ_o of the VCO.

What limits the amplitude of the modulated phase θ_o? Can these limits be extended if necessary? Phase modulation implies some amount of frequency modulation. How large is this FM? Does it exceed the range of the VCO?

1–7 FREQUENCY SYNTHESIS

A frequency synthesizer generates multiples of an accurate reference frequency ω_i. For example, if $\omega_i = 1$ krad/s, then the synthesizer might generate 100, 101, . . . , 200 krad/s. That is, $\omega_o = N\omega_i$, where N varies from 100 to 200. Such a frequency multiplier can be realized with a PLL, as shown in Fig. 1–8. In this application, a frequency divider is included in the feedback path of the PLL. The integer N by which ω_o is divided can be selected by the user. When in lock, the PLL assures that the two frequencies ω_i and ω_o/N at the input to the phase detector (PD) are equal. Then, $\omega_o = N\omega_i$, as desired.

What is the effect of the $\div N$ on the PLL bandwidth? What limits the size of N in practice? How long does it take the PLL to change frequency when N is changed? How do noise in v_i and in the VCO affect the purity of the synthesized frequency?

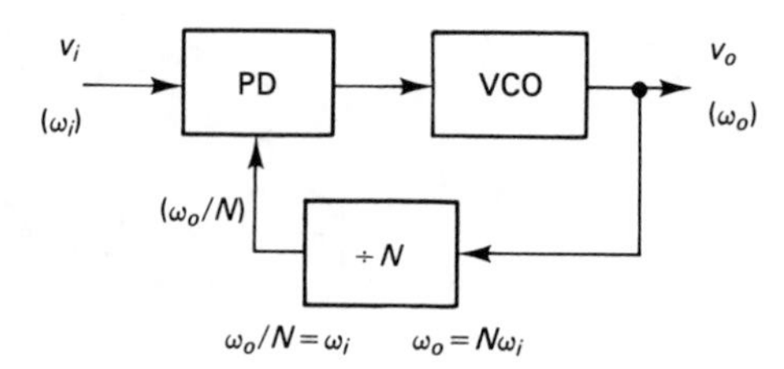

FIGURE 1–8 Frequency synthesis

1–8 ORGANIZATION OF THE TEXT

Chapters 2 through 5 give the mathematical analysis of phase-locked loops and describe their components—loop filter, phase detector, and voltage-controlled oscillator. Various circuit designs for the components are compared.

Chapters 6 through 8 answer some of the questions raised above that are common to many applications. These are questions about the sources of noise and its effects, the time required for lock acquisition as a function of initial frequency error, and the limits of frequency and phase variation that a PLL can track once it is in lock.

Chapters 9 through 11 look at specific PLL applications—phase and frequency modulation and demodulation, clock recovery from data signals, frequency synthesis. These chapters also address the questions raised above that are specific to the application.

1–9 OTHER INFORMATION ON PHASE-LOCKED LOOPS

The purpose of this text is to give the reader an understanding of the fundamentals of phase-locked loops and of their circuit design. From this introduction, the reader should be able to design phase-locked loops for most applications. For specialized and detailed information, the reader will want to refer to some of the literature listed in the Bibliography at the end of this chapter. The basics provided in this text should be a good preparation for such further advances.

BIBLIOGRAPHY

R. E. Best, *Phase-Locked Loops: Theory, Design, and Applications*, McGraw-Hill: New York, 1984.

A. Blanchard, *Phase-Locked Loops: Applications to Coherent Receiver Design*, Wiley: New York, 1976.

F. M. Gardner, *Phaselock Techniques*, Wiley: New York, 1979.

W. C. Lindsey, *Synchronization Systems in Communication and Control*. Prentice-Hall: Englewood Cliffs, NJ, 1972.

W. C. Lindsey and C. M. Chie, Eds., *Phase-Locked Loops*, IEEE Press: New York, 1986.

W. C. Lindsey and M. K. Simon, Eds., *Phase-Locked Loops and Their Application*, IEEE Press: New York, 1978.

V. Manassewitsch, *Frequency Synthesizers*, Wiley: New York, 1987.

U. L. Rohde, *Digital PLL Frequency Synthesizers*, Prentice-Hall: Englewood Cliffs, NJ, 1983.

A. J. Viterbi, *Principles of Coherent Communication*, McGraw-Hill: New York, 1966.

CHAPTER

2

Phase-Locked Loop Basics

2–1 PHASE-LOCKED LOOP CHARACTERISTICS

We have seen that in some applications the PLL should be fast in following the input phase, and in others it should be slow. In other words, the *bandwidth* of the PLL should be either wide or narrow. This is determined by the characteristics of the phase detector (PD), the voltage-controlled oscillator (VCO), and the loop filter, which is introduced in section 2–7.

Another measure of a PLL's performance is the *phase error*—the difference between the input phase θ_i and the VCO phase θ_o. Consider the block diagram of a simple phase-locked loop shown in Fig. 1–1. When it is in lock, the VCO frequency ω_o equals the input frequency ω_i. (How the PLL initially attains frequency lock is dealt with in Chapter 8.) The control voltage v_c necessary to cause $\omega_o = \omega_i$ is provided by the PD output v_d. But the PD requires some phase error between θ_i and θ_o to produce this v_d. We will determine the size of this error in terms of the characteristics of the components of the PLL.

A PLL has other characteristics—frequency range over which it will acquire lock, lock acquisition time, tolerance of modulation without losing lock, output phase noise. These will be discussed in later chapters.

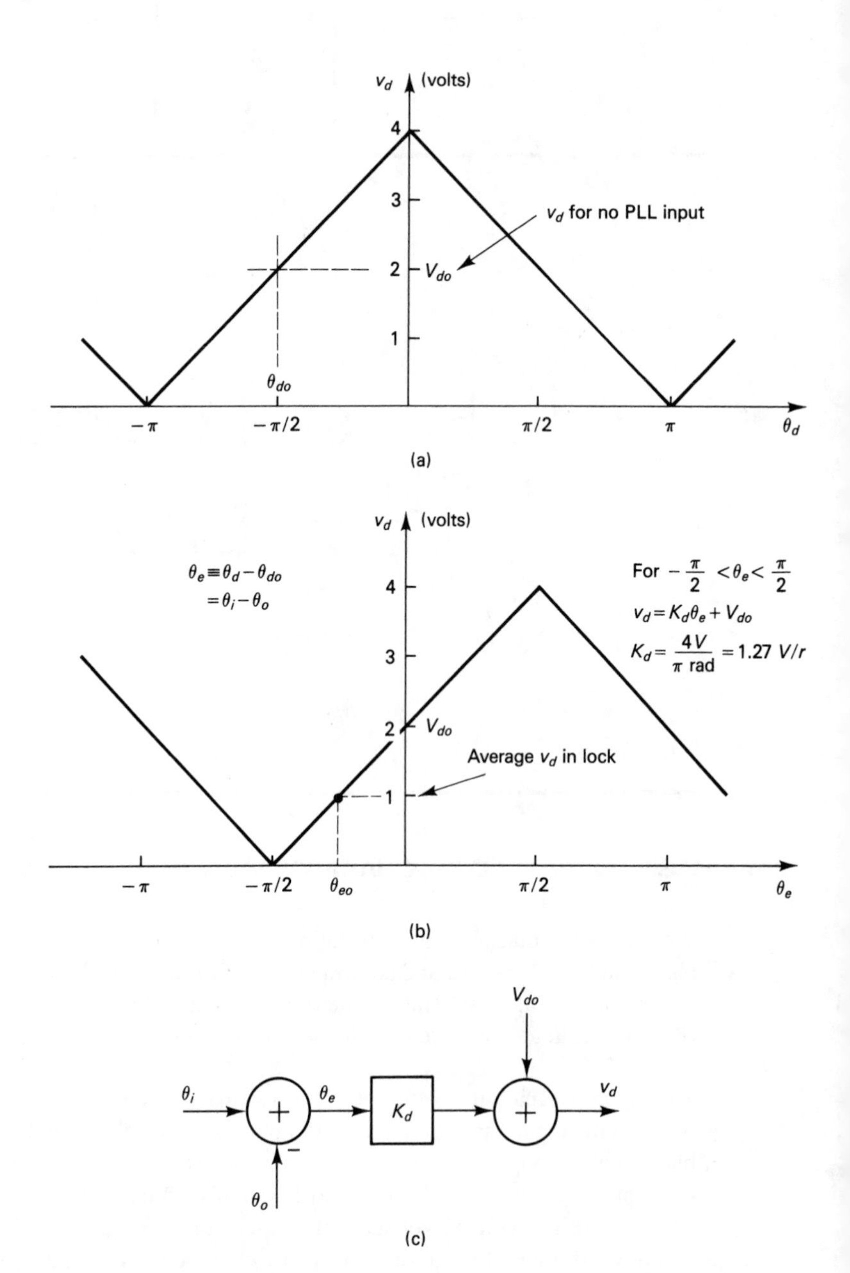

FIGURE 2–1 Phase detector characteristic and model

2–2 PHASE DETECTOR CHARACTERISTICS

Let θ_d represent the *phase difference* between the input phase and the VCO phase. The PD produces a voltage v_d in response to this θ_d; a typical characteristic v_d versus θ_d is shown in Fig. 2–1a. The curve is piecewise linear, and it repeats every 2π radians. This periodicity is necessary since a phase of 2π is generally indistinguishable from a phase of zero. When no signal v_i is applied to the PD, it generates some *free-running voltage* V_{do}, which is shown as 2 V for this case. Corresponding to V_{do} on the curve is some phase θ_{do} (equal to $-\pi/2$ here).

The usual convention is to shift the characteristic so that a phase error of zero corresponds to $v_d = V_{do}$. Therefore, we define the *phase error* to be

$$\theta_e \equiv \theta_d - \theta_{do} \tag{2-1}$$

(see the characteristic in Fig. 2–1b). Because of this shift, $\theta_e = 0$ does not usually correspond to v_i and v_o being in phase, but for analysis purposes it is convenient to define it as zero phase error. We will also use the convention of defining the input phase θ_i and the VCO phase θ_o such that

$$\theta_e = \theta_i - \theta_o \tag{2-2}$$

The plot of v_d versus θ_e in Fig. 2–1b is called the *PD characteristic*. By definition, $v_d = V_{do}$ corresponds to $\theta_e = 0$. In the *range* $-\pi/2 \leq \theta_e \leq \pi/2$ there is a constant slope K_d, where

$$K_d \equiv dv_d/d\theta_e \tag{2-3}$$

In this case, $K_d = (4\text{ V})/(\pi\text{ radians}) = 1.27$ V/rad. In the linear region, the PD can be modeled by

$$v_d = K_d\theta_e + V_{do} \tag{2-4}$$

which is represented by the signal flow graph in Fig. 2–1c. K_d is the *PD gain*, and V_{do} is the free-running detector voltage.

2–3 VCO CHARACTERISTICS

A typical characteristic of a voltage-controlled oscillator is shown in Fig. 2–2a. Here, the VCO frequency ω_o is a linear function of the control voltage v_c. The curve need not be linear, but it usually simplifies the PLL design if the slope is the same everywhere. As v_c varies from 0 to 4 V, the VCO varies over its *range* of 8 Mrad/s to 16 Mrad/s. Outside this range, the performance of the VCO is unacceptable in some way. When the PLL is in

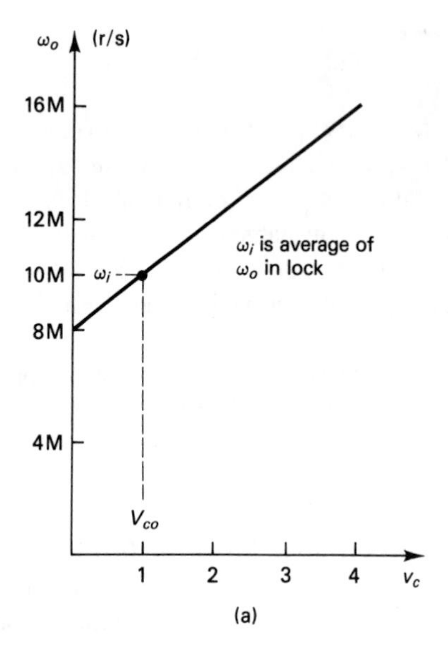

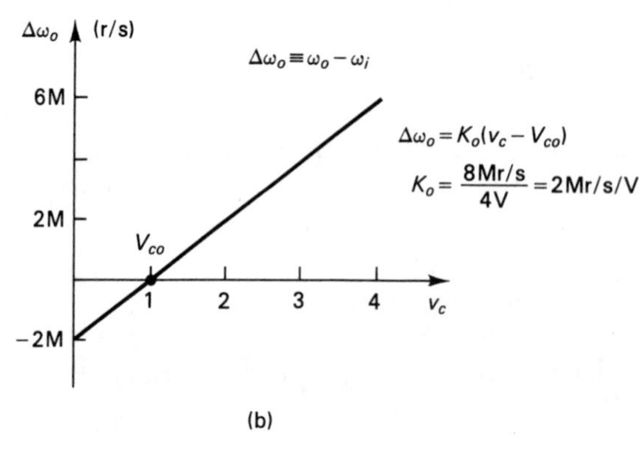

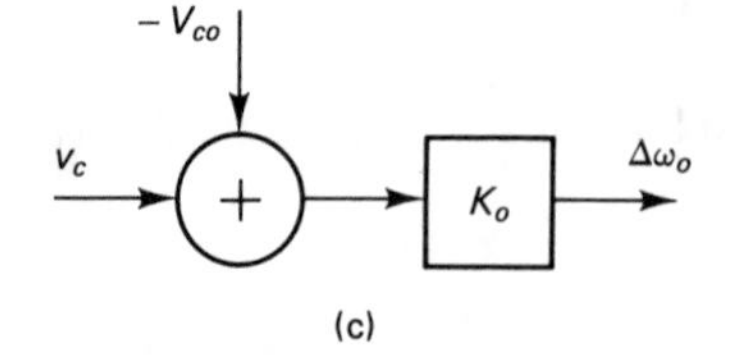

FIGURE 2–2 VCO characteristic and model

lock, $\omega_o = \omega_i$. Suppose that $\omega_i = 10$ Mrad/s. Then according to the characteristic in Fig. 2–2a, $\omega_o = 10$ Mrad/s requires $v_c = 1$ V. This is the *static control voltage* V_{co} corresponding to $\omega_o = \omega_i$. Notice that V_{co} is not a property of the VCO alone; it also depends on the frequency ω_i to which the PLL is locked. (For example, if ω_i were 12 Mrad/s, then according to Fig. 2–2 the V_{co} would be 2 V). This is in contrast to V_{do}, which is a property of the PD alone.

The static operation of the PLL when it is in lock can be found from the PD and VCO characteristics. The lock condition is $\omega_o = \omega_i$. For the case $\omega_i = 10$ Mrad/s, Fig. 2–2a shows $v_c = V_{co} = 1$ V. This voltage is provided in turn by a PD voltage of $v_d = 1$ V. From the PD characteristic in Fig. 2–1b, a phase error $\theta_e = -0.79$ radians is required to produce this v_d. This average θ_e in lock is called the *static phase error* θ_{eo}. It is usually desirable to have θ_{eo} near zero. It certainly must not exceed $\pm\pi/2$ radians, the limits of the linear portion of the PD characteristic. An expression for θ_{eo} in terms of parameters of the PD and VCO characteristics will be developed below.

Sometimes it is convenient to refer to the *output frequency deviation* $\Delta\omega_o$, where

$$\Delta\omega_o \equiv \omega_o - \omega_i \tag{2-5}$$

In lock, the average of ω_o equals ω_i, so $\Delta\omega_o$ is a measure of how far ω_o is from its average in lock. A plot of $\Delta\omega_o$ versus v_c is essentially a shifted VCO characteristic, as shown in Fig. 2–2b. By definition, $\Delta\omega_o = 0$ corresponds to $v_c = V_{co}$.

The slope of the VCO characteristic in the vicinity of the lock frequency is called the *VCO gain* K_o, where

$$K_o \equiv d\omega_o/dv_c = d\Delta\omega_o/dv_c \tag{2-6}$$

Here we have $K_o = (8 \text{ Mrad/s})/(4 \text{ V}) = 2$ Mrad/s/V. Then the frequency deviation can be modeled by

$$\Delta\omega_o = K_o\,(v_c - V_{co}) \tag{2-7}$$

where V_{co} is the control voltage in lock. The signal flow graph in Fig. 2–2c represents Eq. (2-7).

2–4 LINEAR MODEL OF PLL

The descriptions of the PD and the VCO in Eqs. (2-4) and (2-7) are linear, although the linearities hold only for limited ranges. In Chapter 7 we will look at the consequences of these range limitations. In this chapter, we assume that θ_e and ω_o stay in the linear ranges of the PD and VCO. There are several other texts [1-4] that treat this topic and may provide other useful perspectives.

Although the input and output signals of a PLL are often not pure sinusoids, for the moment we will assume they are, for the sake of phase notation:

$$v_i = \sin(\omega_i t + \theta_i)$$

$$v_o = \sin(\omega_i t + \theta_o)$$

where ω_i is a constant, the average input frequency. As the dimension ''radians per second'' implies, frequency is the time derivative of phase, where phase is the argument of the sine function. Thus, the output frequency from the VCO is

$$\omega_o \equiv d(\omega_i t + \theta_o)/dt = \omega_i + d\theta_o/dt \tag{2-8}$$

But as defined in Eq. (2-5), $\Delta\omega_o = \omega_o - \omega_i$. Therefore

$$\Delta\omega_o = d\theta_o/dt \tag{2-9}$$

or

$$\theta_o = \int \Delta\omega_o \, dt \tag{2-10}$$

This relationship between θ_o and $\Delta\omega_o$, together with the signal flow graphs in Fig. 2–1c and Fig. 2–2c, completes a linear model of the PLL (see Fig. 2–3). The VCO model now includes an integrator to provide the phase θ_o as the PLL output. This phase is fed back and compared by the PD with θ_i of the input signal.

We have been referring to ω_i as the input frequency, but it is actually the average input frequency. The full expression for the input frequency is $\omega_i + d\theta_i/dt$. This will be discussed further in section 7–2.

2–5 STATIC PHASE ERROR

By definition, when the PLL is in lock, the average ω_o equals ω_i, and the average $\Delta\omega_o$ is zero. The *static phase error* θ_{eo} is the average value of θ_e in lock. From the signal flow graph in Fig. 2–3, we see that

$$\Delta\omega_o = K_o(K_d\theta_e + V_{do} - V_{co})$$

and taking the time average of both sides,

$$\overline{\Delta\omega_o} = K_o(K_d\overline{\theta}_e + V_{do} - V_{co})$$

In lock, $\overline{\Delta\omega_o} = 0$, and $\overline{\theta}_e \equiv \theta_{eo}$. Therefore

$$\theta_{eo} = (-V_{do} + V_{co})/K_d \tag{2-11}$$

For $K_d = 1.27$ V/rad, $V_{do} = 2$ V, and $V_{co} = 1$ V, Eq. (2-11) gives $\theta_{eo} = -0.79$ radians, as determined graphically before.

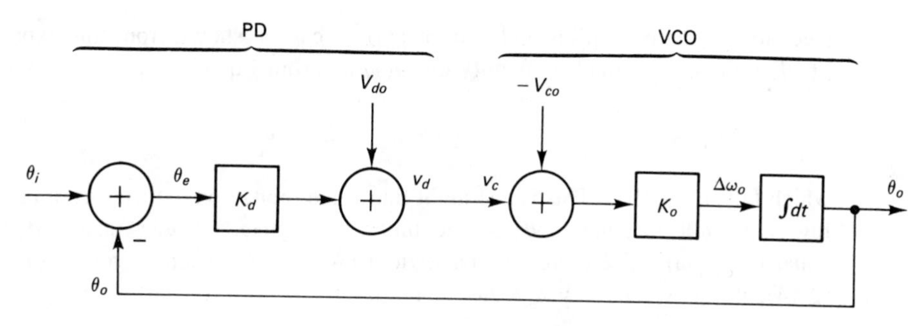

FIGURE 2–3 Linear model of PLL

2–6 PLL BANDWIDTH

In discussing the bandwidth of a PLL, we are concerned with the frequency at which θ_i can vary and still be followed reasonably closely by θ_o. This also holds for the frequency at which ω_i can vary and still be followed by ω_o in the case of FM. Since bandwidth has to do with variations or ac signals, we form an ac model of the PLL by eliminating the dc parameters from the linear model in Fig. 2–3. The resulting ac model is shown in Fig. 2–4. The integration has been replaced by its Laplace transform $1/s$, where s is complex frequency. When finding frequency response, we will replace s by $j\omega$.

Let the *forward gain* of the loop in Fig. 2–4 be $G(s)$:

$$G(s) = K_d K_o/s \tag{2-12}$$

The signal flow graph in Fig. 2–4 is actually a system of equations which can be solved for the phase transfer function θ_o/θ_i. Those familiar with control theory can see by inspection that the transfer function is

$$\frac{\theta_o(s)}{\theta_i(s)} = \frac{G(s)}{1 + G(s)} = \frac{G(j\omega)}{1 + G(j\omega)} \tag{2-13}$$

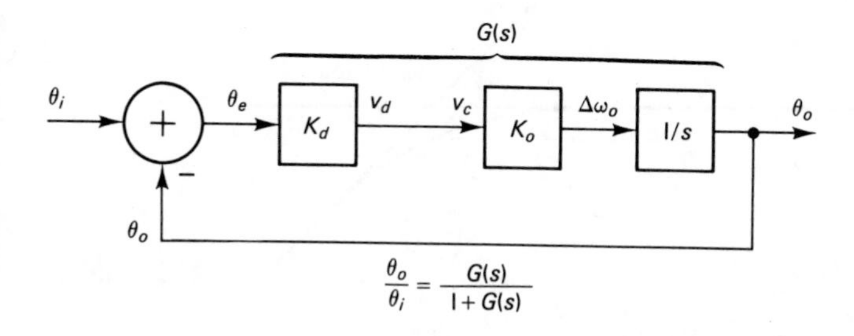

FIGURE 2–4 ac model of PLL

(see, for example, Phillips and Harbor [5]). It can be shown from this expression that $|\theta_o/\theta_i|$ follows the smaller of unity or $|G(j\omega)|$. From Eq. (2-12),

$$|G(j\omega)| = K_d K_o/\omega \tag{2-14}$$

which falls off as $1/\omega$. This is a straight line when plotted on log axes as in Fig. 2–5. For low ω, $|G(j\omega)| > 1$, and $|\theta_o/\theta_i|$ is about unity. For high ω, $|G(j\omega)| < 1$, and $|\theta_o/\theta_i|$ is about equal to $|G(j\omega)|$. Therefore, the bandwidth ω_{3dB} occurs when $|G(j\omega)| = 1$. From Eq. (2-14), this is when $1 = K_d K_o/\omega_{3dB}$, or

$$\omega_{3dB} = K_d K_o \tag{2-15}$$

For the PD and VCO characteristics in Fig. 2–1b and 2–2a, $K_d = 1.27$ V/rad, and $K_o = 2$ Mrad/s/V. Then $\omega_{3dB} = 2.55$ Mrad/s.

Suppose we wished to reduce the bandwidth by a factor of 0.286 to $\omega_{3dB} = 0.73$ Mrad/s. This can be realized by putting a voltage attenuator consisting of R_0 and R_2 between the PD and the VCO, as in Fig. 2–6a. The gain of the attenuator is represented by K_h, where

$$K_h = R_2/(R_0 + R_2) \tag{2-16}$$

For the values $R_0 = 25$ kΩ and $R_2 = 10$ kΩ, we have $K_h = 0.286$. The linear ac model for the PLL now includes K_h in the loop (see Fig. 2–6b), and the forward gain is now

$$G(s) = K_d K_h K_o/s \tag{2-17}$$

As before, the bandwidth is determined by the frequency for which $|G(j\omega)| = 1$. From Eq. (2-17), this is at the frequency

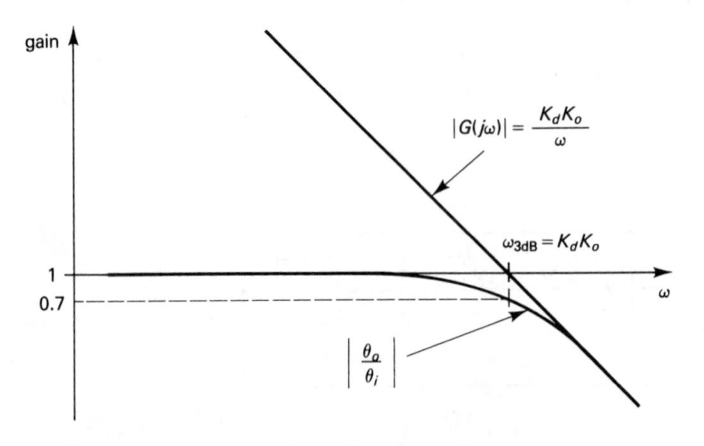

FIGURE 2–5 Frequency response of PLL

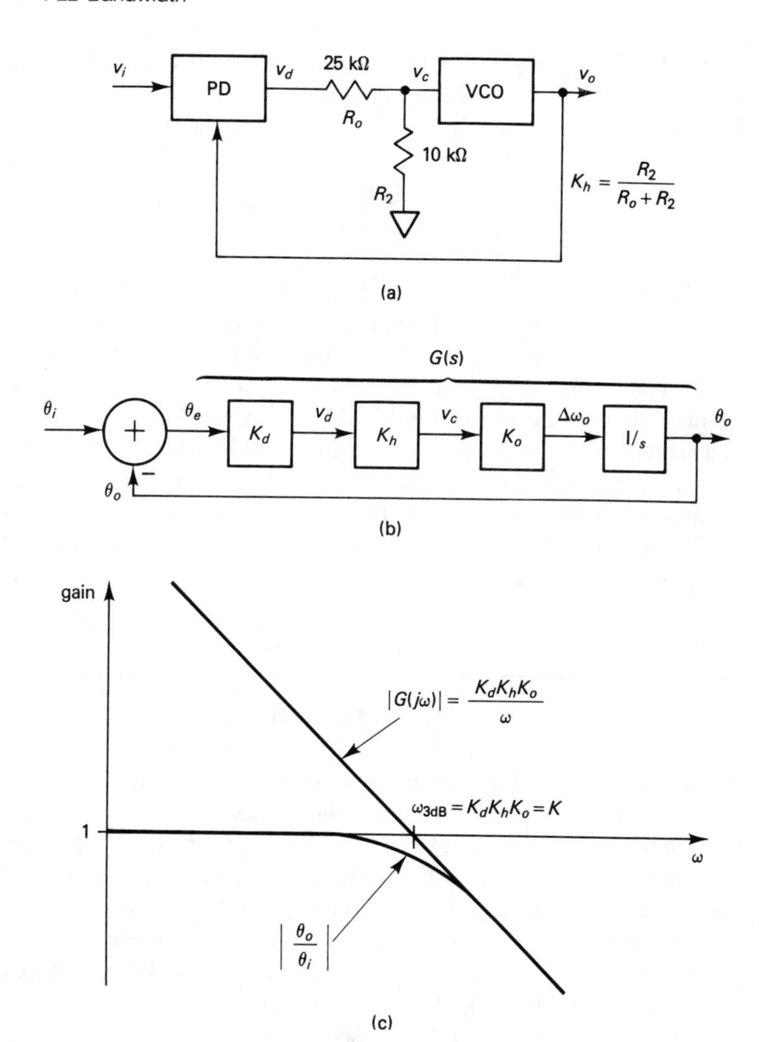

FIGURE 2–6 Narrowed bandwidth PLL

$$\omega_{3dB} = K_d K_h K_o \tag{2-18}$$

(see Fig. 2–6c). For $K_d = 1.27$ V/rad, $K_h = 0.286$, and $K_o = 2$ Mrad/s/V, the bandwidth has been reduced to $\omega_{3dB} = 0.73$ Mrad/s, as desired.

Since the product of these three gains occurs often in the analysis of PLLs, it is standard notation to let

$$K \equiv K_d K_h K_o \tag{2-19}$$

K is called the "loop gain," although it does not include the integration $1/s$, which is also in the loop gain (see Fig. 2–6b). From Eq. (2-18), K is also the 3-dB bandwidth of the PLL. From Eqs. (2-13), (2-17), and (2-19), the transfer function is

$$\frac{\theta_o(s)}{\theta_i(s)} = \frac{K}{s + K} \tag{2-20}$$

A PLL with a simple attenuator (as in Fig. 2–6a) is called a *first-order phase-locked loop* because the transfer function has a first-order polynomial of s in the denominator.

While the attenuator gain K_h has satisfied the ac requirements of the PLL as far as bandwidth, it also affects the static behavior of the PLL. Returning to the complete characteristic of the PD in Fig. 2–1b, we see the maximum v_d is 4 V. Then, after attenuation by $K_h = 0.286$, the maximum v_c is now only 1.14 V. According to the characteristic in Fig. 2–1b, this restricts the VCO to a maximum frequency of 10.3 Mrad/s —barely enough range to let the PLL lock to an input frequency of 10 Mrad/s. In the next section we look at a solution to this problem, but first we will analyze an example involving the PLL in Fig. 2–6a.

EXAMPLE 2–1

A PLL has the VCO characteristic ω_o versus v_c shown in Fig. 2–2a. The input frequency is $\omega_i = 10$ Mrad/s, giving a static control voltage $V_{co} = 1.0$ V. The slope of the characteristic is $K_o = 2$ Mrad/s/V. The PD characteristic is that shown in Fig. 2–1b and reproduced in Fig. 2–7. The slope of the PD characteristic is $K_d = 1.27$ V/rad. There is a sinusoidally modulated phase at the input $\theta_i = 0.3 \sin(\omega_m t)$, where $\omega_m = 5$ Mrad/s.

We will analyze two situations: first with no attenuator ($K_h = 1$), and then with an attenuator with $K_h = 0.286$. In each case we will find the static phase offset θ_{eo}, the bandwidth K, and the amplitude of the output phase swing θ_o.

For $K_h = 1$, $V_{co} = 1.0$ V corresponds to $v_d = 1.0$ V, and from the PD characteristic in Fig. 2–7, the corresponding static phase offset is $\theta_{eo} = -\pi/2 + (1.0\text{ V})/K_d = \underline{-0.79}$ radians. The bandwidth is given by $K = K_d K_h K_o = \underline{2.55\text{ Mrad/s}}$. The magnitude of the frequency response is obtained by taking the magnitude of Eq. (2-20) with $j\omega$ replacing s:

$$\left| \frac{\theta_o(j\omega)}{\theta_i(j\omega)} \right| = \left| \frac{K/j\omega}{1 + K/j\omega} \right| = \frac{K}{\sqrt{\omega^2 + K^2}} \tag{2-21}$$

This response is plotted in Fig. 2–8a. At $\omega = \omega_m = 5$ Mrad/s, Eq. (2-21) gives $|\theta_o/\theta_i| = 0.45$. Since the amplitude of θ_i is 0.3 radians, the amplitude of θ_o is $0.45 \times 0.3 = \underline{0.135}$ radians. Figure 2–8b compares the amplitudes of θ_i and θ_o. Note that there is also a phase shift of 63 deg. due to the phase of the response θ_o/θ_i.

For $K_h = 0.286$, $v_c = 0.286 v_d$, or $V_d = 3.5 v_c$. Then corresponding to $v_c = V_{co} = 1.0$ V, we must have $v_d = 3.5$ V. From the PD characteristic in Fig. 2–7, the correspond-

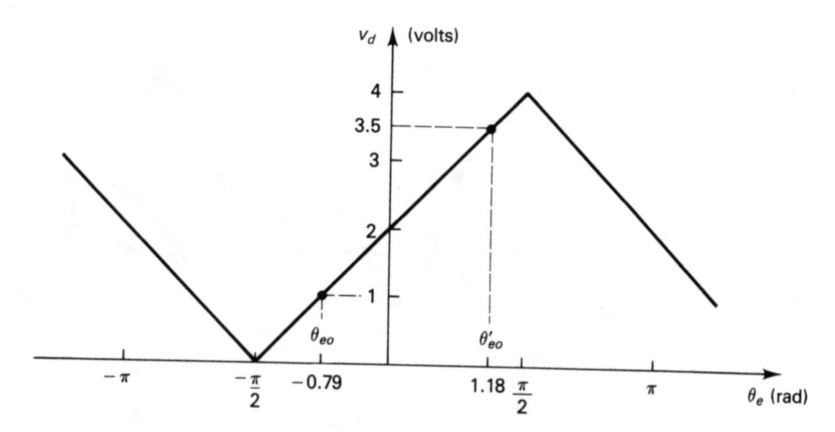

FIGURE 2–7 Phase detector characteristic for Example 2–1

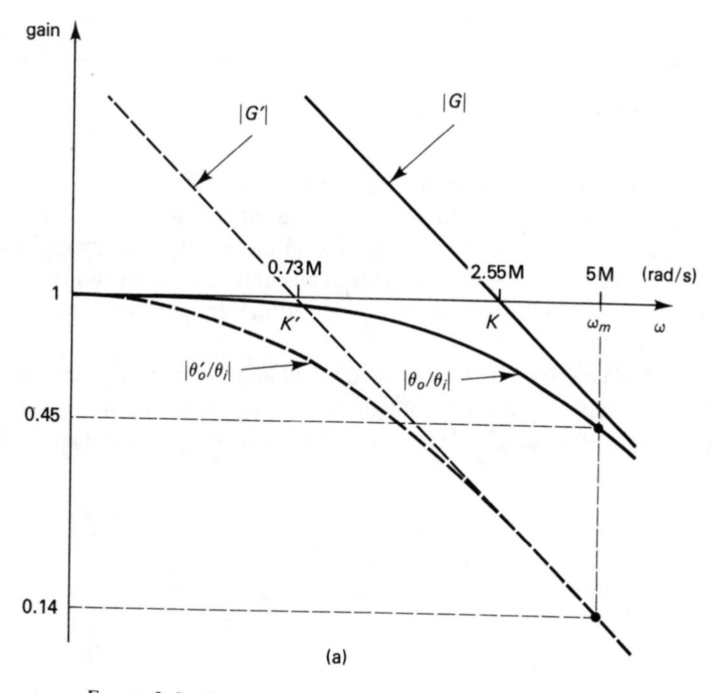

FIGURE 2–8 Response to phase modulation in Example 2–1

ing phase offset is $\theta_{eo}' = -\pi/2 + (3.5 \text{ V})/K_d = \underline{1.18 \text{ radians}}$. The new bandwidth is $K' = K_d K_h K_o = \underline{0.73 \text{ Mrad/s}}$. The new response $|\theta_o'/\theta_i|$ is plotted in Fig. 2–8a. At $\omega = \omega_m = 5$ Mrad/s, Eq. (2-21) gives $|\theta_o'/\theta_i| = 0.14$. Since the amplitude of θ_i is 0.3 radians, the amplitude of θ_o' is $0.14 \times 0.3 = \underline{0.42 \text{ radians}}$. Figure 2–8b compares the amplitudes of θ_i and θ_o'. Note that there is a phase shift of 82 deg. due to the phase of the response θ_o'/θ_i.

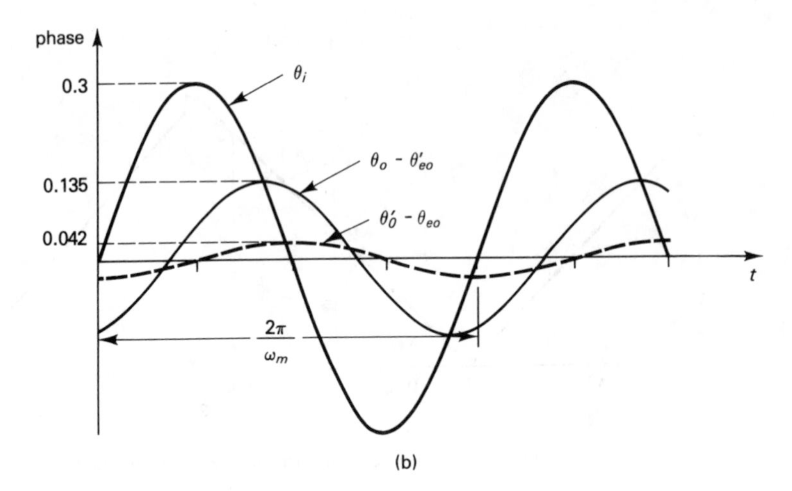

FIGURE 2–8 *(continued)*

2–7 LOOP FILTER

By putting an attenuator in the PLL, we reduced the ac gain K and therefore reduced the bandwidth, as desired. But we also reduced the dc gain and therefore limited the dc voltage V_{co} which the PD can provide to the VCO. This greatly restricts the frequency range of the PLL. The solution is to replace the attenuator in Fig. 2–6a with a *loop filter*. This filter will still act as an attenuator at high frequencies, but it will have unity gain at dc.

A simple loop filter is formed by adding a capacitor to the attenuator, as in Fig. 2–9a. If the capacitor is large enough, the ac attenuation is unaffected. But now the dc path to ground is blocked, and the dc component of v_d is not attenuated. The transfer function of this loop filter is

$$F(s) = K_h \frac{s + \omega_2}{s + \omega_1} \tag{2-22a}$$

where

$$K_h = \frac{R_2}{R_0 + R_2} \tag{2-22b}$$

$$\omega_1 = \frac{1}{(R_0 + R_2)C} \tag{2-22c}$$

$$\omega_2 = \frac{1}{R_2 C} \tag{2-22d}$$

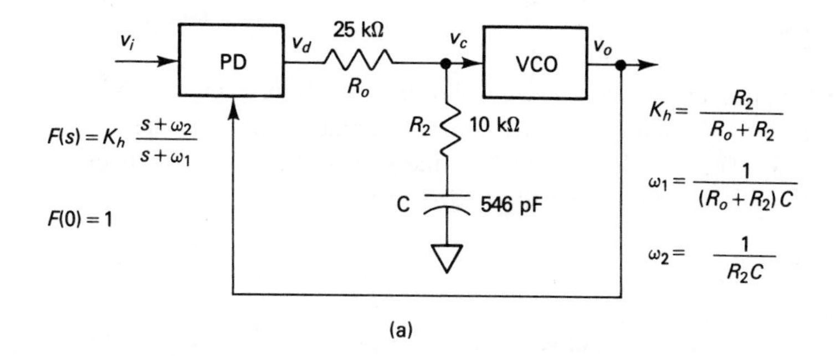

(a)

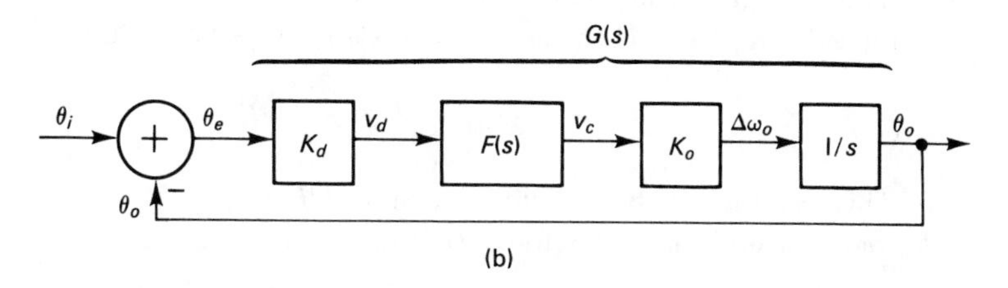

(b)

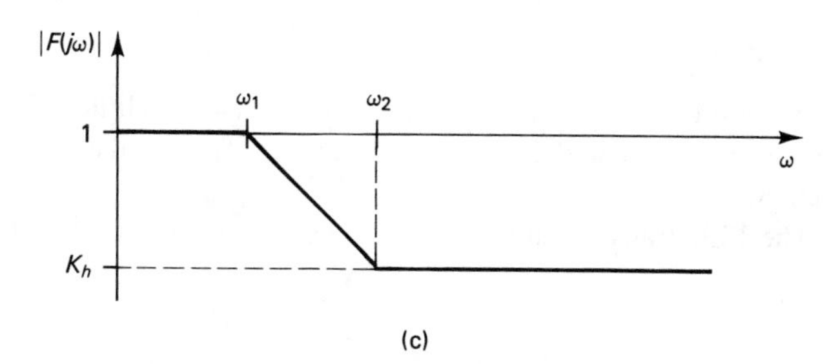

(c)

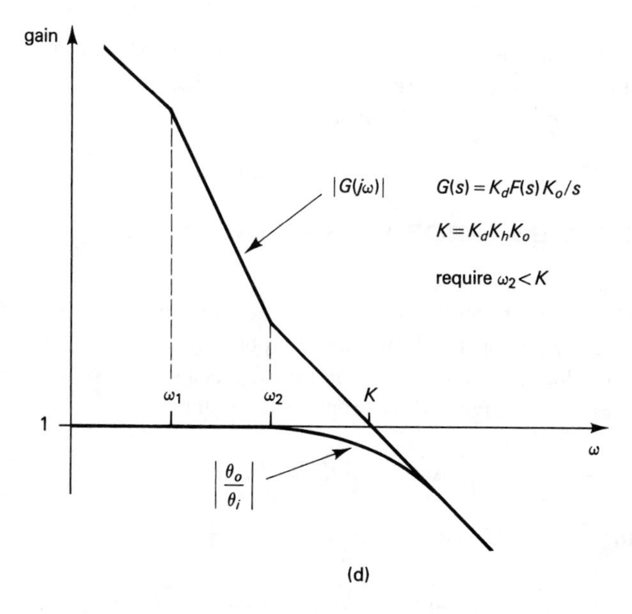

(d)

FIGURE 2–9 Expanded frequency range PLL

The frequency response $|F(j\omega)|$ of the loop filter is plotted in Fig. 2–9c. At dc, the gain is $F(0) = 1$, and at high frequencies (greater than ω_2), the gain is K_h, as desired.

The signal flow graph of the PLL in Fig. 2–9b includes the gain $F(s)$ of the loop filter. The gain of the forward path is

$$G(s) = K_d F(s) K_o / s \tag{2-23}$$

The frequency response of $|G(j\omega)|$ is plotted in Fig. 2–9d. Again, the rule is roughly that $|\theta_o/\theta_i|$ is the lower of unity and $|G(j\omega)|$. At high frequencies, $|F(j\omega)| \approx K_h$, and $|G(j\omega)| = 1$ for $\omega = K_d K_h K_o$. Then, as in Eq. (2-18), the bandwidth is

$$\omega_{3\text{dB}} = K_d K_h K_o \equiv K \tag{2-24}$$

This result for the bandwidth has assumed $|F(j\omega)| = K_h$ when $\omega = K$. But $|F(j\omega)| \approx K_h$ only for $\omega > \omega_2$. Therefore, we require that

$$\omega_2 < K \tag{2-25}$$

as shown in Fig. 2–9d. If ω_2 is less than 4 K, then ω_2 has little effect on the response $|\theta_o/\theta_i|$ (compare Figs. 2–6c and 2–9d). A thorough analysis of the effect of ω_2 is carried out in Chapter 3.

The PLL transfer function is obtained from Eqs. (2-13), (2-22a), and (2-23):

$$\frac{\theta_o(s)}{\theta_i(s)} = \frac{Ks + K\omega_2}{s^2 + (K + \omega_1) + K\omega_2} \tag{2-26}$$

Because the denominator has a second-order polynomial in s, a PLL with a loop filter is called a *second-order phase-locked loop*. In Chapter 3 we look at other loop filters; they also result in a second-order PLL.

2–8 STATIC PHASE ERROR WITH A LOOP FILTER

It is possible to design loop filters with dc gains other than $F(0) = 1$. Therefore, we will need a general expression for the static phase error in terms of $F(0)$. The complete linear model (including dc parameters) in Fig. 2–3 is augmented to include the loop filter in Fig. 2–10. From this signal flow graph, we see that

$$\Delta\omega_o = \theta_e K_d F(s) K_o + V_{do} F(s) K_o - V_{co} K_o$$

and the average (dc or $s = 0$) relation is

$$\overline{\Delta\omega_o} = \bar{\theta}_e K_d F(0) K_o + V_{do} F(0) K_o - V_{co} K_o$$

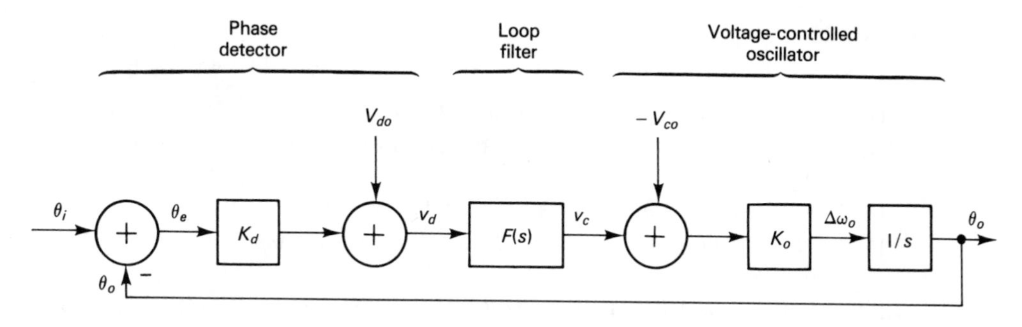

FIGURE 2–10 Full linear model of PLL

But the static phase error θ_{eo} is defined as $\overline{\theta}_e$ when the PLL is in lock (when $\overline{\Delta\omega_o} = 0$). It follows that

$$\theta_{eo} = -V_{do}/K_d + V_{co}/K_dF(0) \tag{2-27}$$

For the particular case of $F(0) = 1$, this reduces to Eq. (2-11). In Chapter 3, we will see how to use an active loop filter to make $F(0)$ essentially infinite so that θ_{eo} is unaffected by V_{co}.

EXAMPLE 2–2

A PLL has a PD with the characteristic in Fig. 2–1b and a VCO with the characteristic in Fig. 2–2a. The input frequency is $\omega_i = 12$ Mrad/s. Design a loop filter to realize a bandwidth of $\omega_{3dB} = 0.73$ Mrad/s. Find the static phase error.

From the characteristics, $K_d = 1.27$ V/rad, and $K_o = 2$ Mrad/s/V, $V_{do} = 2$ V, and $V_{co} = 2$ V (the voltage for which $\omega_o = \omega_i = 12$ Mrad/s). From Eq. (2-24), $K_h = \omega_{3dB}/K_dK_o = 0.286$. From Eq. (2-22b), this is satisfied by $R_0 = \underline{25\ \text{k}\Omega}$ and $R_2 = \underline{10\ \text{k}\Omega}$. First we will try a design with no capacitor; that is, $F(0) = 0.286$. From Eq. (2-27), $\theta_{eo} = 3.93$ radians. But this far exceeds $\pi/2 = 1.57$ radians, the limit for which the linear model holds. Therefore, the solution is false, and the PLL can't lock to the input frequency.

Adding a capacitor to the loop filter will give back the necessary dc gain to let the PLL achieve lock. Then $F(0) = 1$, and Eq. (2-27) gives $\theta_{eo} = \underline{0}$, which puts the PD in the center of its range. The value of the capacitor must be chosen to satisfy Eq. (2-25), which requires that the resulting zero at ω_2 be less than 0.73 Mrad/s. We choose a factor of four less and make $\omega_2 = 0.183$ Mrad/s. Then, from Eq. (2-22d), $C = \underline{546\ \text{pF}}$. The final design is shown in Fig. 2–9a.

REFERENCES

[1] A. J. Viterbi, *Principles of Coherent Communication*, McGraw-Hill: New York, 1966.

[2] A. Blanchard, *Phase-Locked Loops: Applications to Coherent Receiver Design*, Wiley: New York, 1976.

[3] F. M. Gardner, *Phaselock Techniques*, Wiley: New York, 1979.

[4] R. E. Best, *Phase-Locked Loops: Theory, Design, and Applications*, McGraw-Hill: New York, 1984.

[5] C. L. Phillips and R. D. Harbor, *Feedback Control Systems*, Prentice-Hall: Englewood Cliffs, NJ, 1988.

CHAPTER

3

LOOP FILTERS

In Chapter 2 we saw that the PLL bandwidth

$$\omega_{3\text{dB}} = K_d K_h K_o \equiv K \tag{2-24}$$

is determined by the gain K_d of the PD, the high-frequency gain K_h of the loop filter, and the gain K_o of the VCO. Since the PD and the VCO designs are usually less flexible, the design of the loop filter is the engineer's principle tool in determining the bandwidth. We also saw from

$$\theta_{eo} = -V_{do}/K_d + V_{co}/K_d F(0) \tag{2-27}$$

that a large dc gain $F(0)$ of the loop filter is desirable. For a passive filter, the maximum dc gain is unity. In this chapter, we look at active loop filters that achieve an $F(0)$ that is essentially infinite. In some applications it is desirable to add a high-frequency pole to the loop filter. Thus, the engineer has three parameters to choose in designing the loop filter: the high-frequency gain K_h, the placement of the zero that sends $F(0)$ to infinity, and the placement of the pole. This chapter gives a number of circuits for realizing the filter design. It also analyzes the response of the PLL in terms of the loop filter's parameters.

3–1 ACTIVE LOOP FILTER

The design of an active loop filter begins with an amplifier with gain K_h to modify the bandwidth of the PLL. This is realized in Fig. 3–1a by an active circuit with gain $-K_h = -R_2/R_1$. (For a review of op amp circuits, see Kennedy.[1]) When realized by a passive voltage divider (as in section 2–6), K_h was necessarily less than unity. This is no longer so with an active loop filter, but in practice, most PLL designs call for $K_h < 1$.

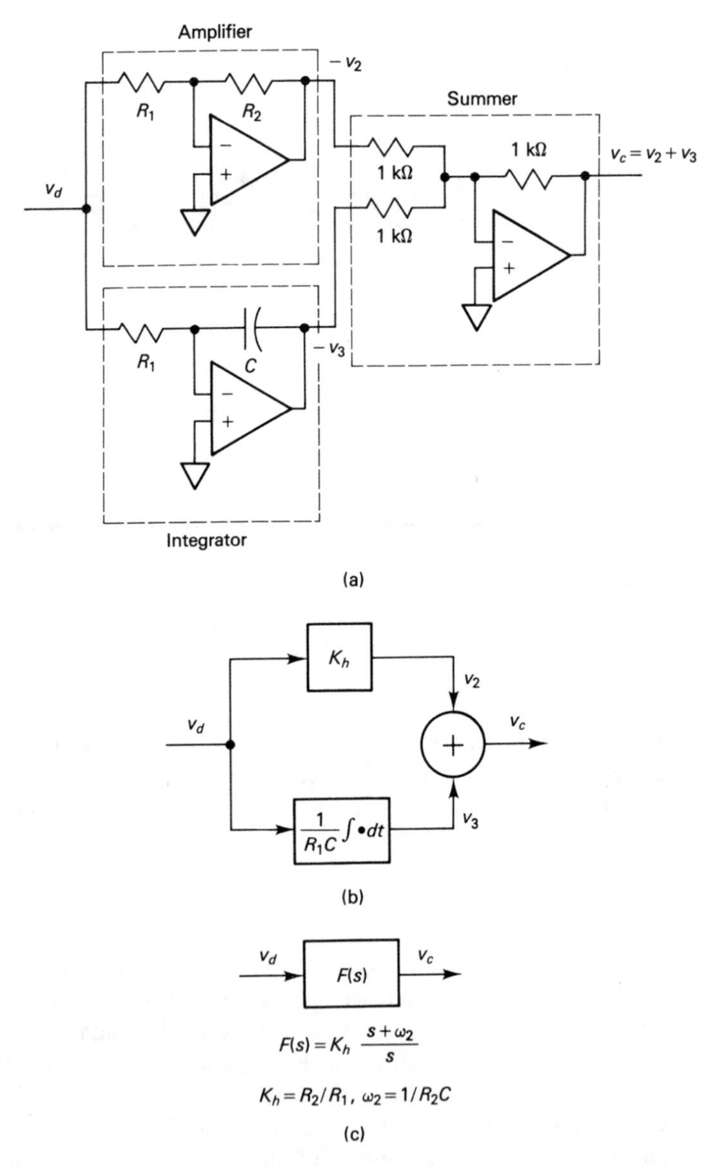

FIGURE 3–1 Model of proportional-plus-integral loop filter (or "active" filter)

The second part of the loop filter design is to realize infinite gain at dc by an integrator, as in the circuit in Fig. 3–1a. By summing the output of the amplifier

$$v_2 = K_h v_d$$

with the output of the integrator

$$v_3 = \frac{1}{R_1 C} \int v_d \, dt$$

we get the complete control voltage

$$v_c = v_2 + v_3 = K_h v_d + \frac{1}{R_1 C} \int v_d \, dt \tag{3-1}$$

as diagrammed by the signal flow graph in Fig. 3–1b.

The corresponding Laplace transform for this relation is

$$v_c = K_h v_d + (1/R_1 C s) v_d \tag{3-2}$$

At high frequencies ($s \rightarrow \infty$) the second term goes to zero, and the gain K_h dominates. At dc ($s = 0$), the second term goes to infinity, and the gain $1/R_1Cs$ of the integrator dominates. The complete gain $F(s)$ of the loop filter is therefore

$$F(s) = K_h + \frac{1}{R_1 C\, s} = K_h \frac{s + \omega_2}{s} \tag{3-3}$$

where

$$K_h = R_2/R_1 \tag{3-4}$$

$$\omega_2 = 1/R_2 C \tag{3-5}$$

The frequency response of the $F(j\omega)$ given in Eq. (3-3) is plotted in Fig. 3–2. As ω goes to zero, the magnitude of $F(j\omega)$ goes to infinity, as desired. For ω greater than the break due to the zero at ω_2, the magnitude is about K_h.

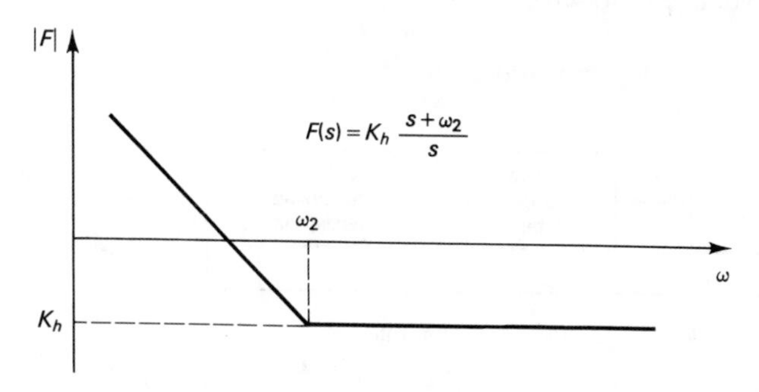

FIGURE 3–2 Frequency response of active filter

3–2 STATIC PHASE ERROR WITH ACTIVE LOOP FILTER

A block diagram of the complete PLL including an active filter is shown in Fig. 3–3. It emphasizes the three main parts to the design of a PLL. This chapter deals with the loop filter, and the next two chapters deal with the phase detector and the VCO.

The linear model of the PLL in Fig. 3–3 has already been given in Fig. 2–10; it is repeated here in Fig. 3–4. In Fig. 2–10, $F(s)$ represented a passive loop filter. Now it represents any form of loop filter, including an active one. An analysis of the linear model showed that the static phase error is

$$\theta_{eo} = -V_{do}/K_d + V_{co}/K_dF(0) \tag{3-6}$$

Now, V_{co} is the VCO control voltage necessary to bring the frequency ω_o equal to the input frequency ω_i in lock. Therefore, V_{co} is not a property of the VCO alone; it depends on ω_i in the particular application. For the designer to be in control of θ_{eo}, it is usually important that θ_{eo} not be a function of V_{co}. For an active loop filter, $F(0) = \infty$, and Eq. (3-6) becomes

$$\theta_{eo} = -V_{do}/K_d \tag{3-7}$$

The disappearance of V_{co} can be explained from a circuit standpoint by considering the loop filter circuit in Fig. 3–1a. During lock acquisition, the integrator accumulates enough charge on the capacitor to provide the control voltage $v_c = V_{co}$ needed by the VCO when in lock. This is the v_3 component of v_c. Once the PLL is in lock, $\overline{v}_d$, the average of v_d, must go to zero so the integrator stops charging. Averaging Eq. (2-4) gives

$$\overline{v}_d = \overline{\theta}_e K_d + V_{do} = \theta_{eo} K_d + V_{do}$$

which leads immediately to Eq. (3-7).

The remaining contributor to θ_{eo} in Eq. (3-7) is V_{do}. This is the free-running voltage of the PD when there is no signal at the PLL input. In Chapter 11, it will be shown that designing for $\theta_{eo} = 0$ improves the purity of a synthesized frequency. From Eq. (3-7), this translates into the desirability of designing for $V_{do} = 0$. We will see in Chapter 8 that acquisition is easier when $V_{do} = 0$. In any case, a basic PLL design consideration is to keep the free-running PD voltage near zero:

$$V_{do} \approx 0 \text{ (desired)} \tag{3-8}$$

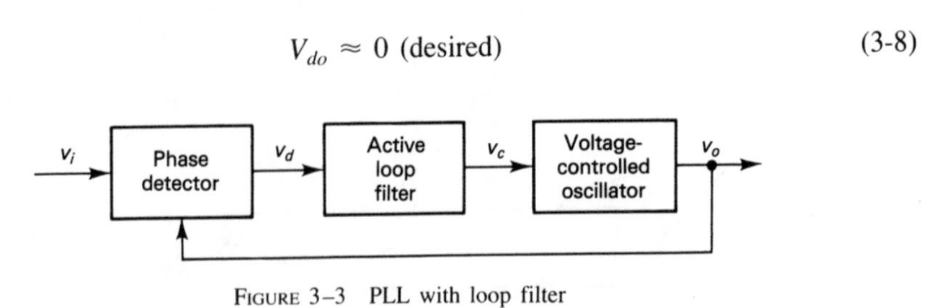

FIGURE 3–3 PLL with loop filter

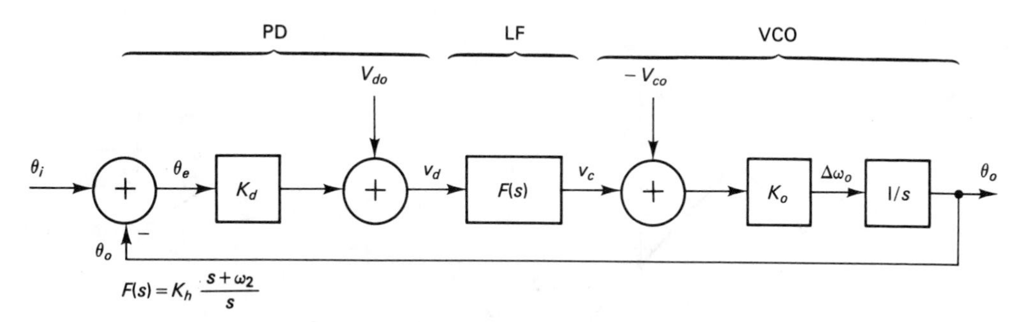

FIGURE 3–4 Linear model of PLL

For this reason, V_{do} is also called the *phase detector offset voltage*. The next section suggests some loop filter designs to help the PD achieve $V_{do} \approx 0$.

3–3 ALTERNATIVE ACTIVE LOOP FILTER DESIGNS

The circuit in Fig. 3–1a is a direct way to realize a proportional-plus-integral loop filter. The signal v_2 is proportional to v_d, and v_3 is the integral of v_d. A simpler way to realize this function is shown in Fig. 3–5a, where v_2 and v_3 are added by placing R_2 and C in series. The only difference is that there is a sign inversion; the transfer function is now $-F(s)$. To be consistent, our convention is that the input to the active loop filter is $-v_d$ so we still have $v_c = F(s)v_d$.

In order to maintain negative feedback for the stability of the PLL, the inversion in an active loop filter must be accompanied by either a negative PD gain $-K_d$ or a negative VCO gain $-K_o$. Since the sign of the PD gain is easily reversed by reversing the v_i and v_o inputs, we will assume throughout this text that an active loop filter is coupled with a PD with gain $-K_d$, where K_d is always positive. Correspondingly, the VCO gain K_o is always assumed to be positive.

The PD characteristic in Fig. 3–5a has $v_d = 0$ for $\theta_e = 0$; that is, $V_{do} = 0$. This is desirable according to Eq. (3-8). In practice, V_{do} will not be exactly zero, of course, but a phase detector used with an active loop filter should have a V_{do} that is nominally zero.

Suppose that a PD produces a nonzero voltage V_{ao} for $\theta_e = 0$, as in Fig. 3–5b. This characteristic can still be used with an active loop filter if the op amp is referenced to V_r rather than ground. Then the PD voltage is effectively

$$v_d = V_r - v_a \tag{3-9a}$$

and the voltage for $\theta_e = 0$ is

$$V_{do} = V_r - V_{ao} \tag{3-9b}$$

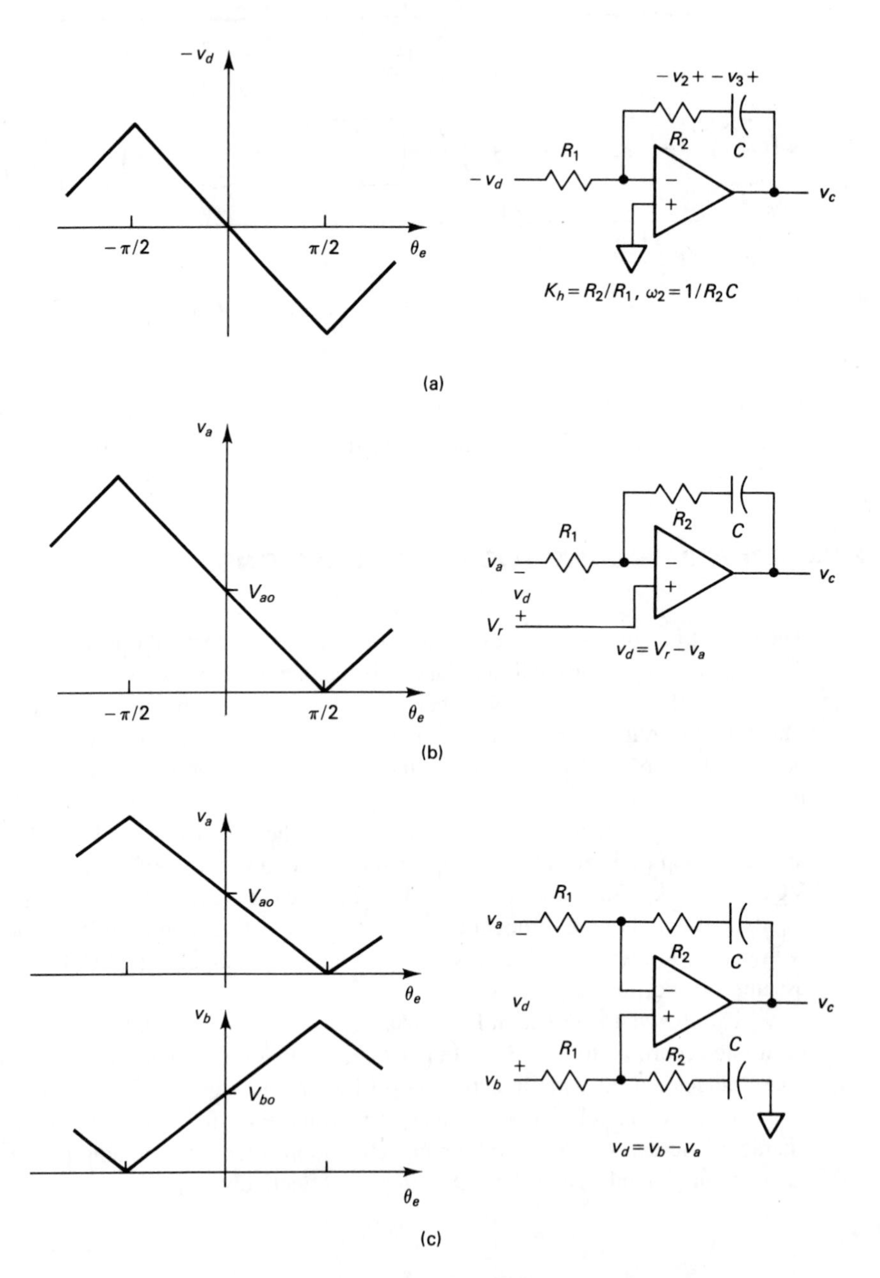

FIGURE 3–5 Active loop filters using one op amp

Now we can set $V_{do} = 0$ by choosing $V_r = V_{ao}$.

There is always some error in realizing $V_r = V_{ao}$. Therefore, it is better to take advantage of a natural balance when it is available. Some phase detectors produce a voltage v_a and its complement v_b, as shown in Fig. 3–5c. Then if the PD voltage is taken as

$$v_d = v_b - v_a \tag{3-10a}$$

the voltage for $\theta_e = 0$ is

$$V_{do} = V_{bo} - V_{ao} \tag{3-10b}$$

The symmetrical design of a PD producing v_a and v_b tends to provide a good match between V_{ao} and V_{bo}, which reduces V_{do}. The transfer function $F(s)$ is the same as that for the filter in Fig. 3–5a when the values of R_1, R_2, and C are the same. The penalty is that the circuit in Fig. 3–5c uses twice as many resistors and capacitors.

3–4 ACTIVE LOOP FILTER OFFSETS

In accord with Eq. (3-8), we try to keep the PD free-running voltage V_{do} as small as possible. Therefore, V_{do} is also referred to as the *phase detector offset voltage*. But the loop filter, now with an active component, contributes its own share of offset to v_d.

Consider the loop filter in Fig. 3–6a. This is the same as the filter in Fig. 3–5b but with an extra resistor R_1 to mitigate the effect of the input bias current of the op amp. Let the input offset voltage of the op amp be V_{IO} (this is the dc voltage that appears across the input terminals of the op amp). Let the input offset current of the op amp be I_{IO} (this is the difference of the dc currents I_{B1} and I_{B2} into the two input terminals of the op amp). It can be shown that these offsets effectively add a dc voltage $V_{IO} + I_{IO}R_1$ to v_d, as the signal flow graph in Fig. 3–6b represents. The dc offset $V_r - V_{ao}$ is contributed by the error in setting $V_r = V_{ao}$ [see Eq. (3-9b)]. Our convention will be to combine the dc sources and label the whole as the PD offset voltage:

$$V_{do} = (V_r - V_{ao}) + V_{IO} + I_{IO}R_1 \tag{3-11a}$$

This is represented by the signal flow graph in Fig. 3–6c. Note that the PD is now considered to be responsible for all dc offsets, even those originating in the loop filter. Although this doesn't correspond with the physical circuit, it simplifies our notation and analysis.

For the balanced loop filter in Fig. 3–5c, the expression for PD offset voltage is similar. Combining Eq. (3-10b) with the op amp offsets:

$$V_{do} = (V_{bo} - V_{ao}) + V_{IO} + I_{IO}R_1 \tag{3-11b}$$

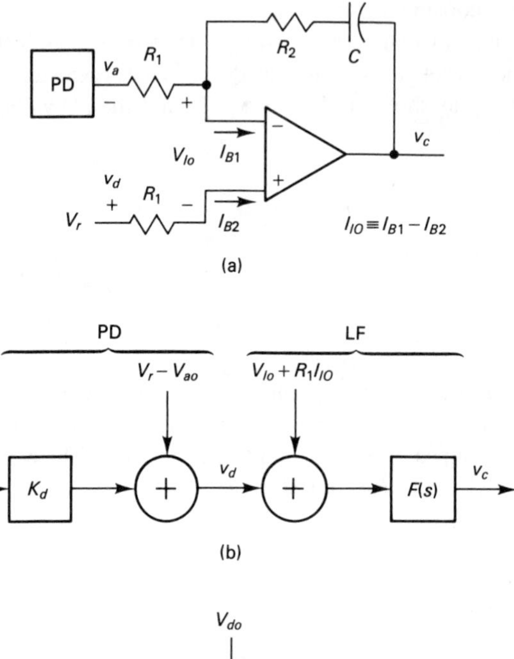

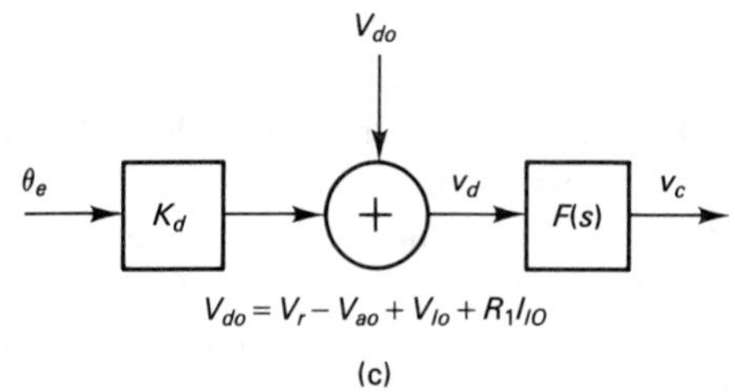

FIGURE 3–6 Offset voltage with an active filter

Typical specifications for an op amp are $|V_{IO}| \leq 5$ mV and $|I_{IO}| \leq 20$ nA. Laser-trimmed op amps are available with $|V_{IO}| \leq 0.1$ mV, and FET-input op amps typically have $|I_{IO}| \leq 1$ nA.

3–5 PLL FREQUENCY RESPONSE

We form the ac model for the PLL (see Fig. 3–7) by eliminating the dc parameters from the linear model in Fig. 3–4. This is the same as the model in Fig. 2–9b, but now $F(s)$ is the response of an active loop filter given in Eq. (3-3): $F(s) = K_h(s + \omega_2)/s$. The forward gain of the PLL is given by

$$G(s) = K_d F(s) K_o / s = K_d K_h K_o (s + \omega_2)/s^2$$
$$= K(s + \omega_2)/s^2 \tag{3-12}$$

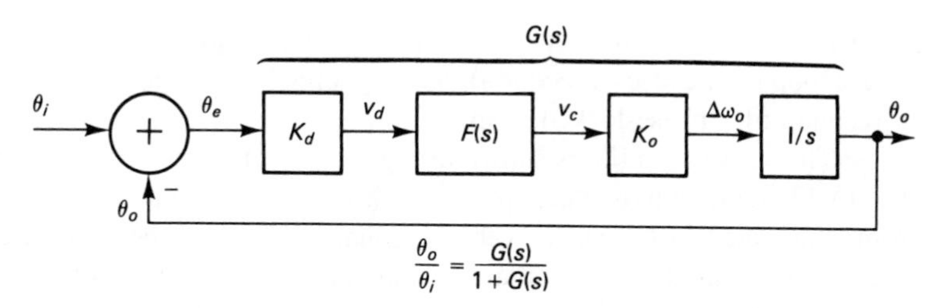

FIGURE 3–7 ac model of PLL

where

$$K \equiv K_d K_h K_o$$

The magnitude of $G(j\omega)$ is plotted in Fig. 3–8. Let the closed-loop phase transfer function be represented by $H(s)$. Then from Eq. 2-13,

$$H(s) \equiv \frac{\theta_o(s)}{\theta_i(s)} = \frac{G(s)}{1 + G(s)} = \frac{Ks + K\omega_2}{s^2 + Ks + K\omega_2} \tag{3-13}$$

For $\omega_2 < K$, it is roughly true that $|H(j\omega)|$ follows the lower of unity and $|G(j\omega)|$, as illustrated in Fig. 3–8. There is some peaking in the response, but this becomes less the

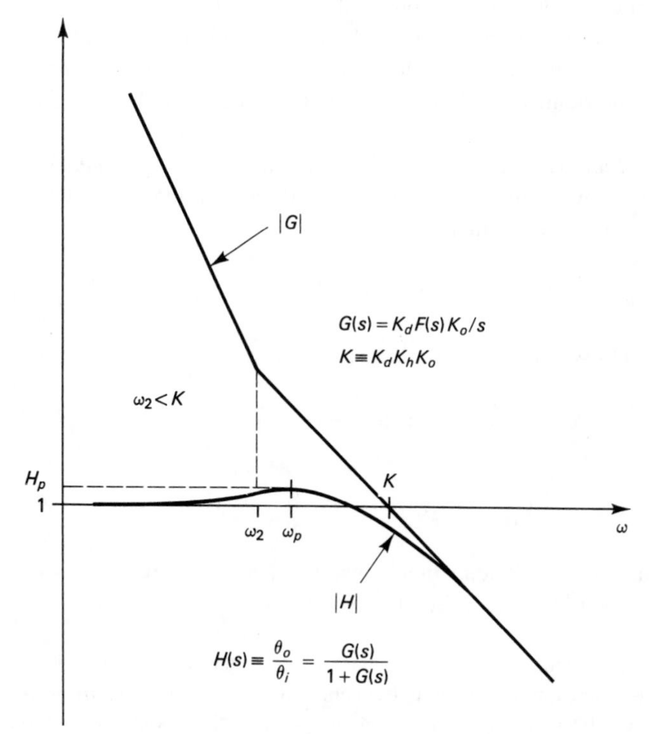

FIGURE 3–8 Frequency response of PLL

farther ω_2 is to the left of K. As we found for the passive loop filter in Chapter 2, the active loop filter causes a second-order polynomial in s in the denominator, and the PLL is a second-order phase-locked loop.

The active loop filters shown in Fig. 3–5 and the corresponding transfer function in Eq. (3-13) serve in most PLL applications. Passive loop filters (see Fig. 2–9a) are used in some integrated PLLs, such as the National NE564 and the Exar XR-210, but a better performance can always be obtained using an active loop filter in a custom PLL design. Therefore, all PLL applications in the following chapters use an active loop filter*.

The expression for $H(s)$ in Eq. (3-13) is in terms of K and ω_2. These two parameters are easily identified in the frequency responses (see Fig. 3–8), and they are easily related to the PLL components [see Eqs. (2-24) and (3-5)]. We mention in passing an alternative expression the reader will often encounter in other PLL literature. It expresses $H(s)$ in terms of a *damping ratio* ζ and a *natural frequency* ω_n. Equation (3-13) can be expressed as

$$H(s) = \frac{2\zeta\omega_n s + \omega_n^2}{s^2 + 2\zeta\omega_n s + \omega_n^2} \tag{3-14}$$

where

$$\zeta = 0.5\sqrt{K/\omega_2}, \quad \omega_n = \sqrt{K\omega_2} \tag{3-15}$$

This notation is derived from common usage in control theory, but it can be misleading. When $\omega_2 << K$ (when $\zeta >> 0.5$), the response $|H|$ hardly depends on ω_2. But ω_n is always a strong function of ω_2, giving the impression that the bandwidth always depends on ω_2. (Note that ω_n, the geometric mean of K and ω_2, is halfway between them on a log frequency scale.)

Let the frequency for which $|H|$ is a maximum be called the *peaking frequency* ω_p (see Fig. 3–8). We can find ω_p by setting $(d/d\omega)|H(j\omega)|^2 = 0$ and solving for ω, where $H(j\omega)$ is found from Eq. (3-13). The result is

$$\omega_p = \omega_2[(2K/\omega_2 + 1)^{1/2} - 1]^{1/2} \tag{3-16}$$

Let the peak value of $|H|$ be $H_p \equiv |H(j\omega_p)|$. From Eq. (3-13) and Eq. (3-16), this gives

$$H_p = [1 - 2\alpha - 2\alpha^2 + 2\alpha(2\alpha + \alpha^2)^{1/2}]^{-1/2} \tag{3-17}$$

where

$$\alpha \equiv \omega_2/K$$

These expressions for ω_p and H_p don't lend much insight. Therefore, we give some approximations that hold for each of three different cases. The overdamped case is ω_2/K

---

* There are other types of active loop filter. In particular, a loop filter with *two* integrators provides zero phase error during a ramp of the input frequency. It makes the PLL a *third-order phase-locked loop*. This specialized application is beyond the scope of this book; "active loop filter" will always mean the loop filter shown in Fig. 3–5.

TABLE 3–1 PEAKING PARAMETER APPROXIMATIONS

Damping	ω_2/K	ω_p	H_p
Over	<0.25	$1.2\omega_2^{3/4}K^{1/4}$	$1 + \omega_2/K$
Critical	0.25	$1.4\omega_2$	1.15
Under	>0.25	$\sqrt{K\omega_2}$	$\sqrt{\omega_2/K}$

< 0.25, the critically damped case is $\omega_2/K = 0.25$, and the underdamped case is $\omega_2/K > 0.25$. The corresponding approximations for ω_p and H_p are given in Table 3–1.

We will be most interested in the overdamped and critically damped cases. The actual values of ω_p/K (the normalized peaking frequency) and $H_p - 1$ (the peaking excess) are plotted in Fig. 3–9. These curves are compared with the approximations for overdamping (see the dashed curves in Fig. 3–9). For $\omega_2/K < 0.1$, it holds within 10% that $\omega_p/K \approx 1.2(\omega_2/K)^{3/4}$, or

$$\omega_p \approx 1.2\omega_2^{3/4}K^{1/4} \tag{3-18}$$

and it holds within 30% that $H_p - 1 \approx \omega_2/K$, or

$$H_p \approx 1 + \omega_2/K \tag{3-19}$$

Equation (3-18) says that ω_p is about a quarter of the way from ω_2 toward K on a log axis (see Fig. 3–8).

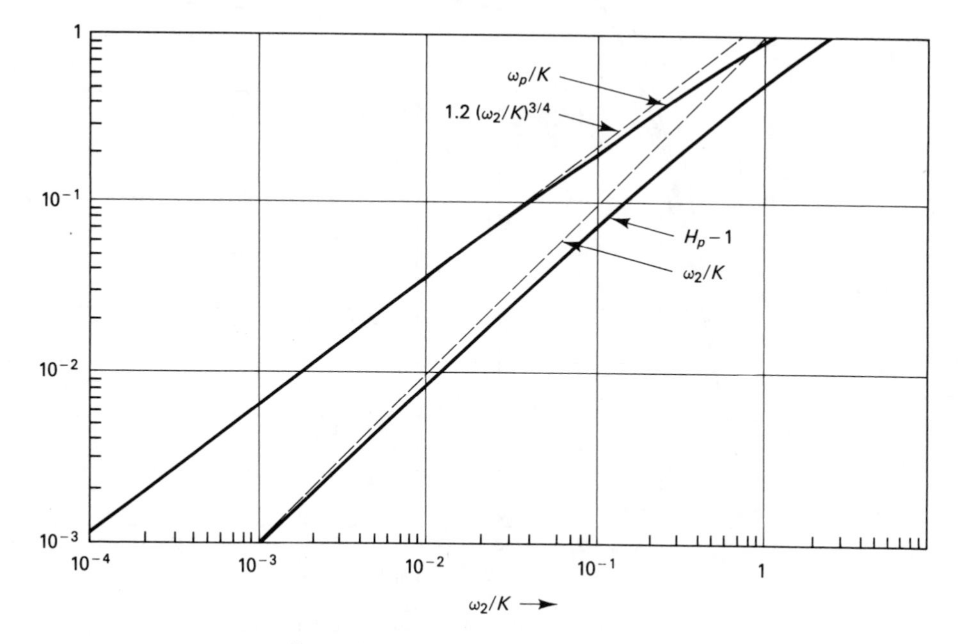

FIGURE 3–9 Peaking parameters H_p and ω_p

3–6 PLL STEP RESPONSE

The response at θ_o to a unit step of phase at θ_i is found by taking the inverse Laplace transform of $H(s)/s$, where $H(s)$ is given in Eq. (3-13). The results for some selected dampings are given in Eq. (3-20).

$$\text{For } \omega_2 = 0, \quad \theta_o = 1 - e^{-Kt} \tag{3-20a}$$

$$\text{for } \omega_2 = K/4, \quad \theta_o = 1 - e^{-0.5Kt}(1 - 0.5Kt) \tag{3-20b}$$

$$\text{for } \omega_2 = K, \quad \theta_o = 1 - e^{-0.5Kt}(\cos 0.866Kt - 0.577 \sin 0.866Kt) \tag{3-20c}$$

Figure 3–10 plots these responses as well as those for some other dampings. Note that as ω_2 gets closer to K, the overshoot increases. The amount of overshoot is plotted as a function of ω_2/K in Fig. 3–11a. For example, the overshoot is 13% for $\omega_2/K = 0.25$. The position of the peak of the step response along the normalized time axis Kt is plotted in Fig. 3–11b as a function of ω_2/K. For example, the peak is at $Kt = 4.0$ for $\omega_2/K = 0.25$.

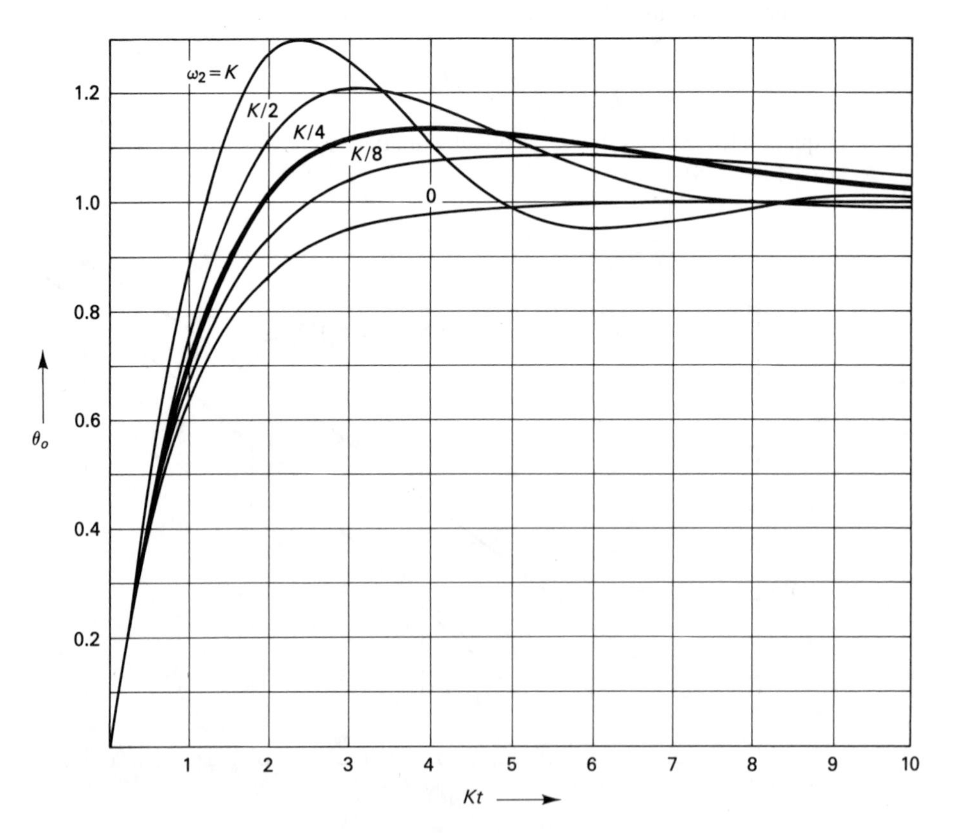

FIGURE 3–10 Step response of PLL

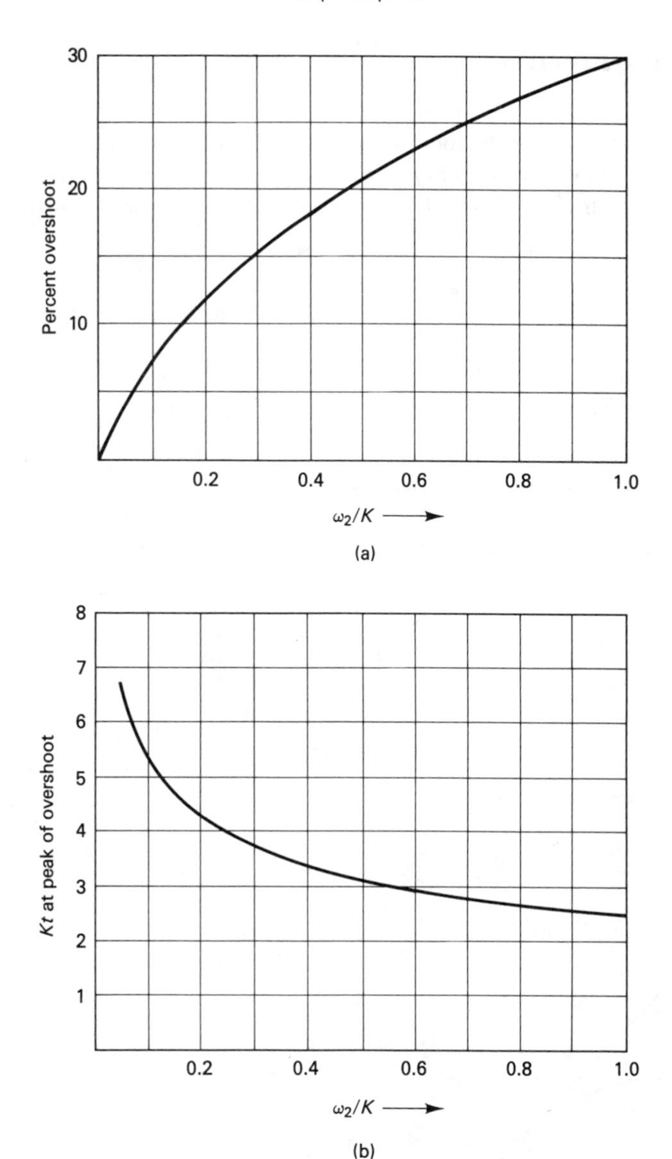

FIGURE 3–11 Overshoot parameters

It would seem from the step responses that it is always best to make ω_2 as low as possible. This slows the response only slightly, and it makes the system very stable, avoiding overshoot. However, from Eq. (3-5) this requires a large capacitor, and we will see that a larger capacitor takes longer to charge during lock acquisition. Therefore, a good rule of thumb is to make $\omega_2 = K/4$ when peaking is not critical; this assures fast acquisition. Chapter 10 looks at an application with many tandem PLLs. In that case, peaking of the frequency response is more of a problem than acquisition time, and ω_2 is selected to limit the peaking to some small value.

3–7 LIMITED LOOP FILTER BANDWIDTH

In Fig. 3–2, we assumed the response of the loop filter has a gain of K_h from ω_2 all the way out to $\omega = \infty$. In practice, the response $|F(j\omega)|$ of an active loop filter rolls off at some frequency ω_3, as shown in Fig. 3–12. The expression for the filter's transfer function is

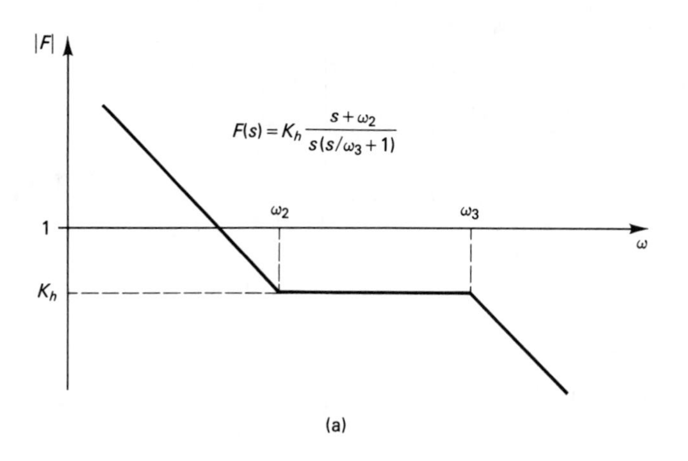

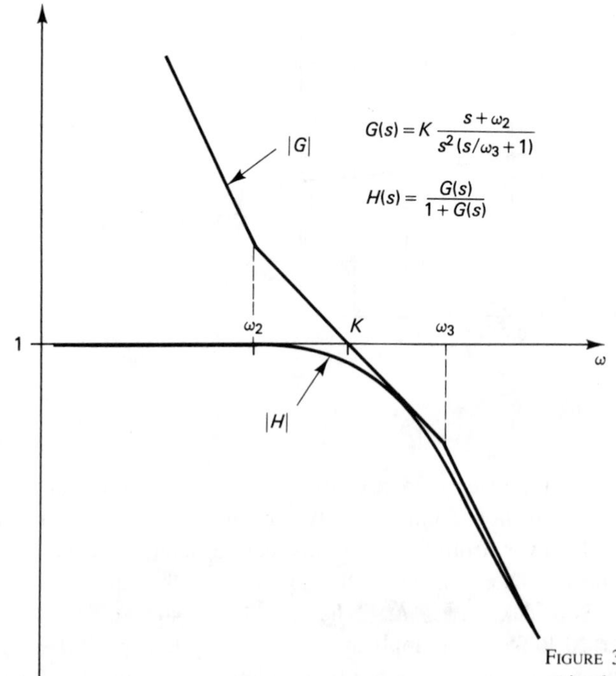

FIGURE 3–12 Frequency response of (a) active loop filter, and (b) corresponding PLL phase transfer function H

$$F(s) = K_h \frac{s + \omega_2}{s(s/\omega_3 + 1)} \tag{3-21}$$

The limited bandwidth of the op amp itself introduces a pole at

$$\omega_3 = 2\pi \frac{R_1}{R_1 + R_2} \text{GBP} = \frac{2\pi}{1 + K_h} \text{GBP} \tag{3-22}$$

where GBP is the gain-bandwidth product. For example, voice-grade op amps typically have GBP = 1 MHz. Then, for $K_h << 1$, $\omega_3 = 2\pi(1 \text{ MHz})$. High-performance op amps such as the Harris HA-2540 are available with a GBP as high as 400 MHz.

How does ω_3 affect the transfer function $H(s)$ of the PLL? The forward gain $G(s) \equiv K_d F(s) K_o/s$ with the cutoff in Eq. (3-21) is then

$$G(s) = K \frac{s + \omega_2}{s^2(s/\omega_3 + 1)} \tag{3-23}$$

$$H(s) = \frac{G(s)}{1 + G(s)} = \frac{Ks + K\omega_2}{s^3/\omega_3 + s^2 + Ks + K\omega_2} \tag{3-24}$$

The responses $|G(j\omega)|$ and $|H(j\omega)|$ are plotted in Fig. 3–12b, where the additional break due to the pole at ω_3 is evident. If $\omega_3 > K$, then $|G(j\omega)|$ still crosses unity at $\omega = K$, and the PLL bandwidth is K. Provided that ω_3 is not too close to K, the step response will still be about that shown in Fig. 3–10. For the case $\omega_2 = K/4$ and $\omega_3 = 4K$, it can be shown the step response is

$$\theta_o = 1 - 3e^{-Kt} + e^{-0.382Kt} + e^{-2.618Kt}$$

This response is plotted versus normalized time Kt in Fig. 3–13. The overshoot is now 18% (compared with 13% for the case $\omega_2 = K/4$ and $\omega_3 = \infty$). Therefore, a good rule of thumb is to keep $\omega_3 \geq 4K$, and then it can be roughly ignored in the analysis of the PLL response. The choice of ω_3 is also influence by the desired acquisition range (see section 8–3).

In some applications, it is desirable to purposely introduce a cutoff at some lower ω_3. One method for implementing this is shown in Fig. 3–14. The resistor R_1 has been split in two, and a capacitor C_3 bypasses to ground the frequencies above ω_3. The relationship is

$$\omega_3 = 4/R_1C_3 \tag{3-25}$$

As before, we still have $K_h = R_2/R_1$ and $\omega_2 = 1/R_2C$

The introduction of a pole at ω_3 may be necessary to suppress high-frequency components that the op amp cannot handle. The op amp, through feedback, must be able to maintain ≤ 0.3 Vp-p at its input to avoid overloading its input stage. In Chapter 11 we will see that ω_3 can be used to reduce phase jitter of a synthesized frequency.

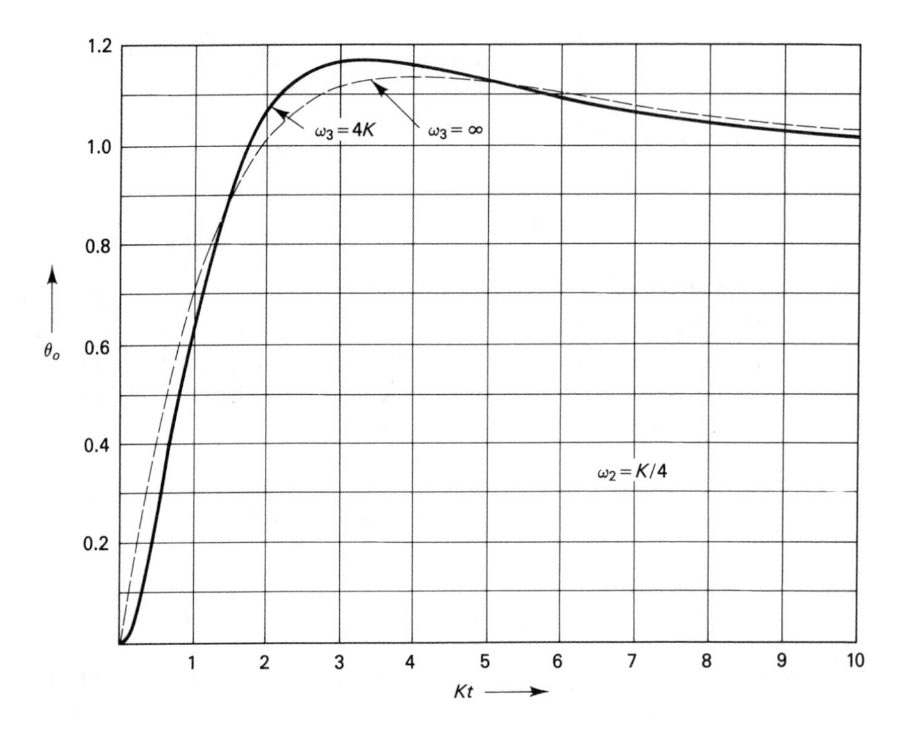

FIGURE 3–13 PLL step response with pole at ω_3

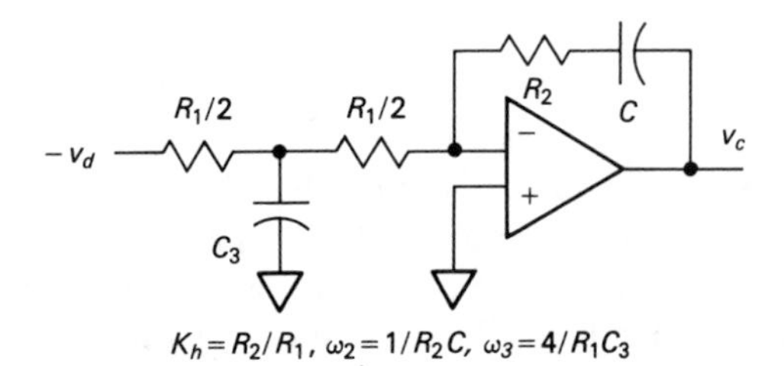

FIGURE 3–14 Active loop filter with lowered ω_3

EXAMPLE 3–1

The input frequency to a PLL is $\omega_i = 52$ Mrad/s. The desired PLL bandwidth is $K = 50$ krad/s, and the peak response is to be $H_p = 1.01$. The PD and VCO characteristics are those shown in Fig. 3–15a and b. The op amp specifications are $V_{IO} = 5$ mV, $I_{IO} = 20$ nA, and GBP = 1 MHz. The capacitor value C is to be kept less than 0.2 μF (small enough not to be polarized). Compare designs using (a) an unbalanced active loop filter, (b) a balanced active loop filter, and (c) a passive loop filter.

(a) The PD has balanced outputs v_a and v_b available, but suppose we use only v_a. Then the op amp must be referenced to $V_r = V_{ao} = 2.5$ V, as in Fig. 3–15c. Since $v_d =$

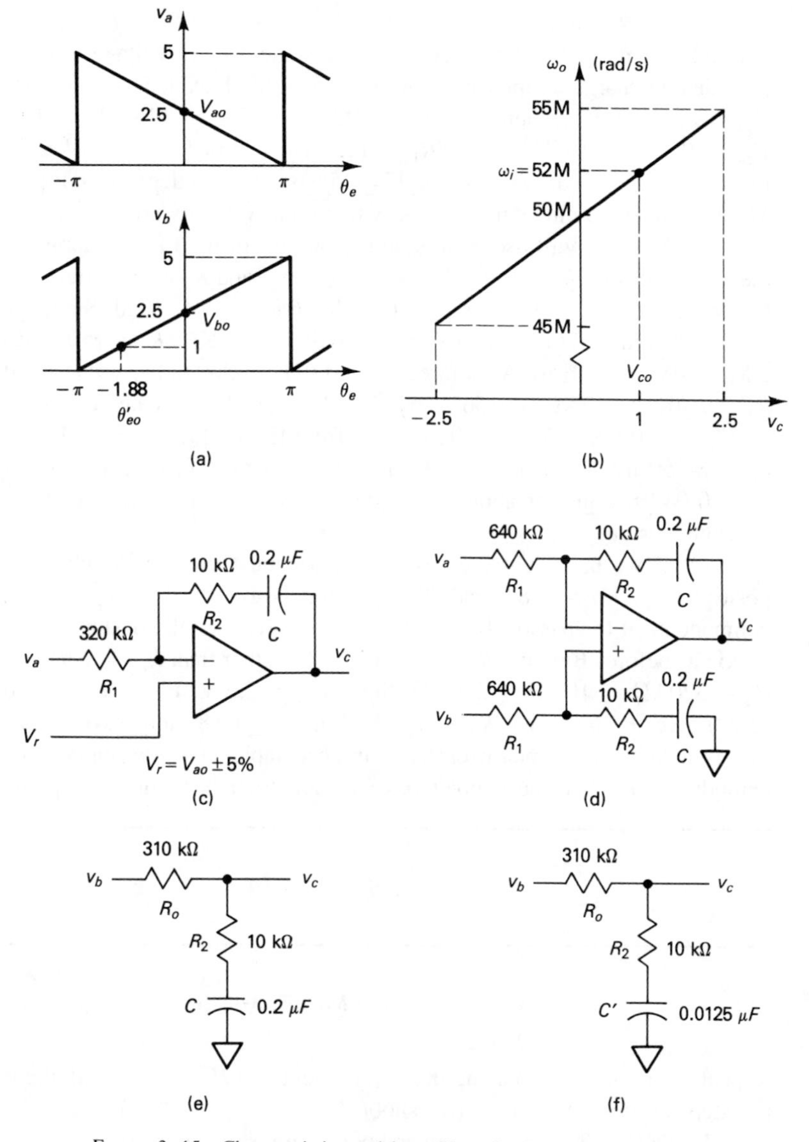

FIGURE 3–15 Characteristics and loop filters for Examples 3–1 and 3–2

2.5 V $-$ v_a, then K_d is the negative of the slope of the v_a curve: $K_d = 5$ V/2π rad $= 0.8$ V/rad. The VCO gain is $K_o = (10$ Mrad/s)/5 V $= 2$ Mrad/s/V. But $K = K_d K_h K_o$. Therefore, $K_h = K/K_d K_o = 0.0313$. From Eq. (3-19), $H_p \approx 1 + \omega_2/K$. Hence

$$\omega_2 = (H_p - 1)K = 0.01\ K = 500 \text{ rad/s}$$

Since $\omega_2 = 1/R_2 C$, choosing C as large as possible will allow R_2 and R_1 to be as small as possible. According to Eq. (3-11a), this will reduce V_{do}. Therefore, we choose $C = \underline{0.2}$

μF. Then $R_2 = 1/\omega_2 C = $ 10 kΩ. Now, $K_h = R_2/R_1$, so $R_1 = R_2/K_h = 10$ kΩ/0.0313 = 320 kΩ. From the VCO characteristic, the static control voltage is $V_{co} = 1.0$ V; this is provided by charge on the capacitor built up during lock acquisition. Suppose V_r matches V_{ao} to within ±5%. Then $V_r - V_{ao} = 0.05 \times 2.5$ V = 125 mV, and from Eq. (3-11a), $V_{do} = (V_r - V_{ao}) + V_{IO} + I_{IO}R_1 = 125$ mV + 5 mV + 6.4 mV = 136.4 mV. From Eq. (3-7), $\theta_{eo} = -V_{do}/K_d = -0.171$ radians, or −10 deg. From Eq. (3-22), $\omega_3 = 6.1$ Mrad/s. This is so much greater than K that it has virtually no effect on the PLL response.

(b) Now we will use both v_a and v_b outputs of the PD and apply their difference to the loop filter in Fig. 3–15d. Then $v_d = v_b - v_a$, and K_d is the difference of the slopes of the two curves: $K_d = 0.8$ V/rad − (−0.8 V/rad) = 1.6 V/rad. Since K_d is now greater, K_h must be reduced to maintain $K = 50$ krad/s: $K_h = K/K_dK_o = $ (50 krad/s)/(1.6 V/rad × 2 Mrad/s/V) = 0.0156. As before, $\omega_2 = 0.01K = 500$ rad/s, and $R_2 = $ 10 kΩ still. Now $R_1 = R_2/K_h = 10$ kΩ/0.0156 = 640 kΩ. Suppose V_{ao} matches V_{bo} within 2%. Then $V_{ao} - V_{bo} = 0.02 \times 2.5$ V = 50 mV, and from Eq. (3-11a), $V_{do} = (V_{ao} - V_{bo}) + V_{IO} + I_{IO}R_1 = 50$ mV + 5 mV + 12.8 mV = 67.8 mV. From Eq. (3-6), $\theta_{eo} = -V_{do}/K_d = -0.042$ radians, or about −2.4 deg. This is about a factor of four better than the previous design.

(c) If we use a passive loop filter, we must use the PD characteristic v_b with a positive slope $K_d = 0.8$ V/rad. Then, as for the filter in Fig. 3–15c, $K_h = 0.0313$. The components of the passive filter in Fig. 3–15e have the relationship $\omega_2 = 1/R_2C$, so $R_2 = $ 10 kΩ as before. But now $K_h = R_2/(R_0 + R_2)$ rather than $K_h = R_2/R_1$. Then $R_0 = R_1 - R_2 = 320$ kΩ − 10 kΩ = 310 kΩ. Since $V_{do} = 2.5$ V, $V_{co} = 1.0$ V, and $F(0) = 1$, Eq. (2-27) gives $\theta_{eo} = -V_{do}/K_d + V_{co}/K_dF(0) = -1.88$ radians, or −107 deg. Although this is a large static phase error, it is acceptable in some applications such as FSK demodulation. Then the simpler passive loop filter here might be preferable.

EXAMPLE 3–2

Repeat Example 3–1 replacing the requirement that $H_p = 1.01$ with the requirement that the step response have 10% overshoot.

From Fig. 3–11, a 10% overshoot requires $\omega_2'/K = 0.16$, or

$$\omega_2' = 0.16\,K = 8 \text{ krad/s}$$

This is 16 times the value of 500 rad/s in Example 3–1. This can be realized by reducing C to $C' = 0.2$ μF/16 = 0.0125 μF in each filter design. Since the bandwidth doesn't change, all other component values stay the same, and the values of θ_{eo} are the same. For example, the passive filter would be that shown in Fig. 3–15f.

3–8 PHASE ERROR RESPONSE

We have developed the transfer function $H(s)$ to find θ_o in response to the input θ_i. It will also be useful to find the phase error θ_e at the PD in response to θ_i. In particular, if θ_e is too large, it will exceed the linear region of the PD characteristic, and the PLL may lose lock.

By definition, the phase error is $\theta_e \equiv \theta_i - \theta_o$. Therefore, the transfer function from θ_i to θ_e is given by

$$\frac{\theta_e(s)}{\theta_i(s)} = 1 - \frac{\theta_o(s)}{\theta_i(s)} \equiv 1 - H(s) \tag{3-26}$$

This transfer function is usually represented by H_e. With Eq. (3-13), this can be expressed as

$$\frac{\theta_e(s)}{\theta_i(s)} \equiv H_e = 1 - H(s) = 1 - \frac{G(s)}{1 + G(s)} = \frac{1}{1 + G(s)} \tag{3-27}$$

where $G(s)$ is given in Eq. (3-23). It can be shown from Eq. (3-27) that $|H_e|$ follows the lower of unity and $|1/G|$. Figure 3–16 shows frequency responses of $|1/G|$ and $|H_e|$. (Note that the $|1/G|$ curve is simply the $|G|$ curve in Fig. 3–12 flipped about the unity-gain line.)

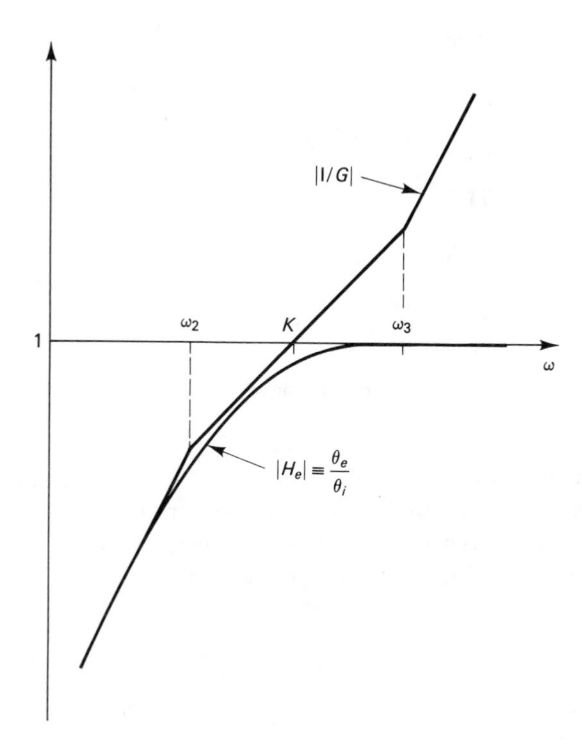

FIGURE 3–16 Phase error response

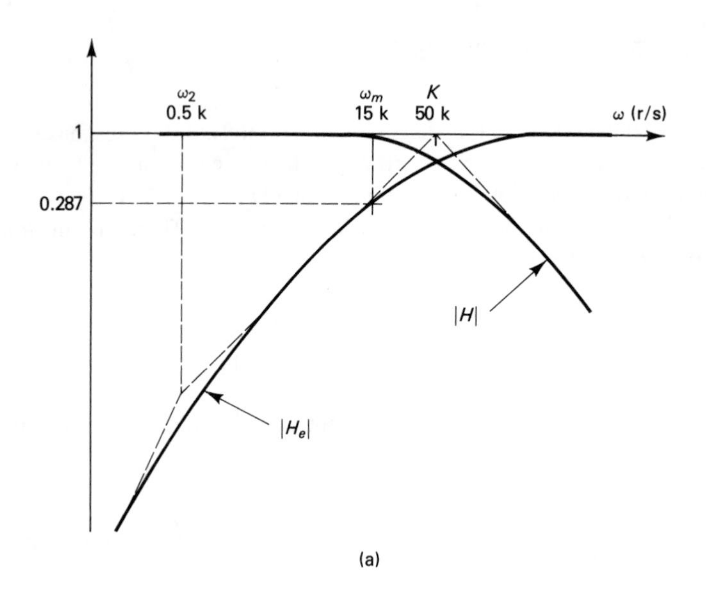

(a)

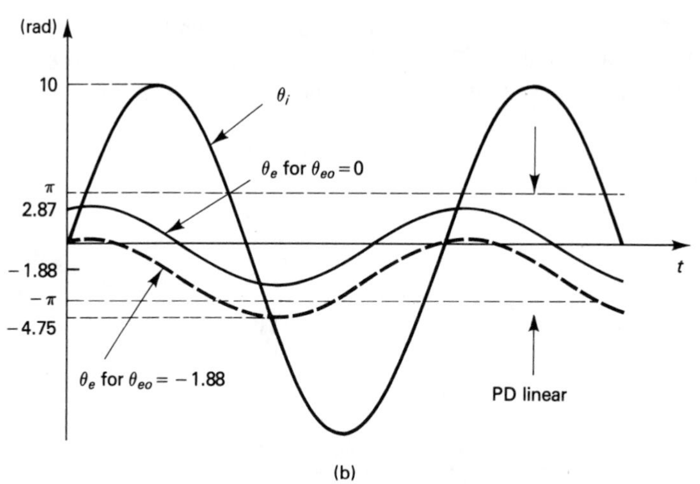

(b)

FIGURE 3–17 Phase error in Example 3–3

It is apparent that $|H_e|$ is a high-pass response with cutoff at $\omega = K$. This amounts to the fact that at low frequencies θ_o tracks θ_i well, and there is little phase error.

If $\omega_3 > 4K$, the pole at ω_3 can usually be ignored in the analysis. Then the expression $G(s) = K(s + \omega_2)/s^2$ in Eq. (3-12) can be used, and

$$H_e(s) \equiv \frac{\theta_e}{\theta_i} = \frac{1}{1 + G(s)} = \frac{s^2}{s^2 + Ks + K\omega_2} \tag{3-28}$$

EXAMPLE 3–3

A PLL has $K = 50$ krad/s and $\omega_2 = 500$ rad/s, and the modulated input phase is $\theta_i = 10 \sin \omega_m t$, where $\omega_m = 15$ krad/s. The PD characteristic is that shown in Fig. 3–15a. Find θ_e, and compare the performance of the PLL when $\theta_{eo} = 0$ with the performance when $\theta_{eo} = -1.88$ radians.

The response $|H_e|$ is sketched in Fig. 3–17a for the given values of K and ω_2. Evaluating Eq. (3-28) at the frequency of the phase modulation yields $H_e(j\omega_m) = 0.287\underline{/73}$ deg. Then the amplitude of θ_e is 0.287×10 radians $= 2.87$ radians. For $\theta_{eo} = 0$, the waveform is centered on the t axis, as shown in Fig. 3–17b. The peak values of θ_e are less than π, so the operation stays in the linear range of the PD (see Fig. 3–15a). If $\theta_{eo} = -1.88$ radians, the waveform is shifted down so the negative peak of θ_e is $-2.87 - 1.88 = -4.75$ radians (see Fig. 3–17b). Since this exceeds $-\pi$, the linear analysis doesn't hold, and the PLL loses lock.

REFERENCES

[1] E. J. Kennedy, *Operational Amplifier Circuits*, HRW: New York, 1988, Chapters 1 and 2.

CHAPTER

4

Phase Detectors

The linear model we have established for a phase detector (PD) is

$$v_d = K_d\theta_e + V_{do} \tag{4-1}$$

where K_d is the PD gain, θ_e is the phase error of the VCO output relative to the input signal, and V_{do} is the offset voltage or "free-running voltage." This linear model breaks down for large enough θ_e. The values of θ_e for which the linear model is valid are called the *range* of the PD.

A variety of devices, both analog and digital, can be used as PDs. We will compare them on the basis of range, offset, and gain. All of them can be thought of as multipliers in some sense.

4–1 FOUR-QUADRANT MULTIPLIERS

A multiplier acts as a PD through the trigonometric identity

$$\sin(A)\cos(B) \equiv 0.5\sin(A - B) + 0.5\sin(A + B) \tag{4-2}$$

Let the inputs to the multiplier be

$$v_i = V_i \sin(\omega_i t) \tag{4-3a}$$

$$v_o = V_o \cos(\omega_i t - \theta_e) \tag{4-3b}$$

The output of the multiplier is

$$\tilde{v}_d = K_m v_i v_o \tag{4-4}$$

where K_m is a constant associated with the multiplier. This is represented by the signal flow graph in Fig. 4–1a. The units of K_m are necessarily volts^{-1} so that $\tilde{v}_d$ will have the dimension of volts. Then by Eqs. (4-2), (4-3), and (4-4),

$$\tilde{v}_d = 0.5K_m V_i V_o \sin(\theta_e) + 0.5K_m V_i V_o \sin(2\omega_i t - \theta_e) \tag{4-5}$$

Figure 4–1 plots $\tilde{v}_d$ for θ_e increasing linearly with time. The two terms in Eq. (4-5) are evident as two sinusoidal components. For a constant θ_e, the output of a PD should be constant according to Eq. (4-1). But the second term of Eq. (4-5) varies with a frequency $2\omega_i$. In most PLL applications, this frequency is high enough that the second term has no effect, or the second term is removed by a filter. In any case, the first term is considered to be the output v_d of the PD:

$$v_d = 0.5K_m V_i V_o \sin(\theta_e) \tag{4-6}$$

Thus, the symbol v_d we have been using for the output of a PD is actually the *average* of the complete output $\tilde{v}_d$. This average is taken over a long enough period to eliminate the $2\omega_i$ component, but not so long as to affect the relationship in Eq. (4-6) when θ_e is a function of time. The accepted convention is to speak of v_d as the PD output voltage, but the designer should not completely lose sight of the second term in Eq. (4-5). Its frequency ω_d—the *detector frequency*—is twice the input frequency ω_i.

The notation in Eq. (4-6) can be simplified as

$$v_d = V_{dm} \sin(\theta_e) \tag{4-7}$$

where the maximum value of v_d is

$$V_{dm} = 0.5K_m V_i V_o \tag{4-8}$$

This sinusoidal characteristic is shown in Fig. 4–1c. For small values of θ_e, $\sin(\theta_e) \approx \theta_e$, and

$$v_d \approx 0.5K_m V_i V_o \theta_e \tag{4-9}$$

Comparing Eqs. (4-1) and (4-9), we see that the PD gain for small values of θ_e is

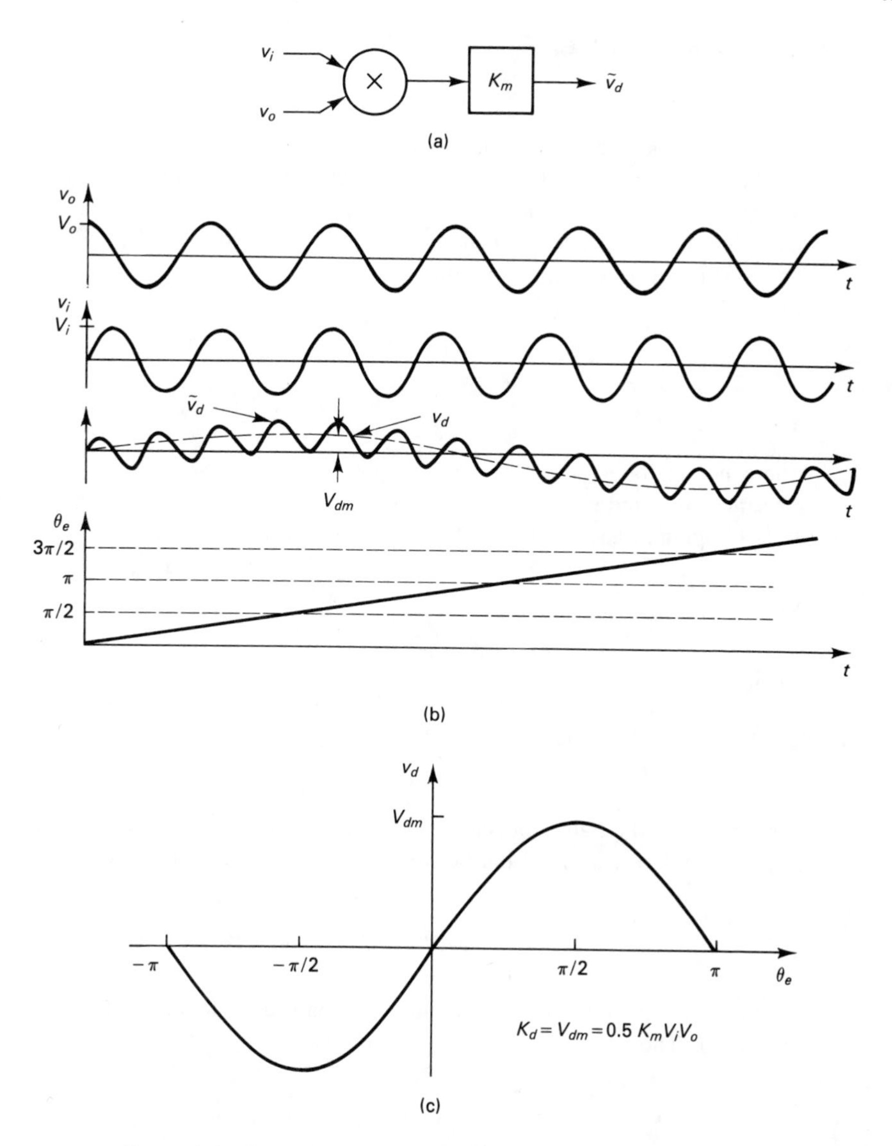

FIGURE 4–1 Four-quadrant multiplier phase detector with sinusoidal inputs

$$K_d = 0.5 K_m V_i V_o \tag{4-10}$$

Note that the PD gain depends on the amplitude of the input signals; it is not a property of the circuit alone.

The waveforms of a four-quadrant multiplier are illustrated in Fig. 4–1b. The adjective *four-quadrant* refers to the ability of the multiplier to handle both positive and negative values at both of its inputs.

4–2 GILBERT MULTIPLIER

One common implementation of a four-quadrant multiplier is the Gilbert multiplier circuit [1] shown in Fig. 4–2a. Here, v_o splits the current I to the left as i_1 or to the right as i_2 according to the characteristic shown in Fig. 4–2b. The current i_1 is split in turn by v_i according to the characteristic shown in Fig. 4–1c. A similar characteristic holds for i_2. The four resulting currents are combined to produce

$$\begin{aligned}\tilde{v}_d &= (i_4 + i_6)R_1 - (i_3 + i_5)R_2 \\ &= (i_4 + i_6 - i_3 - i_5)R\end{aligned}$$

where nominally $R_1 = R_2 = R$. (In practice, of course, R_1 and R_2 are not exactly the same.) If v_i and v_o are kept in the linear region of the characteristics in Figs. 4–2b and 4–2c (amplitude less than 52 mV), it can be shown that

$$\tilde{v}_d = K_m v_i v_o$$

where

$$K_m = RI/(52 \text{ mV})^2 \tag{4-11}$$

A mismatch in the transistors can cause input offsets V_{IO} of a few millivolts that add to v_i and v_o. Similarly, a mismatch between R_1 and R_2 causes an offset

$$V_{OO} = (R_1 - R_2)I/2 \tag{4-12}$$

to be added to the output. These relationships are summarized in the signal flow graph in Fig. 4–2d. The total expression for the output is

$$\begin{aligned}\tilde{v}_d &= K_m(v_i + V_{IO})\,(v_o + V_{IO}) + V_{OO} \\ &= K_m v_i v_o + K_m(V_{IO}v_i + V_{IO}v_o + V_{IO}^2) + V_{OO}\end{aligned}$$

Taking the time average [as when we went from Eq. (4-4) to Eq. (4-7)],

$$v_d = V_{dm}\sin(\theta_e) + K_m V_{IO}^2 + V_{OO}$$

The dc terms can be combined as an effective offset voltage at the PD output:

$$V_{do} = K_m V_{IO}^2 + V_{OO} \tag{4-13}$$

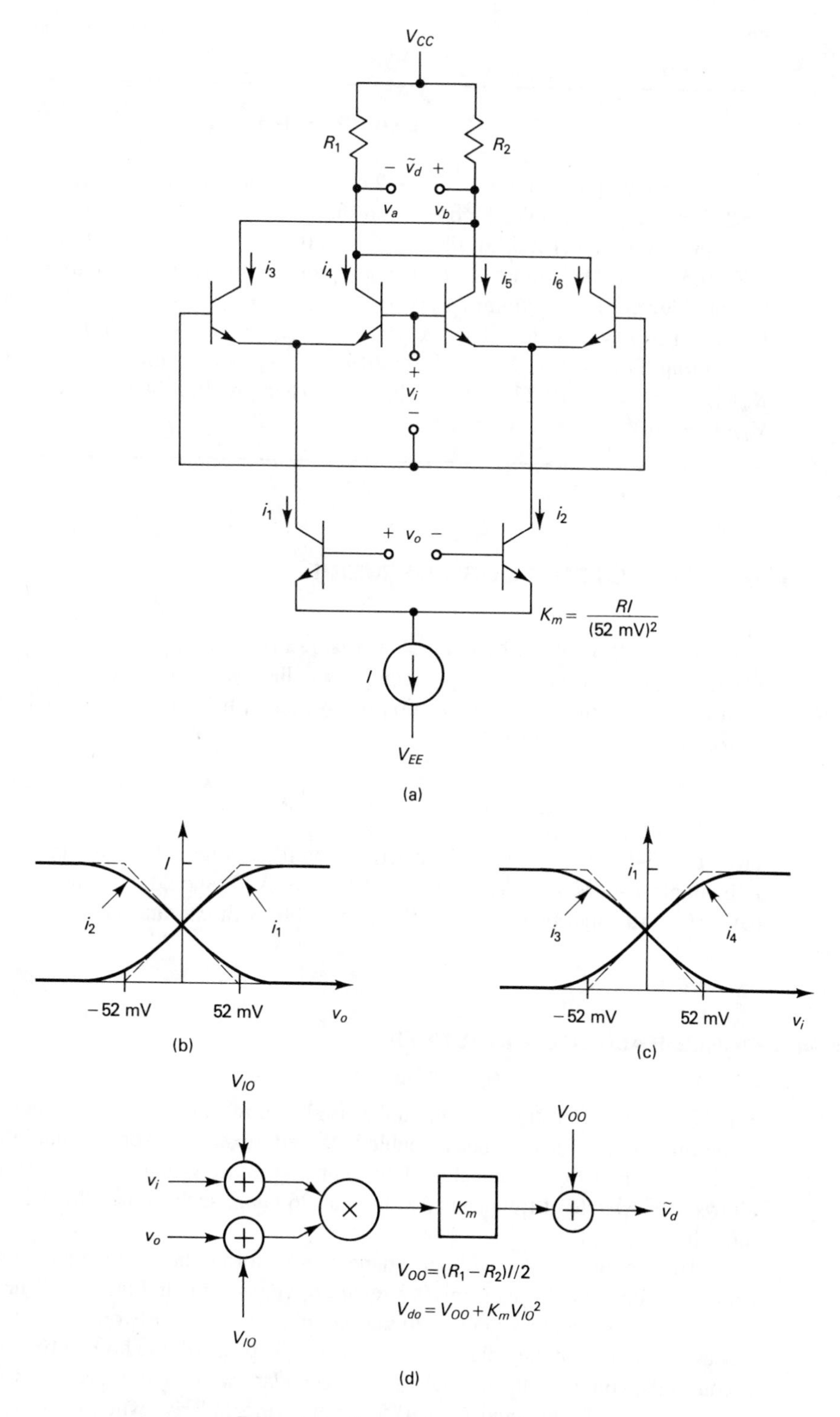

FIGURE 4–2 Four-quadrant multiplier circuit

EXAMPLE 4–1

The Gilbert multiplier circuit in Fig. 4–2a has $I = 2$ mA and $R = 5$ kΩ. R_1 and R_2 differ by 2%, and $V_{IO} = 5$ mV. Find V_{do} and the maximum K_d.

From Eq. (4-11) and Eq. (4-10), $K_m = (10\text{ V})/(52\text{ mV})^2 = 1/(0.27\text{ mV})$ and $K_d = V_iV_o/(0.54\text{ mV})$. Unfortunately, K_d is not a property solely of the circuit but depends on the input levels. The maximum K_d corresponds to $V_i = V_o = 52$ mV, the largest signals for which Eq. (4-8) holds. Then $K_d = (52\text{ mV})^2/(0.54\text{ mV}) = \underline{5.0\text{ V/rad}}$.

From Eq. (4-12), $V_{OO} = (0.02R)I/2 = (100\ \Omega)2\text{ mA}/2 = 100$ mV. Also, $K_mV_{IO}^2 = (5\text{ mV})^2/(0.27\text{ mV}) = 93$ mV. Then by Eq. (4-13), the total offset is $V_{do} = 93\text{ mV} + 100\text{ mV} = \underline{193\text{ mV}}$.

4–3 PHASE DETECTOR FIGURE OF MERIT

Is the $V_{do} = 193$ mV in Example 4–1 a large offset? It depends on how much useful voltage v_d the PD is capable of producing per radian, which is its gain K_d. Therefore, the ratio K_d/V_{do} is a meaningful indication of how small the offset is. We will call this the *figure of merit M* of the PD:

$$M \equiv K_d/V_{do} \tag{4-14}$$

From Eq. (3–7), $M = 1/\theta_{eo}$ with an active loop filter, where θ_{eo} is the static phase offset. In Example 4–1, $M = (5\text{ V/rad})/(193\text{ mV}) = 26$. A PD should reasonably be expected to have $M \geq 15$, and M as high as 500 is possible with careful matching.

4–4 DOUBLE-BALANCED MULTIPLIER

Another form of four-quadrant multiplier is shown in Fig. 4–3a. This circuit is called a diode ring mixer or sometimes a double-balanced mixer (a mixer is a multiplier). Unlike the multiplier in Fig. 4–2a, this circuit consists entirely of passive components. This allows it to operate at frequencies as high as 26 GHz, such as the DMS1-26A manufactured by Anzac. [2]

The circuit operates with any shape of waveforms, but its operation is most easily analyzed if one of the waveforms is a square wave, as the v_o in Fig. 4–3c. Then v_o may be considered a switching voltage, turning on either the bottom two diodes or the top two diodes depending on the polarity of v_o. When v_o is positive, the bottom two conduct, and v_x equals the voltage at the midpoint of the secondary winding of transformer T2, which is ground. Then $v_y = v_i$, and $\bar{v}_d = 0.5v_y = 0.5v_i$. Similarly, when v_o is negative, v_y is

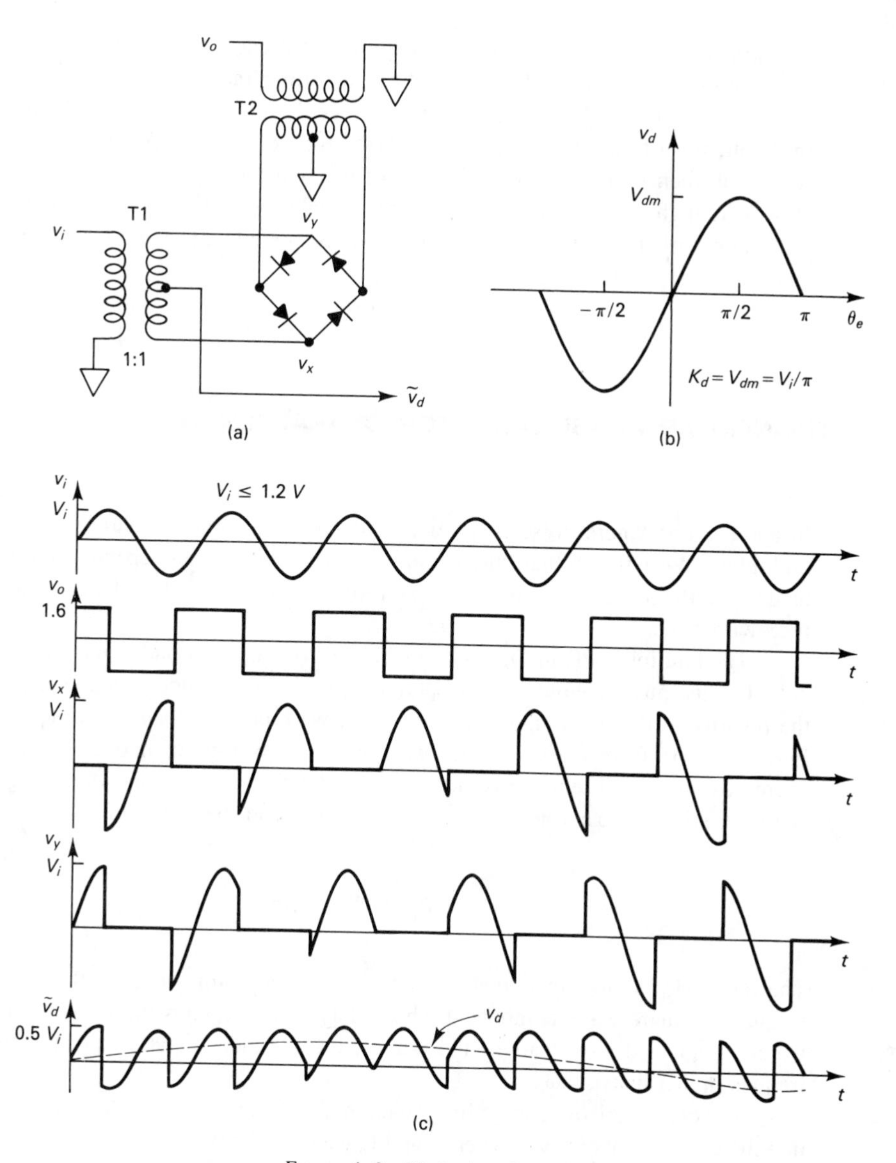

FIGURE 4–3 Diode ring phase detector

effectively grounded, and $\tilde{v}_d = -0.5v_i$. The resulting waveform of $\tilde{v}_d$ is shown in Fig. 4–3c. As θ_e increases with time, the average component v_d varies sinusoidally, corresponding to the PD characteristic in Fig. 4–3b. The maximum of the characteristic equals the PD gain:

$$V_{dm} = K_d = V_i/\pi \tag{4-15}$$

This assumes that both transformers have primary turns equal to secondary turns. For the best figure of merit, the signal amplitude V_i should be kept high to keep V_{dm} high. For the operation as described above, V_i should not be so large as to cause a diode pair to conduct. This corresponds to $V_i \leq 1.2$ V for the transformer ratio here.

With well-matched diodes in an integrated circuit, V_{do} can be kept to less than a millivolt, for a figure of merit $M \geq 400$ (see for example the Anzac MD-158). [3] The weak link then is usually the active loop filter with perhaps 5 mV contribution to V_{do}. However, high-performance op amps are available with V_{IO} as low as 0.1 mV.

For further discussion of double-balanced mixers, see the Anzac catalog [4] and Clarke and Hess. [5]

4–5 TRIANGULAR PHASE DETECTOR CHARACTERISTIC

In applications where noise is not a consideration, it is an advantage to overdrive the multiplier. We will see that this maximizes K_d, eliminates its dependence on the amplitude of the PD inputs, maximizes M, and provides a triangular PD characteristic, which is piecewise-linear.

The multiplier PD in Fig. 4–4a models the overdriven condition by a "slicer" at its output. This causes the output $\tilde{v}_d$ to saturate at $\pm V_{dm}$ for all input signal levels. Since only the polarity of the input signals matters now, we represent v_i and v_o as square waves in Fig. 4–4b. As θ_e increases linearly with time, the average component v_d increases and decreases linearly. The result is the triangular PD characteristic shown in Fig. 4–4c. The characteristic is linear for $-0.5\pi < \theta_e < 0.5\pi$, and the PD gain is

$$K_d = V_{dm}/0.5\pi \tag{4-16}$$

The output $\tilde{v}_d$, shown in Fig. 4–4b, has an average component v_d. It also has a high-frequency square wave component whose duty cycle depends on θ_e. This square wave, which is not a desired part of the output, has a fundamental frequency (the detector frequency ω_d) that is $2\omega_i$.

The characteristics for the four-quadrant multiplier in Fig. 4–2b and Fig. 4–2c show that the circuit is overdriven when v_i and v_o exceed 52 mV. For $v_i > 52$ mV and $v_o > 52$ mV, all of the current I is directed through R_1 in Fig. 4–2a; there is no current through R_2. Therefore $\tilde{v}_d = RI$, which is its maximum value:

$$V_{dm} = RI \tag{4-17}$$

From Eqs. (4-16) and (4-17) we have the PD gain $K_d = 2RI/\pi$, which no longer depends on V_i and V_o. It is also true that mismatches in the transistors do not contribute to V_{IO}. However, as we will see, asymmetry of the square waves effectively causes some V_{IO}.

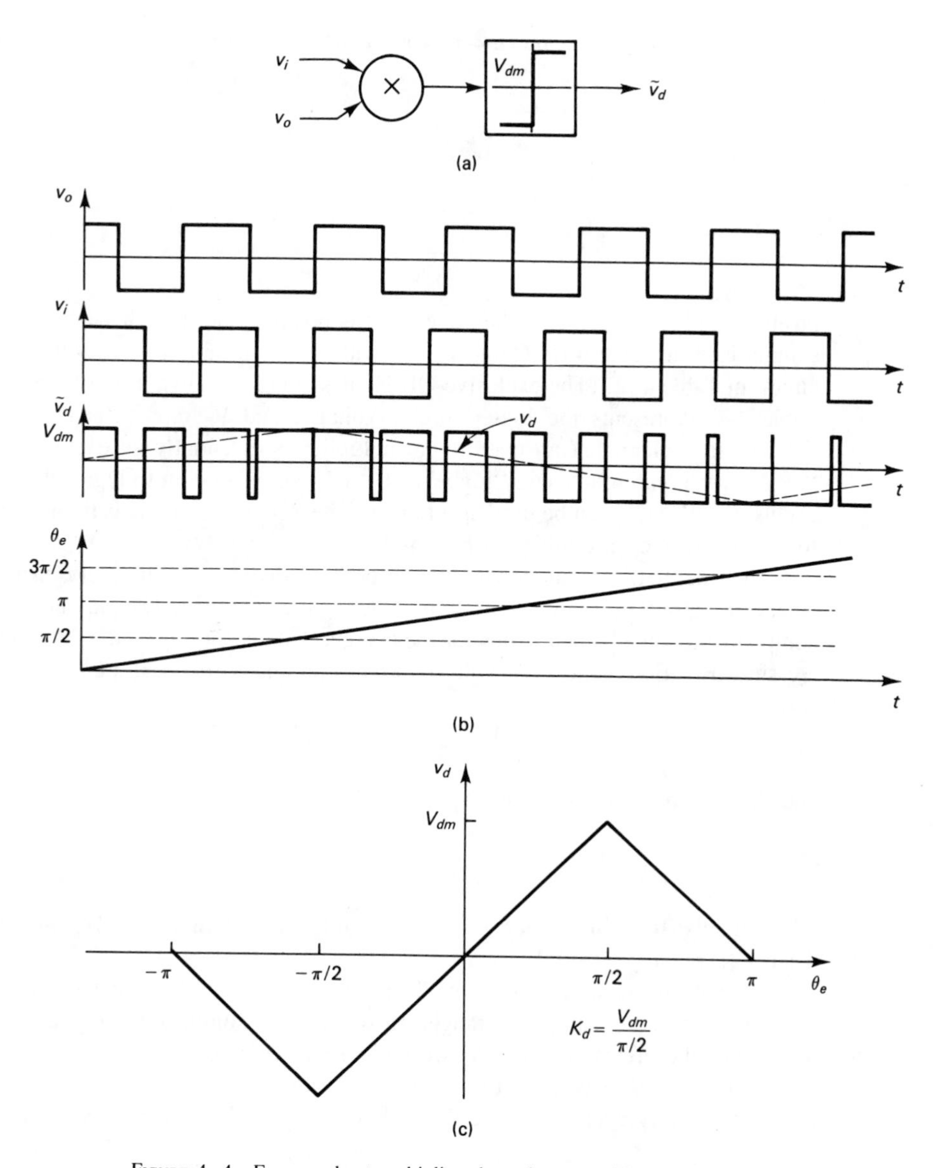

FIGURE 4–4 Four-quadrant multiplier phase detector with over-driven inputs

4–6 EXCLUSIVE-OR PHASE DETECTOR

An exclusive-OR logic circuit is essentially the same as an overdriven multiplier circuit. When overdriven, the multiplier output is saturated at either a positive value, corresponding to a logic "high," or a negative voltage, corresponding to a logic "low." For a multiplier, the output $\tilde{v}_d$ is positive when both inputs v_i and v_o are negative or both are

TABLE 4–1 MULTIPLIER TRUTH TABLE

v_i	v_o	$\tilde{v}_d$
−	−	+
−	+	−
+	−	−
+	+	+

positive, and $\tilde{v}_d$ is negative when one input is positive and the other is negative. This is summarized in Table 4–1. Compare this with the truth table for an exclusive-OR/NOR circuit in Table 4–2. (The exclusive-OR/NOR symbol is shown in Fig. 4–5a.) The V_H in Table 4–2 represents the logic "high" voltage, and V_L represents the logic "low" voltage. It is clear that an overdriven multiplier is essentially an exclusive-NOR with "+" corresponding to logic "high" and "−" corresponding to logic "low." Then an exclusive-OR/NOR can be used as a PD, and the PD characteristic is triangular, as it was for the overdriven multiplier in Fig. 4–4.

In order to obtain an output v_d that is both positive and negative, it is usual to use the balanced output $\tilde{v}_d = \tilde{v}_b - \tilde{v}_a$, where $\tilde{v}_a$ is the exclusive-OR output and $\tilde{v}_b$ is its complement. The characteristics for the average voltages v_a, v_b, and v_d as functions of θ_e are shown in Fig. 4–5c. The voltages corresponding to $\theta_e = 0$ are

$$V_{ao} = V_{bo} = (V_H + V_L)/2 \tag{4-18}$$

and the maximum v_d is given by

$$V_{dm} = V_H - V_L \tag{4-19}$$

Alternatively, the v_a characteristic can be used together with a $V_r = V_{ao}$, as in Fig. 3–5b. Then $V_{dm} = (V_H - V_L)/2$.

Again, the advantages of using the digital exclusive-OR as a PD are greater K_d, less V_{IO}, and greater linear phase range. However, the nonlinearity of the digital circuit aggravates the effect of noise, as we will see in Chapter 6.

So far we have assumed the square waves at the input are symmetrical—that is, they have a 50% duty cycle. Suppose that v_i has a duty cycle of $\delta_i = 0.5$ and v_o has a duty

TABLE 4–2 EXCLUSIVE OR/NOR TRUTH TABLE

v_i	v_o	$\tilde{v}_a$	$\tilde{v}_b$
V_L	V_L	V_L	V_H
V_L	V_H	V_H	V_L
V_H	V_L	V_H	V_L
V_H	V_H	V_L	V_H

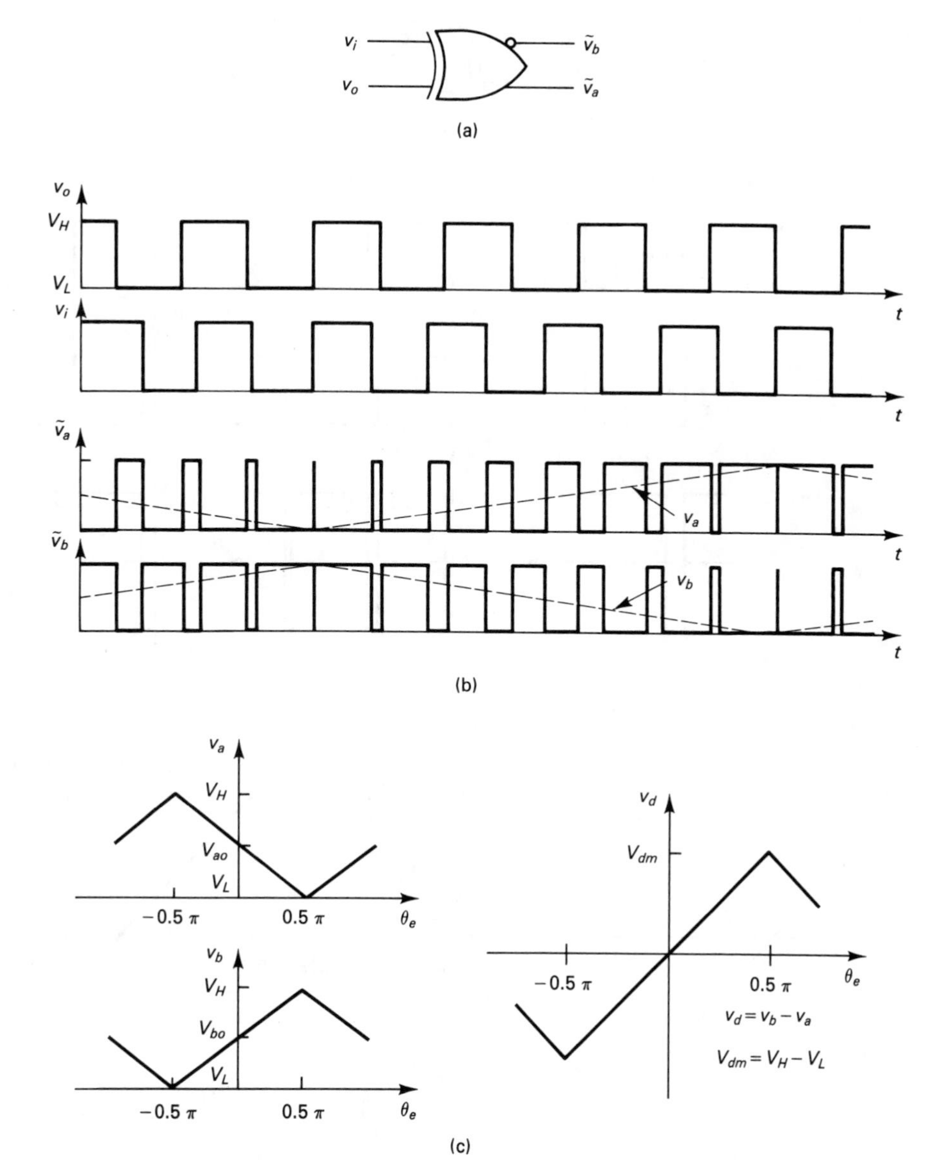

FIGURE 4–5 Exclusive-OR/NOR as a phase detector

cycle of $\delta_o = 0.4$, as in Fig. 4–6b. The effect of $\delta_o \neq 0.5$ is to produce a nonzero free-running voltage and to reduce V_{dm} of the PD characteristic.

The free-running voltage V_{do} is defined as the PD (average) output voltage when there is no input at v_i. For a logic signal, this means that v_i is always "low," or $v_i = V_L$. Then $\tilde{v}_d = V_H - V_L$ when $v_o = V_L$, and $\tilde{v}_d = V_L - V_H$ when $v_o = V_H$. Therefore, the average of $\tilde{v}_d$ is

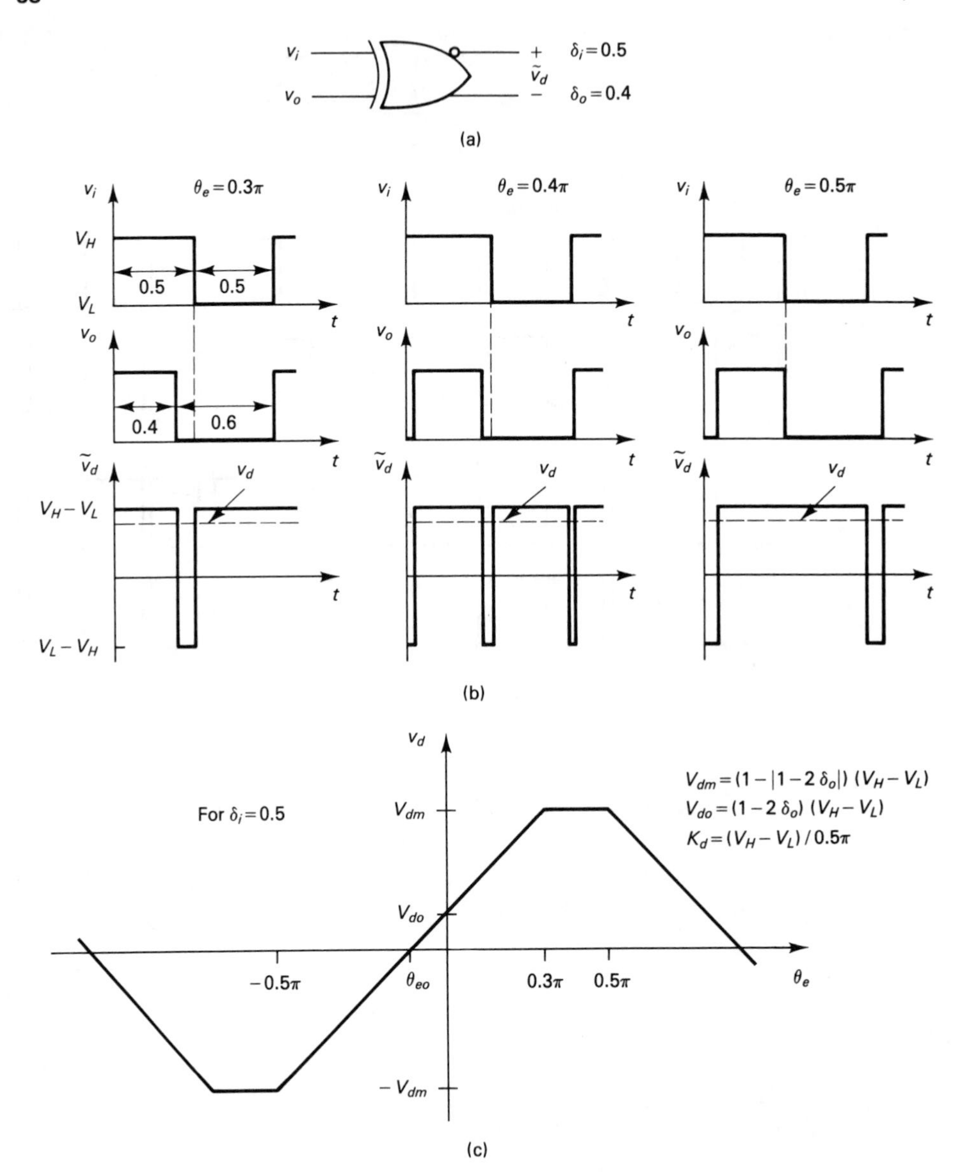

FIGURE 4–6 Exclusive-OR phase detector with $\delta_o \neq 0.5$

$$\begin{aligned} V_{do} &= (1 - \delta_o)(V_H - V_L) + \delta_o(V_L - V_H) \\ &= (1 - 2\delta_o)(V_H - V_L) \end{aligned} \tag{4-20}$$

Note that V_{do} is not zero for $\delta_o \neq 0.5$.

The other effect of $\delta_o \neq 0.5$ is to reduce V_{dm}. Because v_i and v_o no longer have the same waveform, there is no phase for which they exactly match each other and produce $v_d = V_H - V_L$. The best match is when $v_o = V_H$ occurs only during the time $v_i = V_H$.

Figure 4–6b shows three phases for which this is true. In all cases, the average of $\tilde{v}_d$ is the same: $v_d = 0.8(V_H - V_L)$. This is also the maximum of v_d, so $V_{dm} = 0.8(V_H - V_L)$ for $\delta_o = 0.4$. In general,

$$V_{dm} = (1 - |1 - 2\delta_o|)\,(V_H - V_L) \tag{4-21}$$

For $\delta_o \neq 0.5$, this is less than the V_{dm} given by Eq. (4-19). The PD characteristic in Fig. 4–6 shows that the peak of the function has been truncated; for $\delta_o = 0.4$, the function is flat for $0.3\pi \leq \theta_e \leq 0.5\pi$. The PD gain, however, is not a function of δ_o:

$$K_d = (V_H - V_L)/0.5\pi \tag{4-22}$$

This agrees with Eqs. (4-16) and (4-19) for the case $\delta_o = 0.5$.

4–7 TWO-STATE PHASE DETECTOR

A circuit with two states such as a set-reset flip-flop can be used to realize a PD with a characteristic that is linear over a range of $\pm\pi$, as shown in Fig. 4–7d. The two states are represented in Fig. 4–7a, where R and V are the two inputs, and the arrow $\uparrow$ indicates a rising edge on the input. A rising edge on R causes the circuit to go to State 2 ($\tilde{v}_d$ positive), and a rising edge on V causes the circuit to go to State 1 ($\tilde{v}_d$ negative). The timing diagram in Fig. 4–7c shows how the average component v_d varies as θ_e increases linearly. It rises continuously over the whole 2π range, resulting in the "sawtooth" PD characteristic in Fig. 4–7d. The gain is

$$K_d = V_{dm}/\pi = (V_H - V_L)/\pi \tag{4-23}$$

The circuit in Fig. 4–7b realizes an edge-triggered set-reset flip-flop. Input v_i is connected to R, and v_o is connected to V. A rising edge on v_i causes $Q_1 = Q_2$, so $\tilde{v}_d$ is "high" (the "set" state). A rising edge on v_o causes $Q_2 = \overline{Q}_1$, so $\tilde{v}_d$ is "low" (the "reset" state).

One advantage of the two-state PD is the increase in the linear range—double that of an exclusive-OR PD. Another is that the duty cycles of v_i and v_o are not important—only their rising edges.

One disadvantage of the circuit is that it is more sensitive to noise than the exclusive-OR PD. In effect, the flip-flops remember an error due to noise, while the exclusive-OR is memoryless. Another disadvantage is that the high-frequency ω_d of $\tilde{v}_d$ (see Fig. 4–7c) is only ω_i for the two-state PD, rather than the $2\omega_i$ for the exclusive-OR PD. In Chapter 9 it is shown that a lower ω_i can cause more spurious phase modulation because it is closer to the cutoff frequency K of the PLL.

Our standard definition of free-running voltage V_{do} is the v_d when there is no signal at v_i. But for the two-state PD this would result in $v_d = V_{dm}$, which is far from the desired

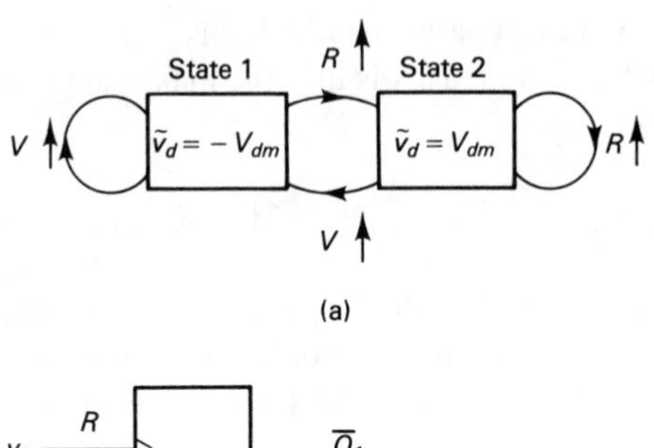

(a)

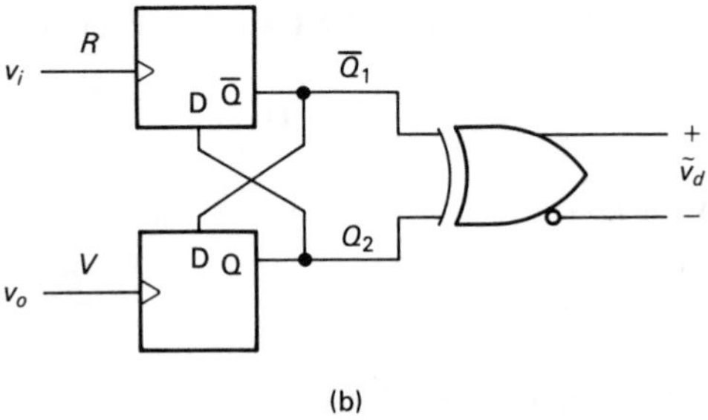

(b)

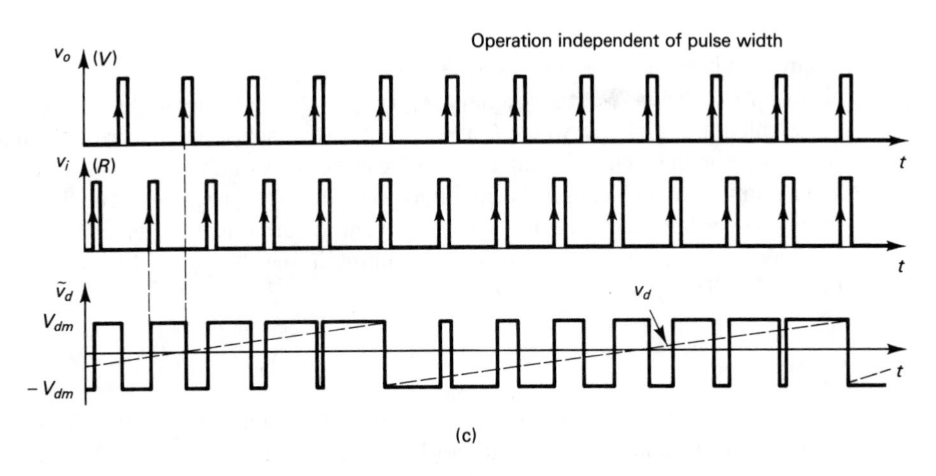

(c)

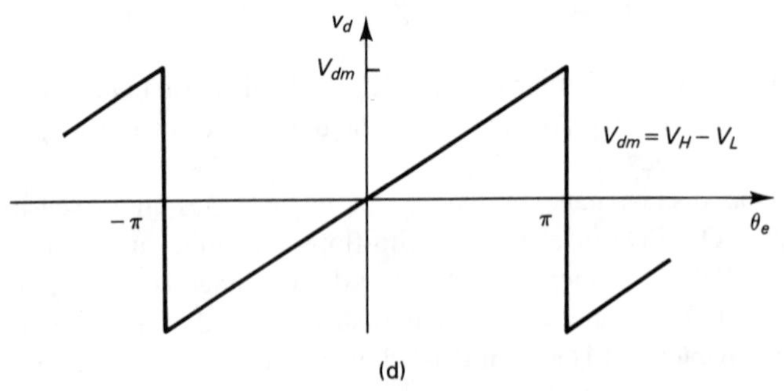

(d)

FIGURE 4–7 Two-state phase detector

value of zero. A more useful definition of V_{do} when studying acquisition is v_d averaged over all θ_e. From Fig. 4–7c, this is given by

$$V_{do} = (V_{dm} - V'_{dm})/2$$

where $V_{dm} = V_{Hb} - V_{La}$ is the maximum, and $-V'_{dm} = V_{Lb} - V_{Ha}$ is the minimum. Then

$$V_{do} = (V_{Hb} - V_{Ha} + V_{Lb} - V_{La})/2 \tag{4-24}$$

where the second subscript a or b refers to the logic voltage at $\tilde{v}_a$ or $\tilde{v}_b$, respectively. Ideally, $V_{do} = 0$ when all logic levels are matched.

4–8 THREE-STATE PHASE DETECTOR

The concept of an n-state PD may be extended to as many states as desired. The three-state PD is widely used because it is simple, has a linear range of $\pm 2\pi$ radians, and can act as both phase detector and frequency detector.

A state diagram for the circuit is shown in Fig. 4–8a. Again, states are changed on rising edges of R and V—R moving to higher states, and V moving to lower states. Suppose the circuit is initially in State 1. Then alternate rising edges on R and V will cycle between states 1 and 2. If V is constantly falling behind R in phase, as in the timing diagram in Fig. 4–8c, then eventually there will be two R rising edges without an intervening V rising edge. This will take the circuit to State 3, and thereafter it will cycle between State 2 and State 3.

The corresponding PD characteristic in Fig. 4–8d grows linearly over a range of 4π radians. Thereafter it remains positive, repeating a cycle every 2π radians. If θ_e decreases, the characteristic decreases linearly over a range of 4π radians. Thereafter it remains negative, repeating a cycle every 2π radians.

The action of a three-state PD as a frequency detector is now clear. For $\omega_i > \omega_o$, θ_e increases with time, and v_d remains positive. For $\omega_i < \omega_o$, θ_e decreases with time, and v_d remains negative. This is a great aid in acquiring lock when the two frequencies are initially different. The details of acquisition with a three-state PD are analyzed in Chapter 8.

The frequency detector action requires that the PD characteristic be multiple-valued (see Fig. 4–8d). This makes it unsuitable in applications where a pulse on v_i may be missed, causing v_d to jump in value by V_{dm}. Therefore, a three-state PD can't be used in high-noise situations or for clock recovery from data. The consequences of a multiple-valued characteristic are examined further in section 4–13.

For $\theta_e = 0$, the rising edges of R and V are coincident, and the PD remains in State 2 almost all the time; there are brief excursions to State 1 or State 3 as V or R comes slightly earlier. Figure 4–8a shows that in State 2 the output is $v_d \equiv v_U - v_D = V_L - V_L = 0$, as desired for $\theta_e = 0$. But in practice, the two V_Ls are not identical, and there is some offset:

$$V_{do} = V_{Lb} - V_{La} \tag{4-25}$$

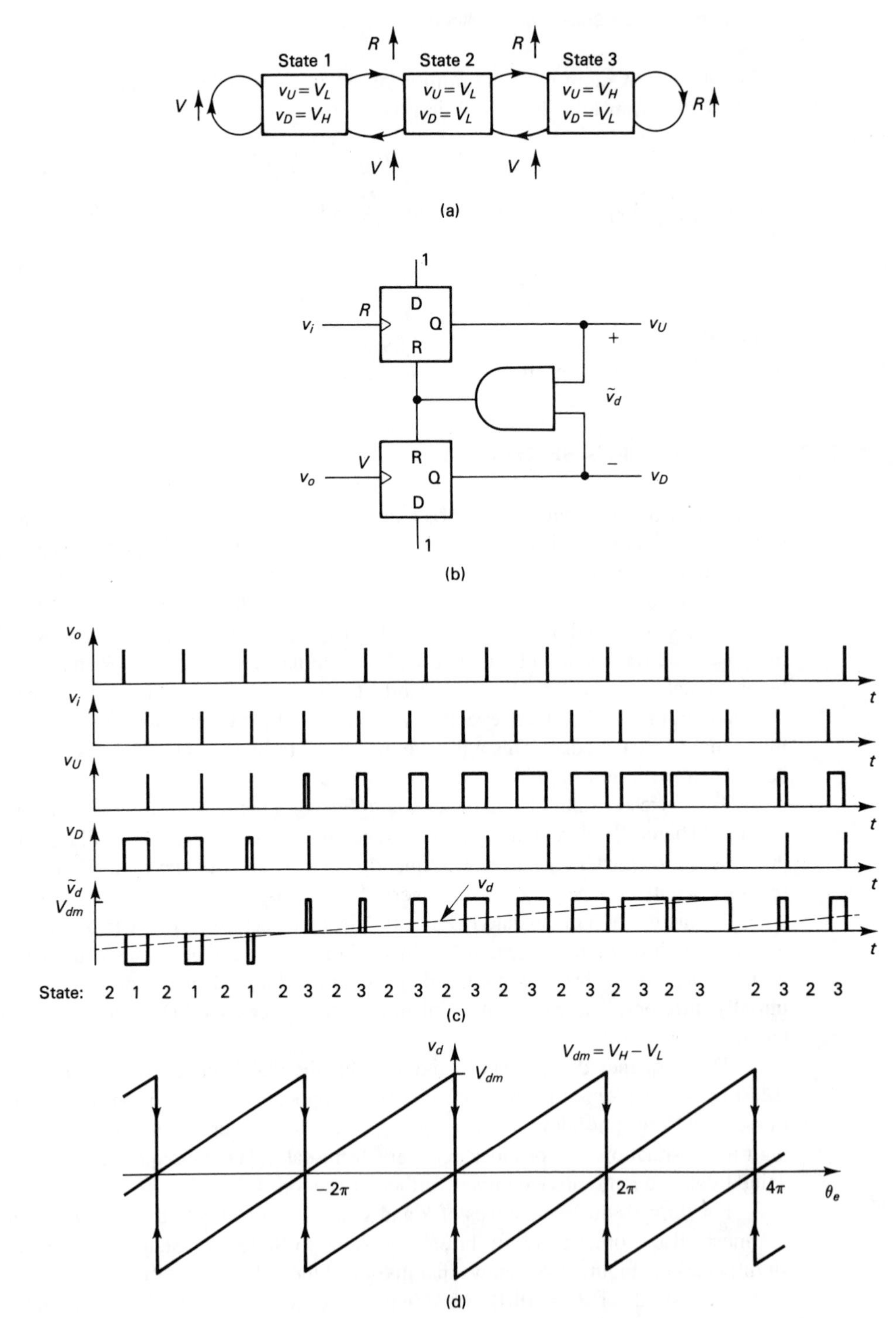

FIGURE 4–8 Three-state phase detector

The maximum v_d corresponds to the PD at State 3 almost all the time. There, $v_d = V_{dm} = V_H - V_L$. Then from Fig. 4–8c the PD gain is

$$K_d = (V_H - V_L)/2\pi \tag{4-26}$$

One realization of the three-state PD is that devised by Shahriary et al. [6] shown in Fig. 4–8b. Suppose v_D and v_U are low (state 2) initially. A rising edge on v_i causes v_U to go high (state 3). Then when a rising edge of v_o occurs, both v_D and v_U are high for an instant. Within a couple propagation delays, the AND gate has reset both flip-flops, and both v_D and v_U are low (State 2 again). This is shown more clearly in the expanded view of the waveforms in Fig. 4–9a. Note that there is a short transient state where both v_D and v_U are high. However, $\tilde{v}_d = 0$ for both State 2 and the transient state, so they can effectively be lumped together as State 2.

The maximum useful frequency for this PD is limited by the minimum duration of State 2. The duration π of State 2 is given by $\pi = \tau_H + \tau v_L$, where τ_H is the duration while $v_U = v_d = V_H$, and τ_L is the duration while $v_U = v_D = V_L$ (see Fig. 4–9a). Then the minimum duration of State 2 is

$$\tau_{\min} = \tau_H + \tau_{L\min}$$

where τ_H is the propagation delay of the AND gate plus that of the flip-flop from R to Q, and $\tau_{L\min}$ is the propagation delay of the AND gate plus the recovery time of the flip-flop from a reset. Therefore, the greatest duty cycle of time spent in State 3 is

$$\delta_{\max} = 1 - \tau_{\min}/T = 1 - \tau_{\min}\omega_i/2\pi$$

where $T = 2\pi/\omega_i$ is the period of the input frequency. Then the maximum v_d is $V_{dm} = \delta_{\max}(V_H - V_L)$, and the maximum phase θ_{em} within the linear range is given by

$$\theta_{em} = 2\pi\,\delta_{\max} = 2\pi - \tau_{\min}\omega_i \tag{4-27}$$

(See Fig. 4–9b) Note that the PD gain remains $K_d = V_{dm}/\theta_{em} = (V_H - V_L)/2\pi$ as in Eq. (4-26).

EXAMPLE 4–2

A three-state PD realized as in Fig. 4–8b has an input frequency $\omega_i = 2\pi \times 20$ MHz, an AND gate propagation delay of 2 ns, a flip-flop propagation delay (R to Q) of 3 ns, and a flip-flop recovery time after reset of 4 ns (this last specification is often not supplied by the manufacturer). Find the usable phase range of the PD.

Adding the propagation delays gives $\tau_H = 2 + 3 = 5$ ns and $\tau_{L\min} = 2 + 4 = 6$ ns. Then $\tau_{\min} = \tau_H + \tau_{L\min} = 11$ ns, and $\theta_{em} = 2\pi - (11\text{ ns})(125.6\text{ Mrad/s}) = 4.9$ radians. This is 78% of the ideal 2π radians phase range.

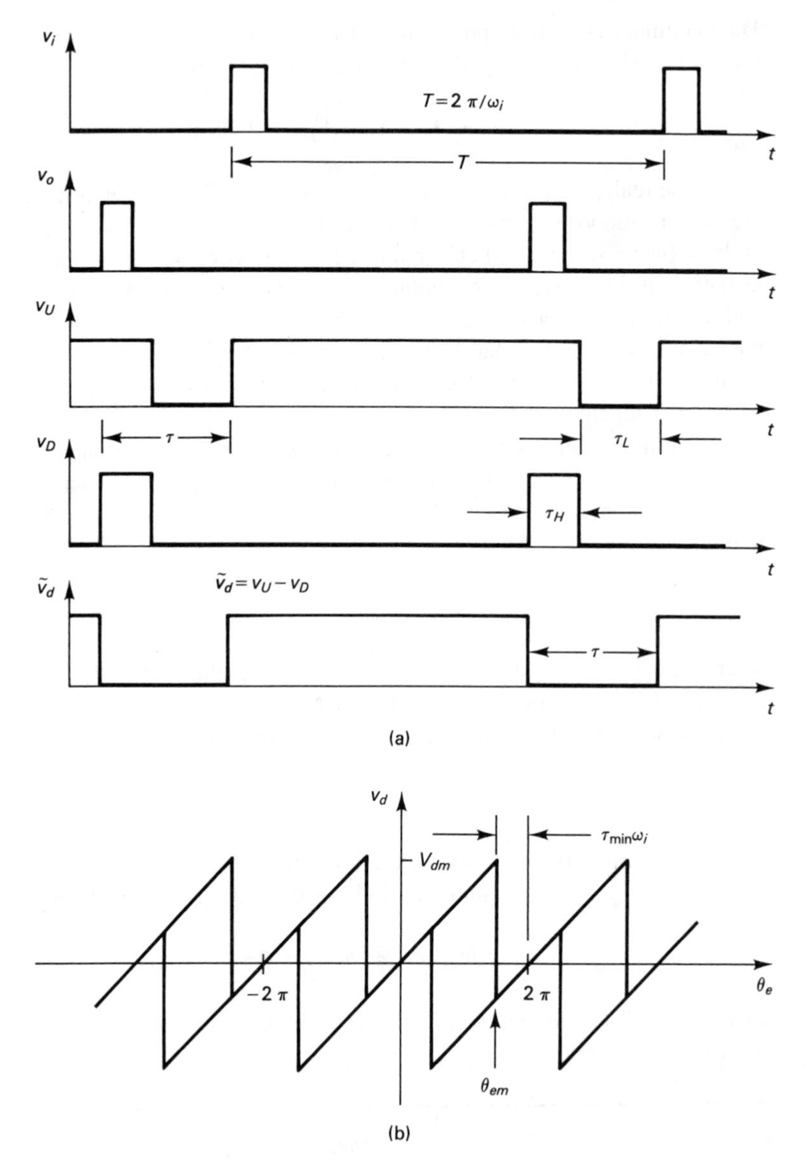

FIGURE 4–9 Three-state phase detector characteristic at high frequencies

Another realization of the three-state PD is manufactured commercially by Motorola [7] in both TTL (the MC4044) and ECL (the MC12040). However, Egan and Clark [8] have shown that these do not make a smooth transition from negative v_d to positive v_d, causing a nonlinearity in the PD characteristic at the origin. Therefore, the realization in Fig. 4–8b is preferred.

4–9 Z-STATE PHASE DETECTOR

A variation of the three-state PD is to have the v_D and v_U outputs drive two CMOS switches, as in Fig. 4–10a. This circuit is intended to be used with a passive filter, as shown. The result is equivalent to an active filter with very low offset. There is no established name for such a PD; we suggest *Z-state PD*, with ''Z'' referring to the high-impedance state. A commercial version of such a PD is the 74HC4046 manufactured by Harris Semiconductor. [9]

In State 1, v_D is high, and the lower switch connects $\tilde{v}_d$ to ground. In State 2, neither v_D nor v_U is high, and neither switch is closed. Since no current flows through R_0, $\tilde{v}_d$ equals the v_3 across the capacitor. In State 3, v_U is high, and the upper switch connects $\tilde{v}_d$ to V_{DD}. The resulting $\tilde{v}_d$ waveform as θ_e increases is shown in Fig. 4–10b. Note that $\tilde{v}_d$ is not determined solely by the PD circuit; at times it equals the voltage v_3 in the loop filter.

The result is the PD characteristic shown in Fig. 4–10c. The slope K_d depends on v_3, and in general the K_d for $\theta_e > 0$ is different from that for $\theta_e < 0$. But the important feature is that with both switches open, the free-running voltage V_{do} equals v_3. But in the steady state, $v_3 = \tilde{v}_c \equiv V_{co}$. Therefore

$$V_{do} = V_{co}$$

For a passive filter $F(0) = 1$, and the static phase error given by Eq. (2–27) is

$$\begin{aligned}\theta_{eo} &= -V_{do}/K_d + V_{co}/K_dF(0) \\ &= (V_{co} - V_{do})/K_d = 0\end{aligned}$$

Thus, there is no static phase error, which is usually not true of a passive filter. In practice, though, the bias current I_B to the VCO is not zero. The resulting voltage drop I_BR_O across R_O leads to a small static phase offset

$$\theta_{eo} = I_BR_O/K_d \tag{4-28}$$

I_B can be kept to picoamperes with FET buffering at the VCO input. The result is a PD with an effective figure of merit M in the thousands!

Such a tremendous M is sometimes worth the nonlinearity that occurs in the PD characteristic at the origin. Another limitation, though, is that it takes time for $\tilde{v}_d$ to settle to v_3 when both switches are open—a result of some 25 pF of stray capacitance at the $\tilde{v}_d$ node that must discharge through R_0. This restricts the circuit to rather low-frequency applications when variation in θ_e is expected.

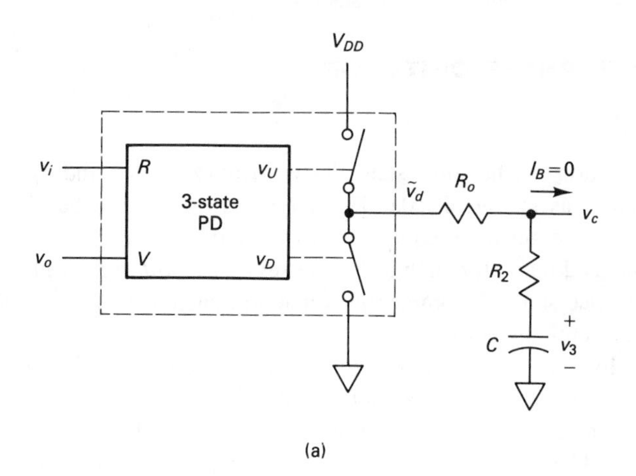

(a)

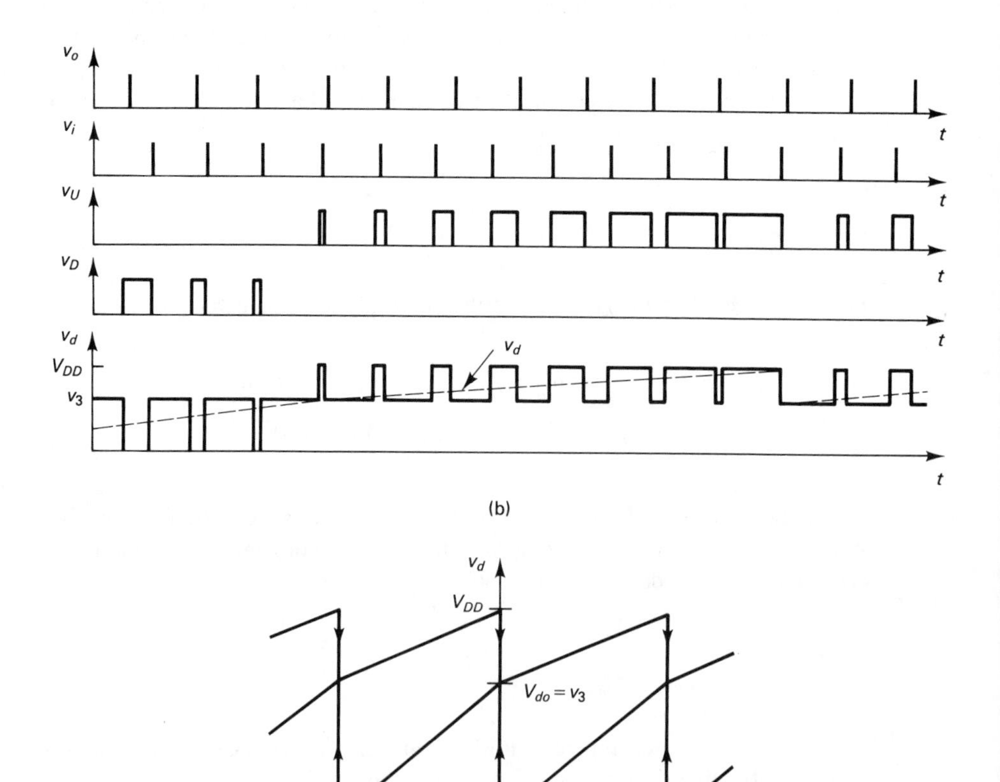

(b)

(c)

FIGURE 4–10 Z-state phase detector

4–10 SAMPLE-AND-HOLD PHASE DETECTOR

The sample-and-hold circuit shown in Fig. 4–11a can act as a phase comparator. Pulses on v_o cause a switch to close momentarily, charging the capacitor to the current values of v_i. The following buffer keeps the charge from leaking off the capacitor while the switch is open. As θ_e increases, successive values of v_i are sampled, and $\tilde{v}_d$ has the form of v_i but at a lower frequency. (This is the effect of aliasing.) Therefore, the form of the PD characteristic is sinusoidal if v_i is sinusoidal, and the characteristic is triangular if v_i is triangular.

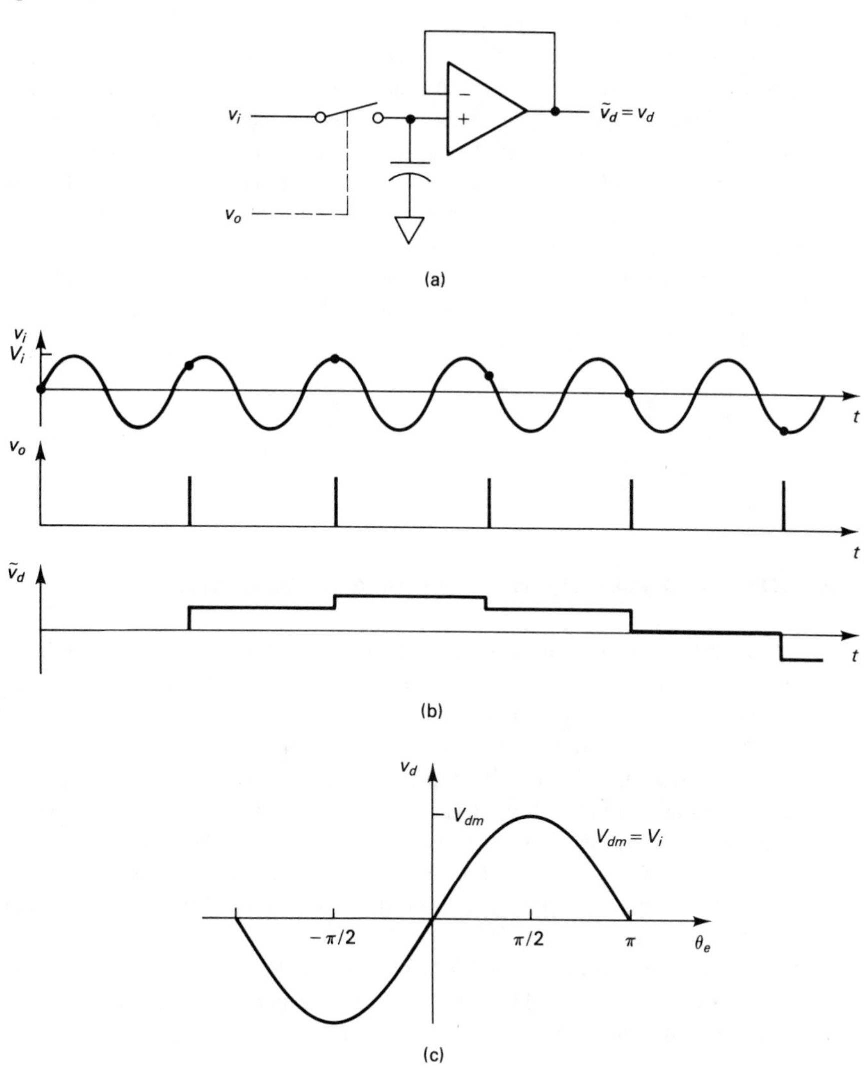

FIGURE 4–11 Sample-and-hold phase detector

The one advantage of a sample-and-hold PD is that $\bar{v}_d$ contains no high frequency at ω_i or $2\omega_i$. If the phase is constant, then $\bar{v}_d$ is a straight dc line. This is desirable in applications such as frequency synthesis, where any spurious modulation by $\bar{v}_d$ degrades the spectral purity of the synthesized frequency.

4–11 EXTENDED RANGE: FREQUENCY DIVISION

In PM demodulation applications and phase jitter smoothing applications, it is sometimes necessary for the PD to handle large values of θ_e. The largest range we have seen so far is that of the three-state PD with $\pm 2\pi$ radians. With more complex designs, there is no limit as to how wide the range can be made, but there are some tradeoffs.

The circuit in Fig. 4–12a uses the same two-state PD we analyzed in Fig. 4–7. But now there is a frequency divider in front of each input, so the frequencies at v_i' and v_o' are 1/3 those at v_i and v_o. The phase at v_o must slip three cycles relative to v_i in order for the phase at v_o' to slip one (long) cycle. This extends the range of the two-state PD from $\pm\pi$ to $\pm 3\pi$ radians (see Fig. 4–12c).

The high-frequency ω_d that appears at $\bar{v}_d$ (see Fig. 4–12b) is now $\omega_i/3$ rather than the ω_i for the two-state PD by itself. This makes it more difficult to keep ω_d out of the passband of the PLL. In general

$$\omega_d = \omega_i/N \tag{4-29}$$

for an extended range PD with $\div N$ frequency dividers.

4–12 EXTENDED RANGE: *n*-STATE PHASE DETECTOR

Another method of extending the range of a PD is to increase the number of states in an *n*-state PD beyond the three-state we have already seen. Oberst [10] refers to this as a "generalized phase comparator." The method for realizing four or more states involves combining a three-state PD with a shift register. [11]

Consider the six-state PD represented in Fig. 4–13a. The number of each state here refers to the number of output variables ($v_{\bar{D}}$, v_U, v_1, v_2, and v_3) in Fig. 4–13b that are high. For simplicity of notation, let 1 stand for logic high and 0 stand for logic low. Then the condition $v_{\bar{D}}v_Uv_1v_2v_3 = 10000$ and the condition $v_{\bar{D}}v_Uv_1v_2v_3 = 00100$ are both State 1. Rising edges on R move the circuit to a higher state, and rising edges on V move the circuit to a lower state.

The connection of the PD to the passive loop filter is shown in Fig. 4–13b. the analysis can be simplified by replacing the PD and the five resistors of value $5R_0$ with the Thevenin equivalent in Fig. 4–13c. The Thevenin voltage is

$$\bar{v}_d = k\, V_H/5 + (5 - k)V_L/5 \tag{4-30}$$

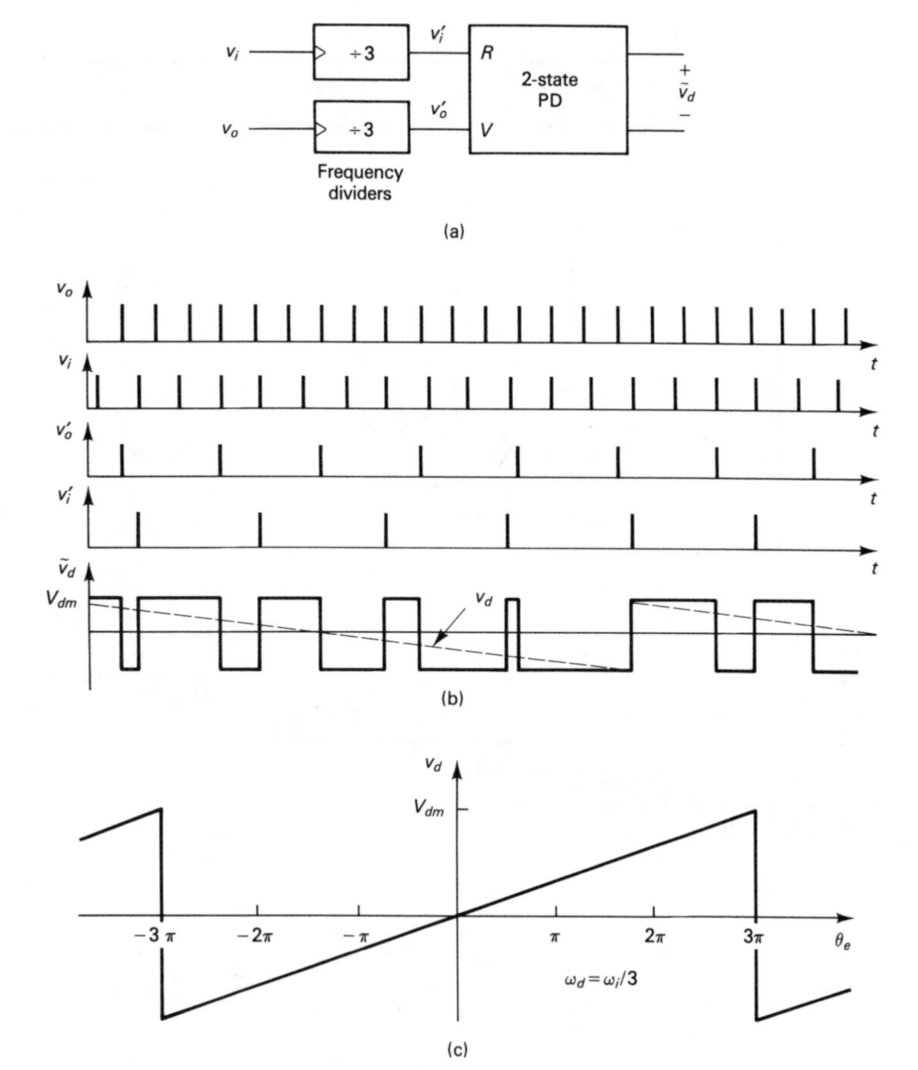

FIGURE 4–12 Extended range: frequency division

where k is the state (0 through 5) of the circuit, and V_H and V_L are the high and low logic voltages. The Thevenin resistance is R_0, which completes the loop filter. Fig. 4–13d shows $\tilde{v}_d$ as θ_e increases with time for the case $V_H = 5$ V and $V_L = 0$. The PD first alternates between states $k = 0$ and 1, then between 1 and 2, etc. The average value v_d correspondingly increases from 0 to 5 V as k goes from 0 to 5.

The resulting PD characteristic, shown in Fig. 4–13e, has a linear range of $\pm 5\pi$ radians. Note that as θ_e increases beyond 5π radians, v_d stays between 4 and 5 V, and as θ_e decreases beyond -5π radians, v_d stays between 0 and 1 V. This is similar behavior to the three-state PD, and it allows all n-state PDs for $n \geq 3$ to act as frequency detectors.

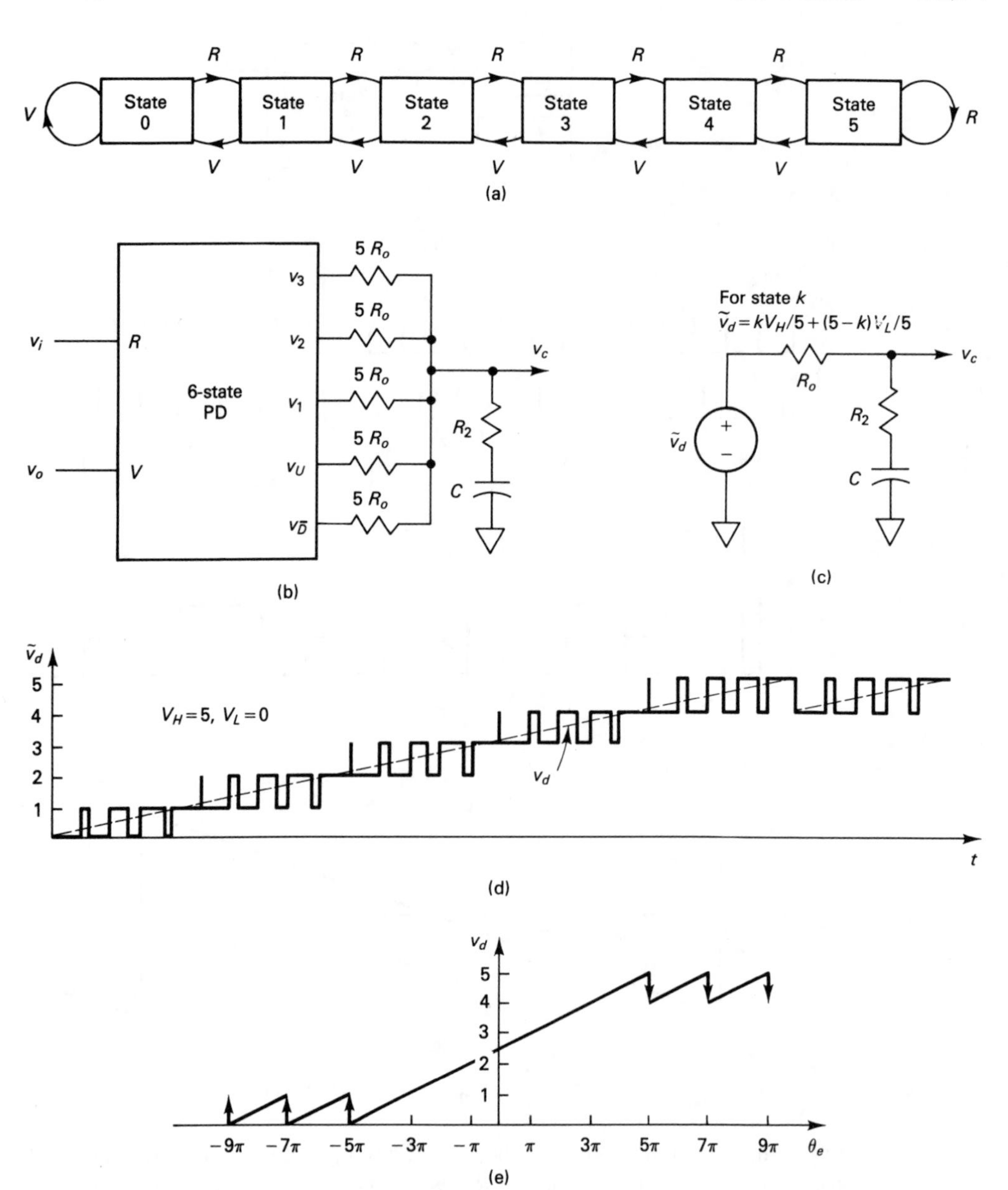

FIGURE 4–13 Extended range: n-state phase detector

The use of a six-state PD with an active loop filter is shown in Fig. 4–14a. Here, the five summing resistors with value $5R_1$ act in parallel as R_1 of the loop filter. The op amp is referenced to a V_r halfway between the maximum Thevenin voltage of V_H and the minimum Thevenin voltage of V_L:

$$V_r = (V_H - V_L)/2 \tag{4-31}$$

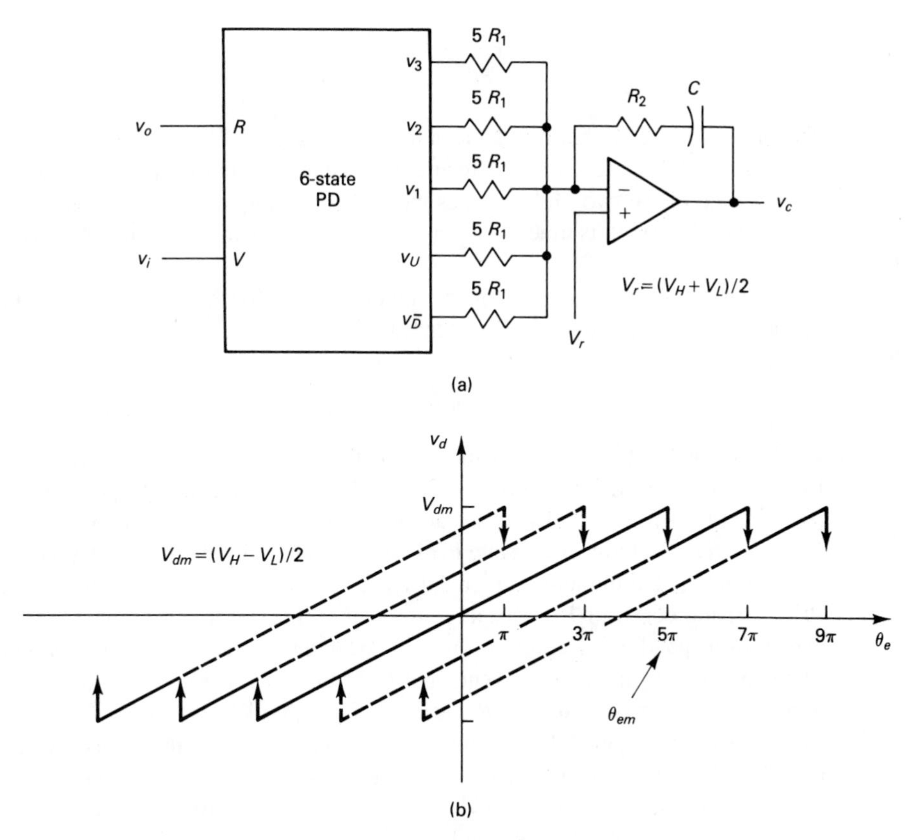

FIGURE 4–14 Six-state phase detector with active loop filter

The PD output is effectively

$$-\ \bar{v}_d = k\ V_H/5 + (5 - k)V_L/5 - V_r \tag{4-32}$$

[compare Eq. (4-30)]. The sign inversion introduced by the active loop filter has been compensated by reversing v_i and v_o at the inputs R and V to the PD. Otherwise the negative feedback would have become positive, and the PLL would be unstable.

The PD characteristic is shown in Fig. 4–14b. The maximum value is given by

$$V_{dm} = (V_H - V_L)/2 \tag{4-33}$$

The maximum value of the phase range is $\theta_{em} = 5\pi$. In general,

$$\theta_{em} = (n - 1)\pi \tag{4-34}$$

where n is the number of states. The PD gain is $(V_H - V_L)/10\pi$. In general,

$$K_d = \frac{V_{dm}}{\theta_{em}} = \frac{V_H - V_L}{(n - 1)2\pi} \tag{4-35}$$

The offset voltage V_{do} really has meaning only for n odd. Then it is defined as the value of v_d for the middle state, as it was for the three-state PD. For n even, the PD spends equal time at each of the two middle states when there is no phase modulation. For n odd, the PD spends almost all its time at the one middle state, greatly reducing the ac component of $\tilde{v}_d$.

The unwanted high-frequency component of $\tilde{v}_d$ for the n-state PD (see Fig. 4–13d) has a frequency (the detector frequency) of

$$\omega_d = \omega_i \tag{4-36}$$

as is true for the two-state and three-state PDs. This is a great improvement over the frequency divider method in Fig. 4–12, for which ω_d is inversely proportional to the range [see Eq. (4-29)]. The tradeoff is that the n-state PD has more complex circuitry.

A circuit realizing a six-state PD is shown in Fig. 4–15a. It includes a three-state PD with output v_U and output complement $v_{\overline{D}}$. Two "slip detectors" monitor the three-state PD for rising edges of R and rising edges of V that don't cause a state change. These events correspond to the end loops in Fig. 4–8a; the three-state PD slips 2π radians here. The slip detectors note these events and record them in the up-down (or left-right) shift register. When a rising edge on R causes a slip, a slip detector causes the shift register to shift up, shifting a logic "1" into the bottom stage. When a rising edge on V causes a slip, the other slip detector causes the shift register to shift down, shifting a logic "0" into the top stage. Thus, a slip either adds or subtracts a "1" from the contents of the shift register, which appear at v_1, v_2, and v_3. The "1" makes up for the 2π radians either lost or gained by the slip.

A state diagram of the six-state PD is shown in Fig. 4–15b. There are actually twelve states, but we lump together states that have the same number of "1's" at $v_{\overline{D}}$, v_U, v_1, v_2, and v_3. For example, State 2 comprises states 11000, 10100, and 00110. Movement between rows corresponds to a state change of the three-state PD, and movement along the top or bottom row corresponds to a slip changing the state of the shift register. Eventually the six-state PD itself will have a slip—the end loops at State 0 and State 5. These slips can be put off further by adding stages to the shift register; a four-stage shift register would raise the circuit to a seven-state PD with a linear range of $\pm 6\pi$ radians.

The realization of the slip detectors in Fig. 4–15a needs to be addressed. When a rising edge on R causes no state change of the three-state PD, this is defined as a slip. The state diagram of the three-state PD, shown in Fig. 4–8a, is redrawn in Fig. 4–16a. The transient State 2′ is shown explicitly here; it lasts for only a couple propagation delays and reverts spontaneously to State 2. Note that when R changes the state from 1 to 2 (via 2′) or when it changes the state from 2 to 3, it causes a rising edge on v_U (from V_L to V_H). When R occurs during State 3, it causes no state change (a slip) and no rising edge on v_U. Therefore, a slip is equivalent to a rising edge on R with no corresponding rising edge on v_U.

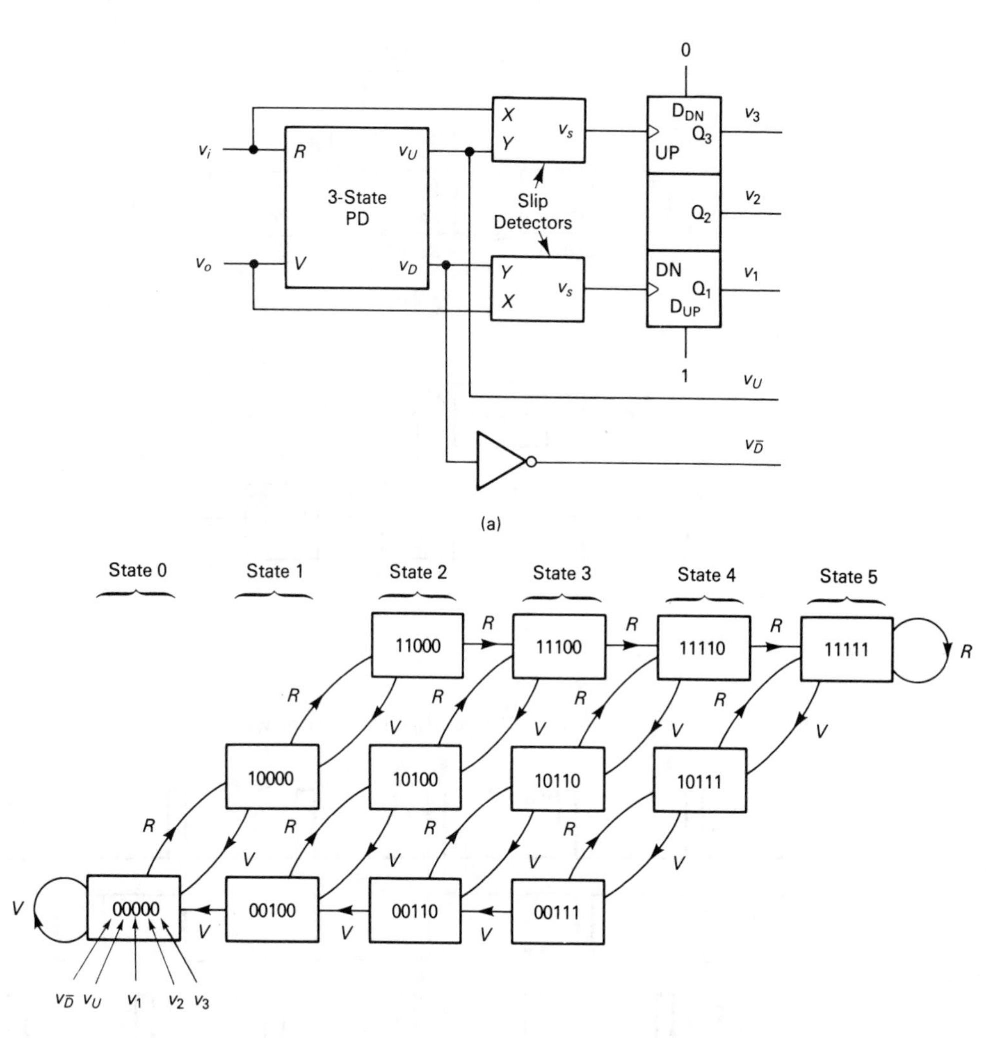

FIGURE 4–15 Realization of six-state phase detector

A circuit realizing the slip detector is shown in Fig. 4–16b. The first flip-flop records a rising edge on v_U as Z high. The rising edge on R (that caused the rising edge on v_U) samples Z after a delay of τ. (The delay can be either the propagation delay of a few gates, a delay line, or a monostable multivibrator.) It finds Z high, making v_0 low, and immediately resets Z (see Fig. 4–16c). If X', the delayed R, samples Z and finds it low, v_0 goes high, indicating no rising edge on v_U since the last time Z was reset. This is a slip, and it causes a pulse on v_s.

The delay τ must be long enough to allow a state change to propagate through the three-state PD and through the first flip-flop. (It must not be longer than the interval between rising edges on R.) This necessary delay causes a delay of τ between the occurrence of a slip and its detection at v_s. The effect is a τ-wide transient in v_d. In wide-

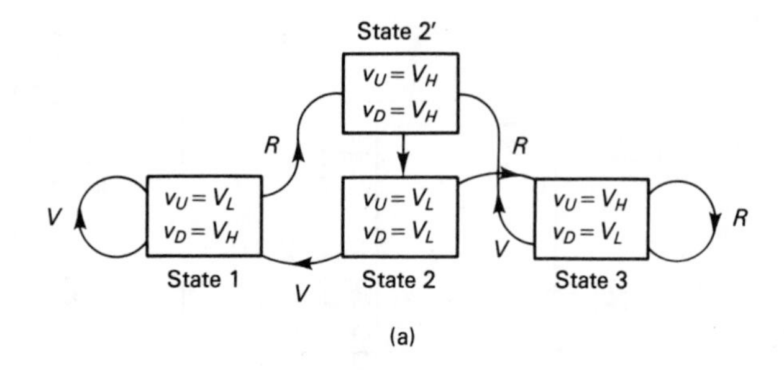

(a)

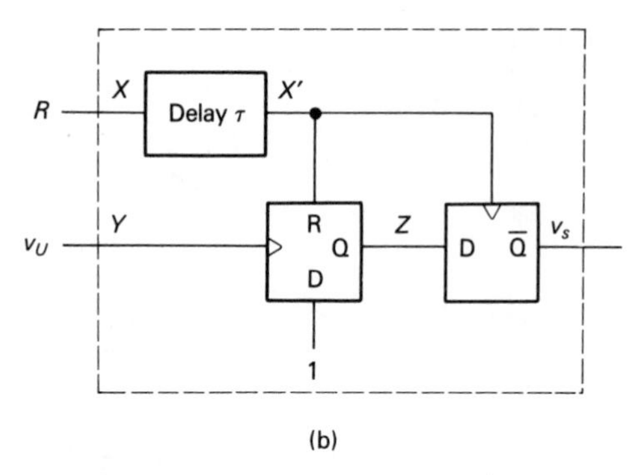

(b)

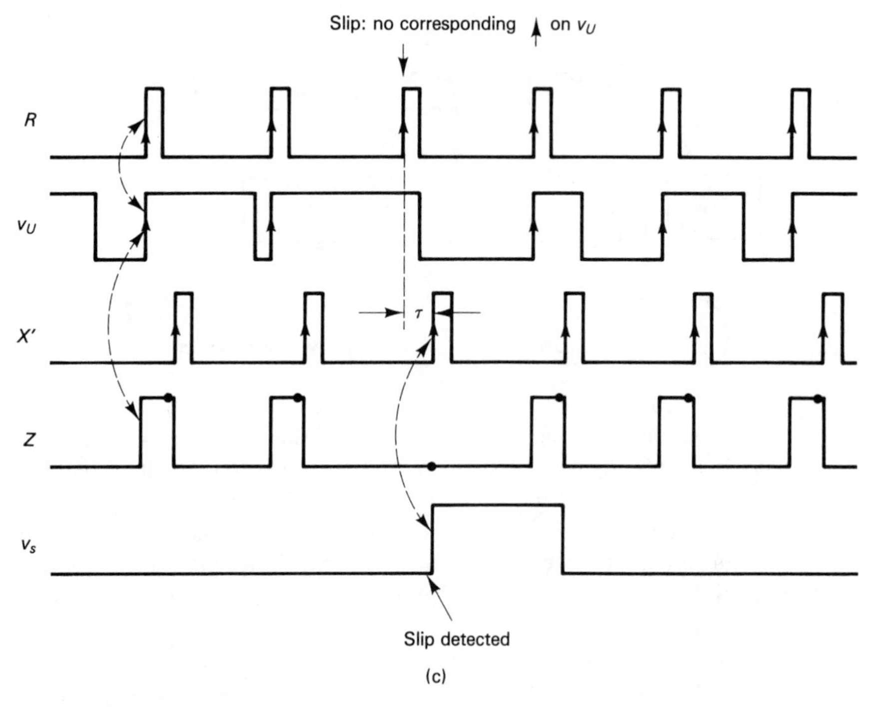

(c)

FIGURE 4–16 Slip detector circuit

band applications where it is not negligible, the transient can be eliminated by inserting delays of τ in the v_U and $v_{\overline{D}}$ lines in Fig. 4–15a.

4–13 MODIFIED PHASE DETECTOR CHARACTERISTIC

In Chapter 6 we study the PLL response to noise, in Chapter 8 the transient response during acquisition, and in Chapter 9 the response to modulation. In some situations, two or more of these conditions are present simultaneously. With a linear system this would be easily handled by superposition. But because of the nonlinearity of the PD, superposition is generally not valid for a PLL. In this section we analyze how the PD characteristic $v_d(\theta_e)$ behaves when there are two independent components to θ_e. Let the total phase error be

$$\theta_e = \theta_e' + \theta_e'' \tag{4-37}$$

where θ_e' is the component of interest and θ_e'' is the "interfering" component. For instance, θ_e' could be the phase error due to θ_i, and θ_e'' the phase error due to noise. We will see that the presence of θ_e'' effectively presents a modified PD characteristic $v_d'(\theta_e')$ to the other component θ_e'.

We will assume that θ_e'' has zero mean and that it has an even probability density function $p(\theta_e'')$. Then θ_e' can be thought of as the average of θ_e over the variable θ_e''. If the PD output $v_d(\theta_e)$ is averaged over θ_e'', then this average v_d' as a function of θ_e' is the *modified PD characteristic*:

$$v_d'(\theta_e') = \int_{-\infty}^{\infty} p(\theta_e'')\, v_d(\theta_d' + \theta_e'')\, d\theta''_e \tag{4-38}$$

If $v_d(\theta_e)$ were linear, the presence of θ_e' and θ_e'' together wouldn't affect the characteristic, and we would have $v_d'(\theta_e') = v_d(\theta_e')$, an unchanged PD characteristic. But for large enough θ_e'', the nonlinearity causes $v_d'(\theta''e) \neq v_d(\theta_e')$.

Because we have assumed that $p(\theta_e'') = p(-\theta_e'')$, Eq. (4-38) can be seen to be the correlation of p and v_d:

$$v_d'(\theta_e) = p(\theta_e) * v_d(\theta_e)$$

and therefore

$$V_d'(\omega) = P(\omega) \times V_d(\omega) \tag{4-39}$$

where V_d', P, and V_d are the Fourier transforms of v_d', p, and v_d. Use of the Fourier transform usually simplifies the calculation of $v_d'(\theta_e)$.

Figure 4–17a shows an example of obtaining the modified PD characteristic $v_d'(\theta_e')$ from a sinusoidal PD characteristic

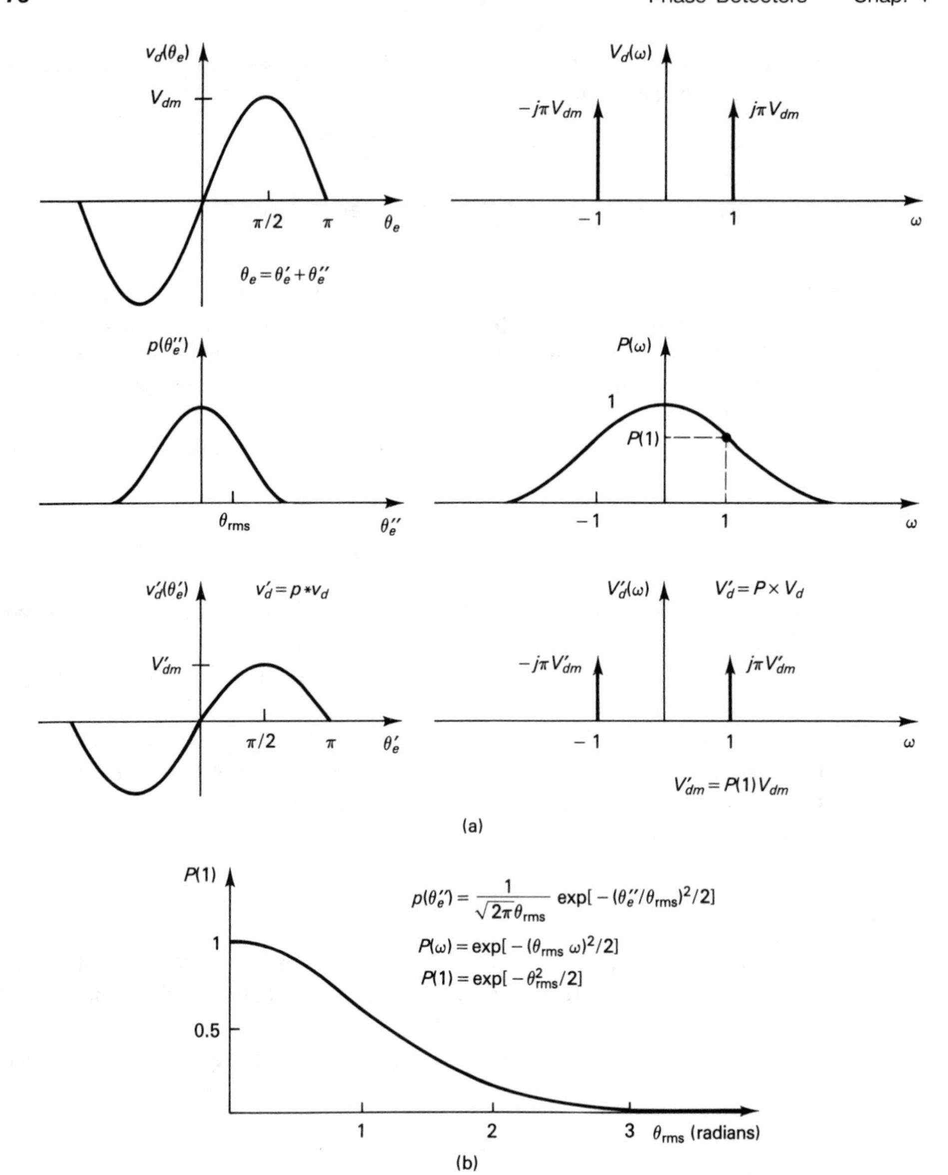

FIGURE 4–17 Modified Sinusoidal phase detector characteristic

$$v_d(\theta_e) = V_{dm}\sin(\theta_e) = V_{dm}\sin(\theta_e' + \theta_e'') \tag{4-40}$$

Suppose that noise is the cause of θ_e'' (see Chapter 6) so its probability density function is Gaussian with standard deviation θ_{rms}:

$$p(\theta_e'') = \frac{1}{\sqrt{2\pi}\theta_{rms}} \exp\left[\frac{-(\theta_e'')^2}{2(\theta_{rms})^2}\right] \tag{4-41}$$

The Fourier transform of $v_d(\theta_e)$ is

$$V_d(\omega) = -j\pi V_{dm}\delta(\omega + 1) + j\pi V_{dm}\delta(\omega - 1) \qquad (4\text{-}42)$$

where δ is the Dirac-delta function (see Fig. 4–17a). The Fourier transform of $p(\theta_e'')$ is

$$P(\omega) = \exp(-\omega^2\theta^2_{\text{rms}}/2) \qquad (4\text{-}43)$$

Then from Eq. (4-39),

$$V_d'(\omega) = P(\omega)V_d(\omega) = -j\pi V_{dm}'\delta(\omega + 1) + j\pi V_{dm}'\delta(\omega - 1) \qquad (4\text{-}44)$$

where the amplitude of the modified PD characteristic is

$$V_{dm}' = P(1)\, V_{dm} = \exp(-\theta^2_{\text{rms}}/2)\, V_{dm}$$

Taking the inverse Fourier transform of Eq. (4-44) gives the modified PD characteristic

$$v_d'(\theta_e') = V_{dm}'\sin(\theta_e') \qquad (4\text{-}45)$$

with a modified gain (slope at the origin) of

$$K_d' = K_dP(1) = K_d\exp(-\theta^2_{\text{rms}}/2) \qquad (4\text{-}46)$$

The reduction factor $P(1)$ is plotted in Fig. 4–17b for our Gaussian example. If the rms value of θ_e'' is 1.0 radian, then V_{dm}' is about 60% of V_{dm}. K_d is also reduced to 60% of its original value. The consequence of θ_e'' is also seen in the acquisition of lock (see Chapter 8) when θ_e' is due to the transient and θ_e'' is due to θ_i or noise.

The same technique can be used to analyze the effect of θ_e'' on triangular and sawtooth PD characteristics. The result is always a rounding of the corners of $v_d'(\theta_e')$ as $P(\omega)$ attenuates the "harmonics" of $v_d(\theta_e)$. See Pouzet [12] for the modified characteristics of these other phase detectors.

The three-state (and higher-state) PDs have characteristics with double-value functions (see Fig. 4–18a). This hysteresis complicates the analysis somewhat in determining the modified PD characteristic. Let θ_e'' be limited to some maximum magnitude $\theta_{\max}$; the uniform probability density function $p(\theta_e'')$ in Fig. 4–18b is an example. If θ_e remains entirely on a linear portion of the characteristic (see cases 1 and 2 in Fig. 4–18a), then the characteristic is unchanged (see cases 1 and 2 in Fig. 4–18c). But if θ_e ever touches a discontinuity in the characteristic, the operation jumps to a new linear portion (see case 3 in Fig. 4–18a). Therefore, v_d' can't have the higher value that v_d does for that average phase error θ_e'.

The modified PD characteristic in Fig. 4–18c plots v_d' as a function of θ_e' for the case $\theta_{\max} = 0.5\pi$. The effect is to reduce the modified phase range of the PD to $\pm 1.5\pi$. In general, the modified range is reduced by $\theta_{\max}$ on each end.

For $\theta_{\max} > \pi$, another phenomenon occurs, as shown in Fig. 4–18d. There

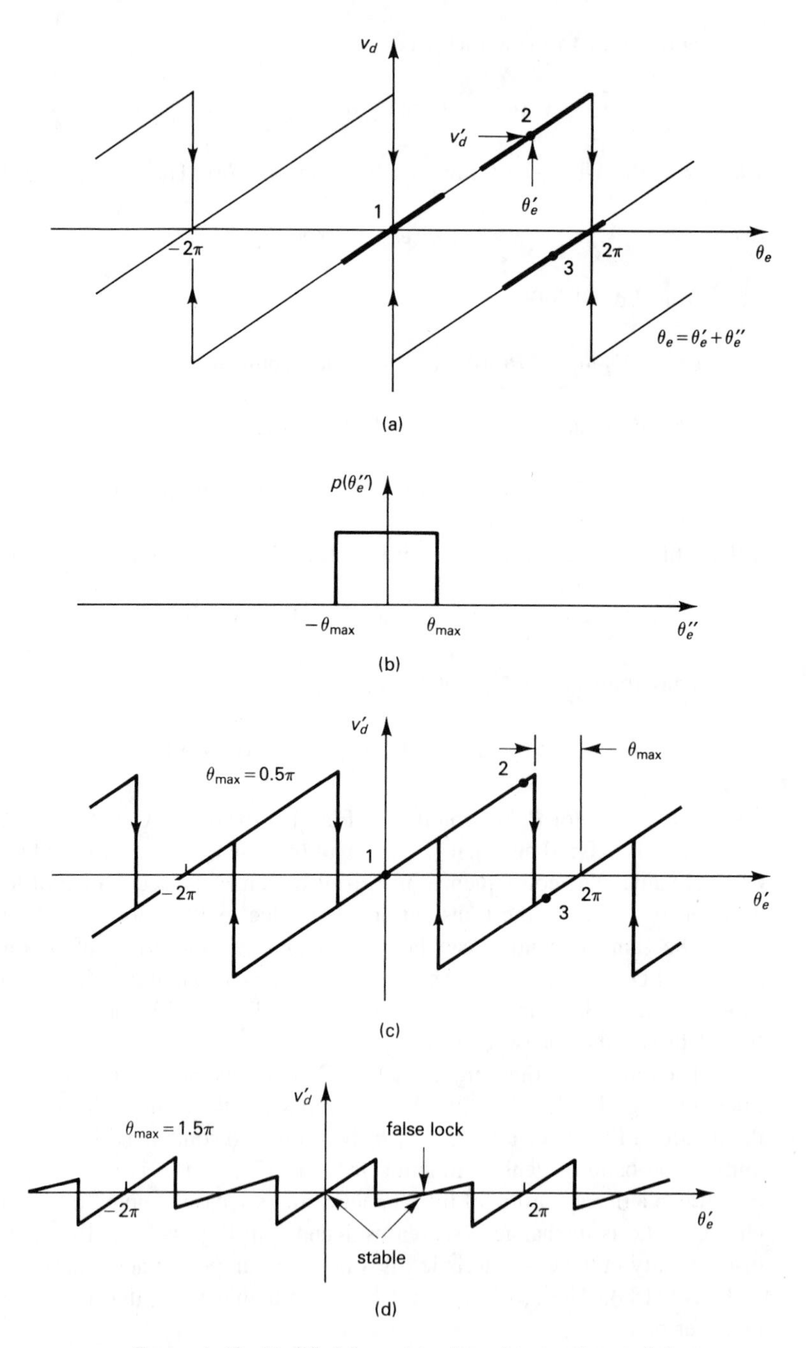

FIGURE 4–18 Modified three-state phase detector characteristic

continues to be a usable linear range about the origin, but a spurious linear range appears at $\theta_e' = \pi$. This provides a stable point of operation for the PLL, but with $\theta_{eo} = \pi$ rather than the desired $\theta_{eo} = 0$. One situation in which this can occur is when θ_e'' is due to modulation of θ_i while the PLL is acquiring lock, and θ_e' is the phase transient during acquisition. Then to avoid a false lock at $\theta_e' = \pi$, the modulation of θ_i must be such that

$$\theta_e'' < \pi \qquad (4\text{-}47)$$

If θ_e'' is unbounded, as in the case of noise-generated phase error, then the analysis of three-state PD behavior is intractable. The double-valued function $v_d(\theta_e)$ makes it impossible to know even the probable value of v_d. For this reason, three-state PDs can't be used in applications where noise is significant.

REFERENCES

[1] P. R. Gray and R. G. Meyer, *Analysis and Design of Analog Integrated Circuits*, Wiley: New York, 1984, Section 10.3.

[2] *Anzac RF & Microwave Signal Processing Components*, Adams-Russell: Burlington: Mass., 1989, p. 341.

[3] *Anzac Components*, p. 272.

[4] *Anzac Components*, pp. 234–41.

[5] K. K. Clarke and D. T. Hess, *Communication Circuits: Analysis and Design*, Addison-Wesley: Reading, Mass., 1978, Section 8.4.

[6] I. Shahriary, G. Des Brisay, S. Avery, and P. Gibsan, ''GaAs Monolithic Phase/Frequency Discriminator,'' *IEEE GaAs Symposium*, 1985, pp. 183–86.

[7] *Motorola MECL Device Data*, Motorola, Inc.: Phoenix, Ariz., 1989, Section 6.

[8] W. Egan and E. Clark, ''Test Your Charge-Pump Phase Detectors,'' *Electronic Design*, 26, no. 12 (June 7, 1978), pp. 134–37.

[9] *Data Book: RCA High-Speed CMOS Logic ICS*, Harris Semiconductor: Melbourne, Florida, 1989, pp. 493–509.

[10] J. F. Oberst, ''Generalized Phase Comparators for Improved Phase-Locked Loop Acquisition,'' *IEEE Trans. on Communication Technology*, v. COM-19, pp. 1142–48, December, 1971.

[11] D. H. Wolaver, ''Extended Range Phase Detector,'' Patent 4,920,902, owned by General Signal/Tau-tron, Inc., February 20, 1990.

[12] A. H. Pouzet, ''Characteristics of Phase Detectors in Presence of Noise, *Proc. 8th Int. Telemetry Conf.*, Los Angeles, Calif., 1972, pp. 818–28.

CHAPTER

5

VOLTAGE-CONTROLLED OSCILLATORS

5–1 PROPERTIES OF VCOS

In Chapter 2 we introduced the VCO characteristic as a linear function of the VCO frequency ω_o with respect to the control voltage v_c. The linear model is

$$\omega_o = \Delta\omega_o + \omega_i = K_o(v_c - V_{co}) + \omega_i \tag{5-1}$$

where ω_i is the average input frequency and V_{co} is the value of v_c such that $\omega_o = \omega_i$. K_o is the VCO gain. Since V_{co} depends on the input to the PLL, it is not considered a property of the VCO itself. In the simple examples so far, K_o has a well-defined value because the VCO characteristic has been linear. In practice, this is only an approximation at best.

Figure 5–1a shows an example of a VCO characteristic more like one would encounter in practice. Because the slope of the curve is not constant, the VCO gain K_o has a range of values from zero to 5 Mrad/s/V. Suppose for some particular application an acceptable range is 4 Mrad/s/V $\leq K_o \leq$ 5 Mrad/s/V. Then the VCO can't be used for $\omega_o <$ 1 Mrad/s, where $K_o <$ 4 Mrad/s/V. In this example, if $v_c >$ 2.5 V is applied (corresponding to $\omega_o >$ 10 Mrad/s), the VCO stops oscillating. Thus, the *range* of the VCO is 1 Mrad/s $< \omega_o <$ 10 Mrad/s. When the range is small, it is sometimes expressed as some

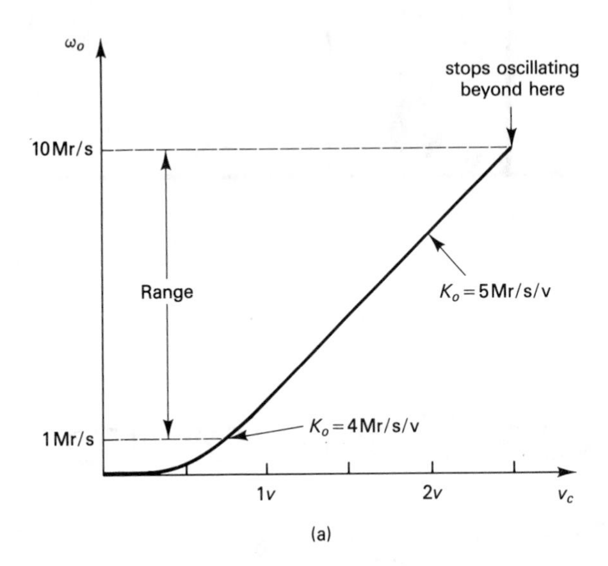

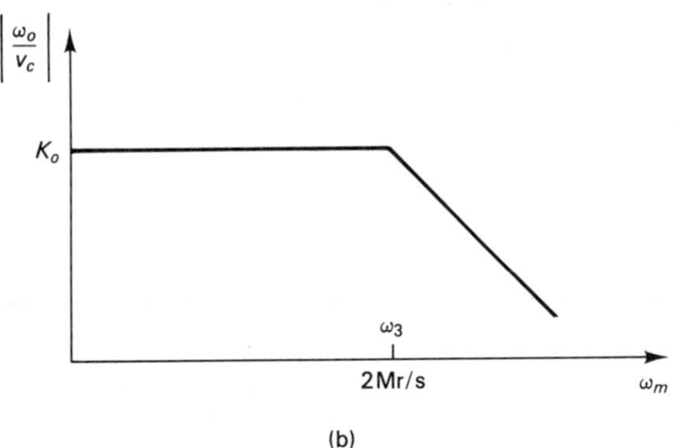

FIGURE 5–1 VCO characteristics

deviation about a center frequency, such as 5 Mrad/s $\pm 10\%$. Similarly, K_o can be expressed as a center value with a deviation, such as $K_o = 4.5$ Mrad/s/V $\pm$ 11%. The percent deviation of K_o is a measure of the *linearity* of the characteristic. There is obviously a tradeoff between range and linearity.

The VCO characteristic implies that there is a fixed relationship between v_c and ω_o no matter how quickly v_c changes. But if v_c is modulated at too high a frequency, the modulation of ω_o is less than that indicated by K_o. Fig. 5–1b shows a typical modulation response; the ac gain $|\omega_o/v_c|$ is K_o for low frequencies of modulation ω_m, but it falls off for $\omega > \omega_3$. This reflects a pole at ω_3 in the VCO transfer function:

$$\frac{\omega_o(s)}{v_c(s)} = \frac{K_o}{1 + s/\omega_3} \tag{5-2}$$

where ω_3 is called the *modulation bandwidth*. As shown in section 3–7, a pole at ω_3 in the forward loop gain $G(s)$ can cause instability if $\omega_3 < K$. If ω_3 is to have little effect on the PLL response, a good rule is to keep $\omega_3 \geq 4\ K$ (see Fig. 3–13).

In this chapter, we look at three kinds of VCOs—astable, multivibrators, L-C oscillators, and crystal oscillators. Each has advantages and disadvantages in terms of range, linearity, modulation bandwidth, and immunity to outside influence. This last property is seen in the VCO phase noise, which is analyzed in section 6–5. It is also seen in susceptibility to being "pulled" in frequency and phase by other signals. This important practical consideration, called *injection locking*, is the subject of the last third of this chapter. In general, VCOs less sensitive to noise and injection are also less sensitive to temperature and to variations in power supply voltage.

5–2 VOLTAGE-CONTROLLED MULTIVIBRATORS

There are two general classes of oscillators: relaxation oscillators (or astable multivibrators) and resonant oscillators (or Vanderpole oscillators). Figure 5–2 shows a circuit for a voltage-controlled multivibrator. A current i_x charges and discharges a capacitor C_x between two values of the threshold voltage v_t—between 2 V and 3 V in this case (see the waveform for v_1). The output v_o is a square wave. The frequency of oscillation is determined by the rate of charging C_x, which is proportional to $|i_x| = i_c$, which increases with v_c. The relationships are $i_c = (v_c - 0.6\text{ V})/R$, and $T/2 = V_1C_x/i_c$, where V_1 is the difference between the thresholds and $T/2$ is the time to charge the capacitor by V_1. But T is the period of the oscillation, so $\omega_o = 2\pi/T$. Therefore

$$\omega_o = \frac{\pi}{V_1 R C_x}(v_c - 0.6\text{ V}) \tag{5-3}$$

In commercially available oscillator chips, the capacitor C_x is left as an external component so the user can select the frequency of operation. For the circuit here, $R = 1\text{ k}\Omega$ and $V_1 = 1$ V. If $C_x = 628$ pF, then Eq. (5-3) gives $\omega_o = (5\text{ Mrad/s/V})(v_c - 0.6\text{ V})$, and the VCO characteristic is that shown in Fig. 5–1a.

It is usually not necessary to construct a circuit like that in Fig. 5–2; commercial integrated circuits are available, such as the Motorola MC4024. The published characteristic [1] for the MC4024 (see Fig. 5–3a) gives the product f_oC_x as a function of v_c. Corresponding to $v_c = 4.3$ V (about the center of the linear range) is $f_oC_x = 330$ MHz-pF. Selecting $C_x = 110$ pF, for example, gives $f_o = 330/110 = 3.0$ MHz, or $\omega_o = 2\pi f_o =$ 18.8 Mrad/s. Then as v_c varies from 3.7 V to 4.9 V (the linear portion of the characteristic), f_o varies from 1.8 MHz to 4.2 MHz, or 3.0 MHz ± 40% (see Fig. 5–3c). The VCO gain is $K_o = 2\pi(4.2 - 1.8\text{ MHz})/(4.9 - 3.7\text{ V}) = 12.6$ Mrad/s/V.

In some applications, the VCO range needs to be constrained to keep the PLL from locking to the wrong frequency component (as in the example in Fig. 1–2) or to reduce the time required to attain lock.

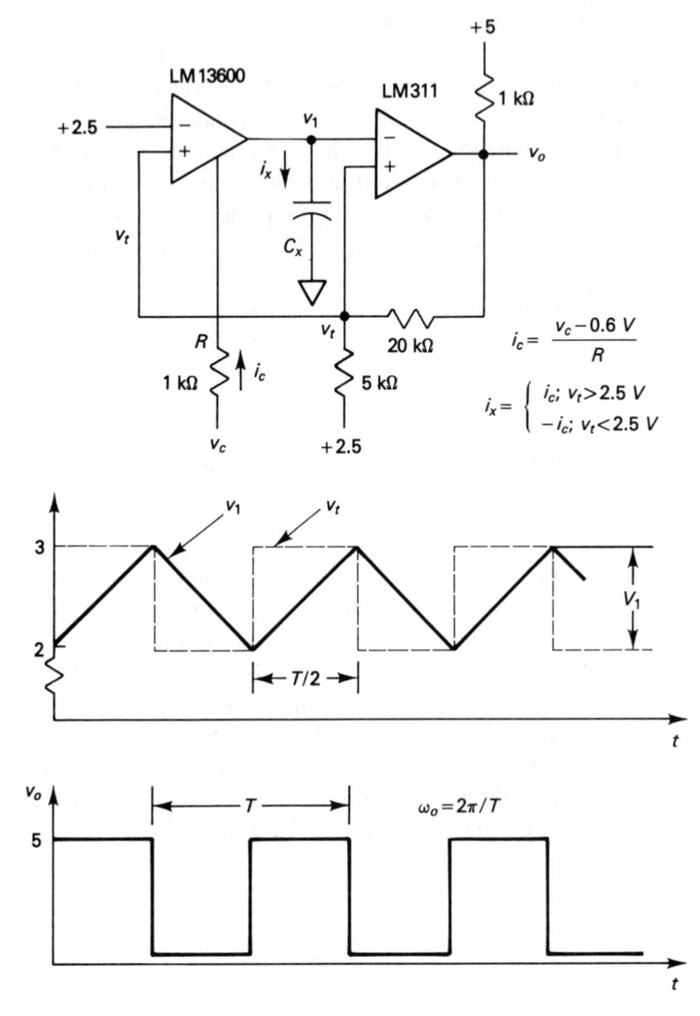

FIGURE 5–2 Voltage-controlled multivibrator

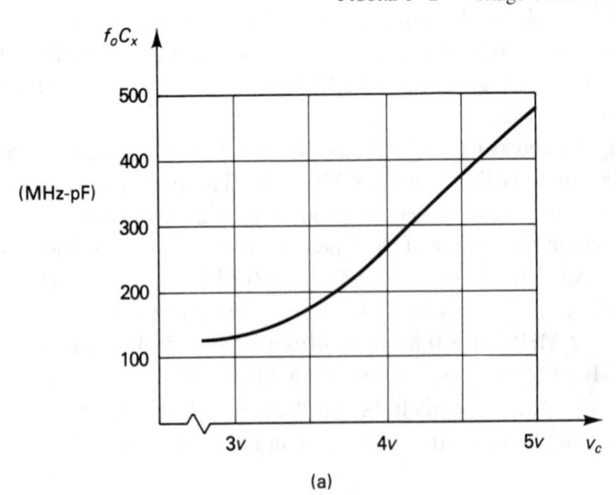

FIGURE 5–3 Reducing VCO range

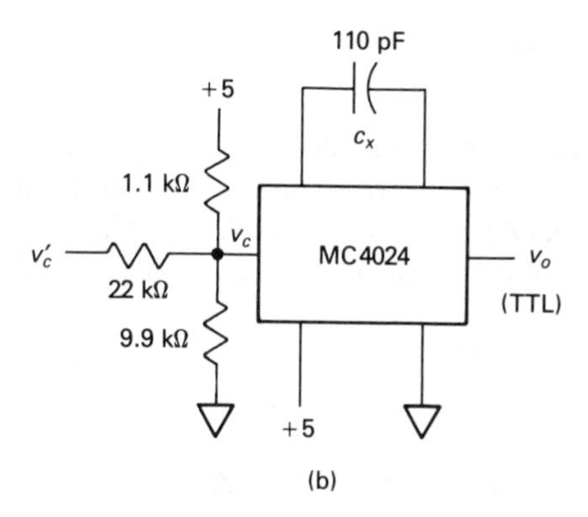

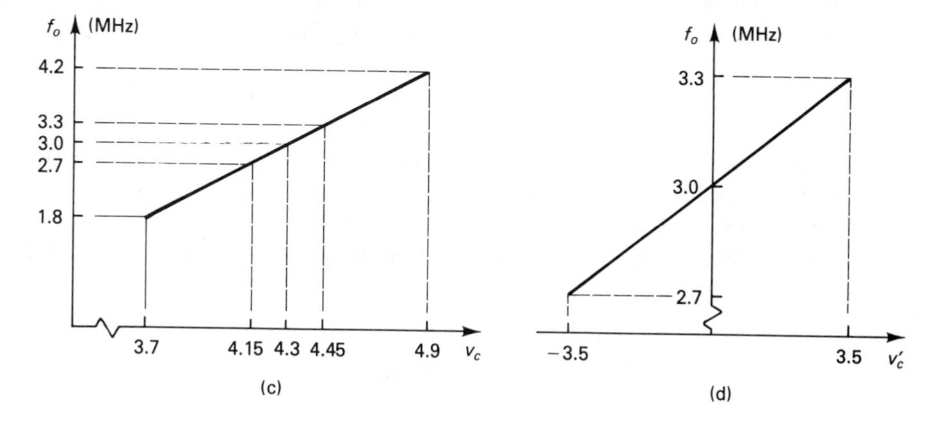

FIGURE 5–3 *(continued)*

EXAMPLE 5–1

The voltage $v_{c'}$ from the loop filter is constrained to $-3.5 \leq v_{c'} \leq 3.5$. Design a VCO whose range is constrained to 3.0 MHz ±10%. By choosing $C_x = 110$ pF, we get the characteristic in Fig. 5–3c with f_o centered in the linear range. According to this characteristic, we constrain 2.7 MHz $< f_o <$ 3.3 MHz by constraining v_c to the range 4.15 V $\leq v_c \leq$ 4.45 V. Then we need a resistor network to attenuate the swing of $v_{c'}$ by $(4.45 - 4.15)/(3.5 + 3.5) = 0.043$ and shift the O-V center of v_c' to the 4.3-V center of v_c. The resistor network shown in Fig. 5–3b provides the necessary relationship:

$$v_c = 4.3 \text{ V} + 0.043\, v_c'$$

Specifically, $v_c' = -3.5$ V results in $v_c = 4.15$ V, and $v_c' = 3.5$ V results in $v_c = 4.45$ V. The modified VCO characteristic is that shown in Fig. 5–3d with the range limited to 2.7 MHz $\leq f_o \leq$ 3.3 MHz. With the voltage divider now part of the VCO, the gain is reduced to $K_o' \equiv \Delta\omega_o/\Delta v_c' = 2\pi(3.3 - 2.7 \text{ MHz})/7 \text{ V} = 0.54$ Mrad/s/V.

5–3 RESONANT VCOS

Resonant oscillators can operate at higher frequencies than multivibrators, and they are less influenced by noise, stray signals, temperature, and supply voltage. Figure 5–4a illustrates the principle of a resonant oscillator. A resonant circuit—a parallel L-C tank—converts the current i_1 from a current source to a voltage v_1. At resonance, that is at the frequency

$$\omega_o = 1/\sqrt{LC} \tag{5-4}$$

the admittances of L and C cancel, and the tank has the impedance r_p. This resistance is usually not an actual resistor but rather a model of effective resistance due to losses in the current source and the inductor. [2] The quality factor or Q of the tank is

$$Q = r_p/\omega_o L \tag{5-5}$$

Then at resonance, $v_1 = r_p\, i_1$. But the source of i_1 is dependent on v_1: $i_1 = g_m\, v_1$, where g_m is the transconductance of the amplifier. Then it follows that for positive feedback with unity loop gain we must have $g_m = 1/r_p$. This relation is often maintained by an automatic gain control. The result is oscillation at the frequency ω_o given by Eq. (5-4).

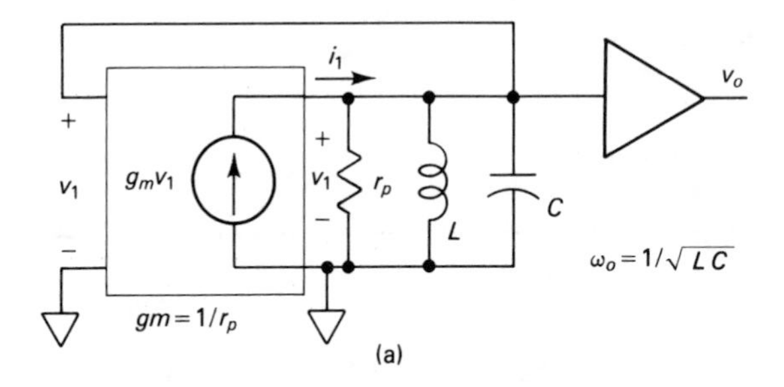

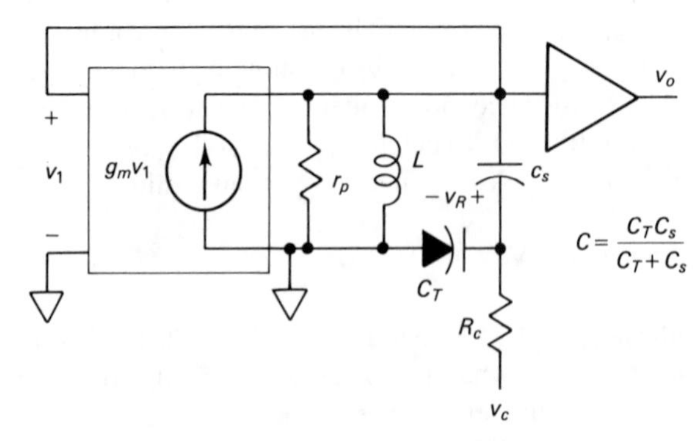

FIGURE 5–4 Resonant oscillator

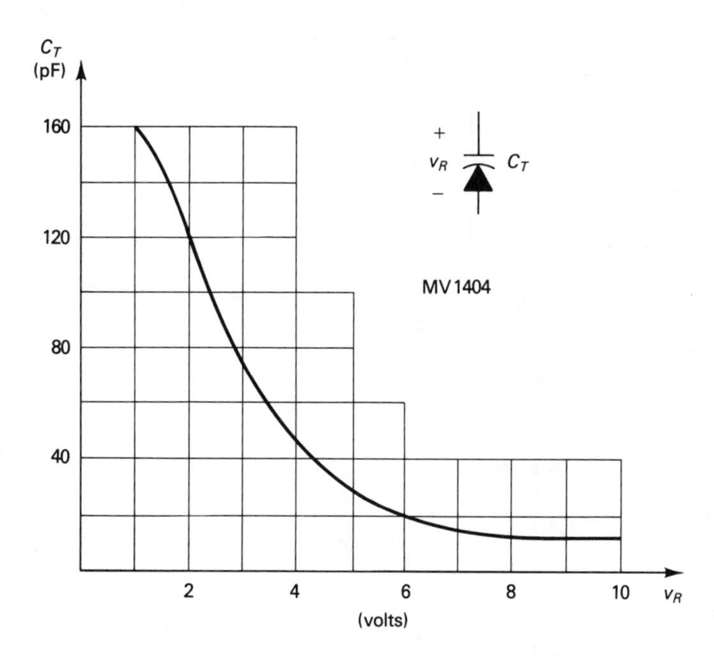

FIGURE 5–5 Varactor characteristic

The frequency ω_o may be controlled by varying C electronically, converting the oscillator to a VCO. The oscillator in Fig. 5–4b includes a varactor diode as part of the tank capacitance. A varactor is a reverse-biased diode whose junction capacitance C_T is a function of the reverse bias v_R; a typical varactor characteristic is shown in Fig. 5–5. In the circuit in Fig. 5–4b, v_R is applied by the control voltage v_c through a buffer resistor R_c, which keeps the source of v_c from loading down the tank. A series capacitor C_s blocks dc current that would otherwise flow through R_c and L. Therefore, $v_R = v_c$ for slow variations in v_c. The total capacitance is given by $C = C_sC_T/(C_s + C_T)$.

EXAMPLE 5–2

Design a resonant VCO with a range from 67 Mrad/s to 134 Mrad/s. (This factor-of-two range is called an *octave* range.) Make the VCO gain K_o vary as little as possible over the range. The circuit has the form shown in Fig. 5–6a, where the transconductance amplifier (with AGC to maintain the proper gain) has been provided by the Motorola MC1648. The varactor is a Motorola MV1404, with the characteristic shown n Fig. 5–5. The series capacitor C_s and the parallel capacitor C_p provided flexibility in designing the VCO characteristic.

The relationships governing the design are Eq. (5-4) relating ω_o to C

$$C = \frac{C_sC_T}{C_s + C_T} + C_p \tag{5-6}$$

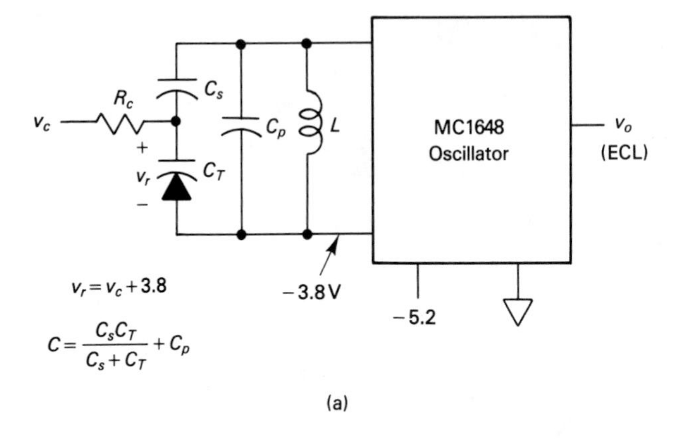

(a)

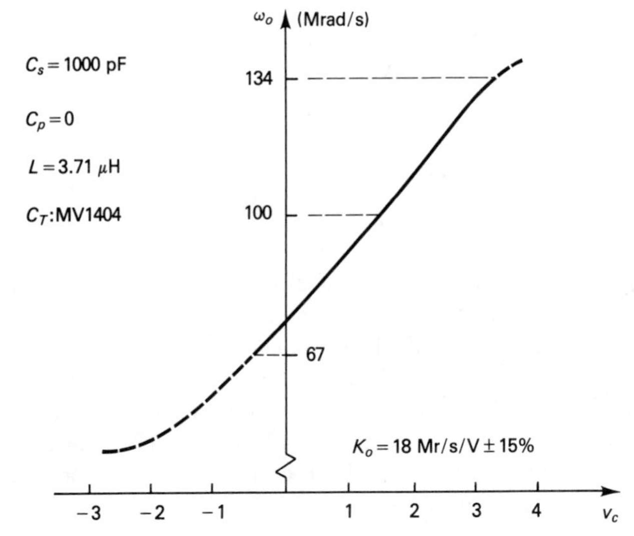

FIGURE 5–6 Resonant VCO example

relating C to C_T, and Fig. 5–5 relating C_T to v_R. The largest possible range is achieved when $C \approx C_T$ by choosing $C_p = 0$ and C_s large compared to C_T—say, $C_s = 1000$ pF. From Fig. 5–5, C_T has a 10-to-1 range, and from the square root in Eq. (5-4), ω_o therefore has a maximum $\sqrt{10}$-to-1 range. This is greater than the required 2-to-1 range, but linearity is improved if we are not forced to use all of the VCO range.

Table 5–1 shows the computed parameter values for the choice $C_p = 0$ and $C_s = 1000$ pF. Because the anode of the varactor is at −3.8 Vdc, $v_R = v_c + 3.8$ V. The values of ω_o versus v_c in Table 5–1 are plotted in Fig. 5–6b. $L = 3.71\ \mu$H was selected to put the inflection point in the center of the desired range 67 Mrad/s to 134 Mrad/s. This achieves the greatest linearity. The gain K_o is a maximum in the center and a minimum on the two ends: $K_o = 18$ Mrad/s/V ±15%. For ω_o to be limited to the desired range, v_c needs to be limited to the range $-0.5\text{ V} \leq v_c \leq 3.1$ V. This can be done with a resistor network, as in Fig. 5–3b, or diodes can be used to limit v_c, as in the next example.

TABLE 5–1

v_c	v_R	C_T	C	ω_o
−2.8 V	1 V	160 pF	138 pF	44.2 Mrad/s
−1.8	2	120	107	50.2
−0.8	3	77	71	61.3
0.2	4	47	45	77.5
1.2	5	30	29	96.2
2.2	6	20	20	117.2
3.2	7	15	15	135.0

EXAMPLE 5–3

Design a resonant VCO with a frequency range of 100 Mrad/s ±5% (that is, 95 Mrad/s to 105 Mrad/s). The VCO gain K_o is to be as small as possible while maintaining a ±10% tolerance on K_o over the frequency range.

One solution would be to use the same C_s and C_p values in Example 5–2 and constrain the control voltage to 1.4 V ± 275 mV. But this is a very small range for v_c, corresponding to a high value of K_o. This makes the VCO more sensitive to noise and dc offsets.

To reduce K_o, we need to make the total capacitance C, calculated in Eq. (5-6), be less sensitive to C_T. This can be done by both increasing C_p and decreasing C_s. Increasing C_p tends to flatten the upper end of the VCO characteristic by swamping out C_T when C_T becomes small. Decreasing C_s tends to flatten the lower end of the VCO characteristic by dominating C_T when C_T becomes large. Trial and error is required to find values of C_p and C_s so that the slope (or VCO gain) K_o is as small as possible while maintaining less that a ±10% variation in K_o over the desired ±5% range of ω_o. A spreadsheet computer program is useful here.

Two designs are shown in Fig. 5–7a. In Design #1, C_p = 300 pF, C_s = 200 pF, and L = 0.286 μH. This achieves low gain; the whole available range from v_c = −2.8 V to 3.2 V is used to go from ω_o = 95 Mrad/s to 105 Mrad/s. But because the inflection point is too far to the left, K_o varies a great deal over the range—from 2.6 Mrad/s/V to 0.7 Mrad/s/V. The excessive flattening of the curve at high frequencies indicates C_p is too large. In Design #2, C_p is decreased to 120 pF, and C_s is 60 pF to again require a 6-V change in v_c for a 10% change in ω_o. The last step is to choose L = 0.68 μH for the desired frequency range about 100 Mrad/s. The inflection point is now centered, and K_o varies from 2.45 Mrad/s/V to 1.05 Mrad/s/V. This is less variation, but it still exceeds the specified ± 10% variation. We need to increase K_o and discard the ends of the curve.

Two more designs are shown in Fig. 5–7b. In Design #3, the gain is increased by decreasing C_p to 17 pF. For C_s = 22 pF, there is a quite linear range over the desired 10% change in ω_o. Choosing L = 3.40 μH centers this range on 100 Mrad/s. The variation of K_o over the range $95 \le \omega_o \le 105$ Mrad/s is only ±10% about a value of 3.5 Mrad/s/V. To constrain ω_o to this range, we need to constrain $-0.5 \le v_c \le 3.1$ V. This is easier to do if the voltage range were centered on 0 V.

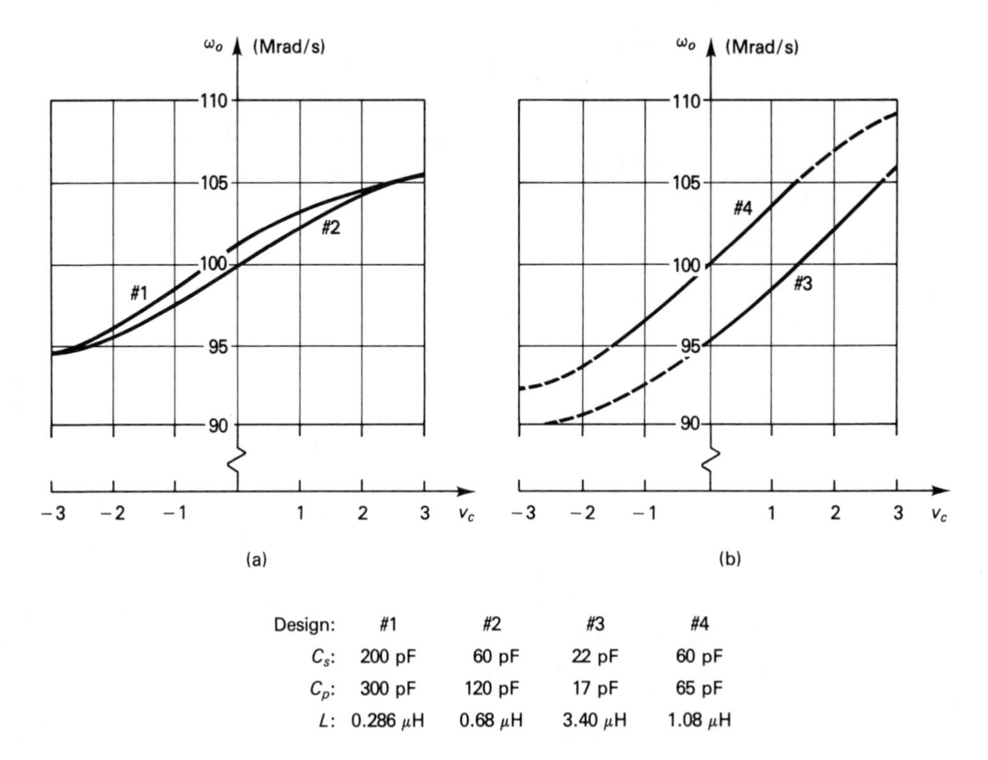

FIGURE 5–7 VCO characteristic for Example 5–3

In Design #4, the linear portion of the curve is moved to be centered horizontally on $v_c = 0$ V. C_s is increased to 60 pF to make the lower part of the curve more linear. $C_p = 65$ pF provides a 10% change in ω_o for a 2.8-V change in v_c, and $L = 1.08$ μH centers the range vertically on 100 Mrad/s. The gain is again $K_o = 3.5$ Mrad/s/V $\pm$ 10%. Table 5–2 gives the parameter values for this final design.

To constrain $95 \leq \omega_o \leq 105$ Mrad/s, we need to constrain $-1.4 \leq v_c \leq 1.4$ V. This can be done with diodes, as shown in Fig. 5–8. Since v_d develops little voltage across R_2, v_c is not more than two diode drops away from ground.

TABLE 5–2

v_c	v_R	C_T	C	ω_o
−2.8 V	1 V	160 pF	109 pF	92.3 Mrad/s
−1.8	2	120	105	93.8
−0.8	3	77	99	96.8
0.2	4	47	91	100.6
1.2	5	30	85	104.3
2.2	6	20	80	107.5
3.2	7	15	77	109.6

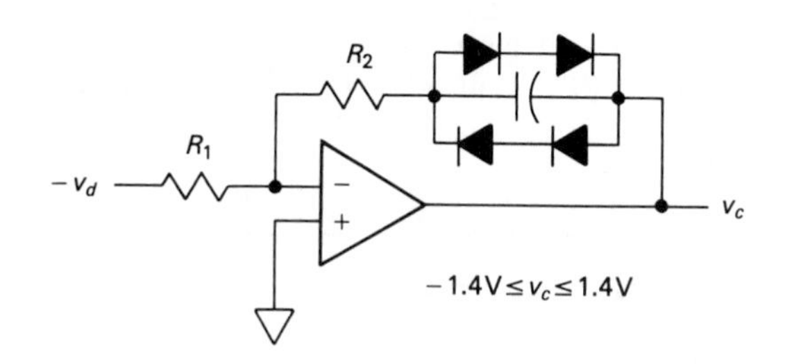

FIGURE 5–8 Limiting v_c

Further information on the theory of resonant oscillators can be found in Clarke and Hess. [3] More on the design of resonant VCOs is available in Rohde. [4]

Note that we still need to chose a value for R_c in Fig. 5–6a. This value depends on the desired modulation bandwidth.

5–4 MODULATION BANDWIDTH

To change ω_o, v_c must change v_R by charging C_T and C_s through R_c (see Fig. 5–6a). C_p is essentially shorted out by the inductor since v_c changes slowly compared to the oscillation frequency. A model is shown in Fig. 5–9 to find the transfer function from v_c to v_R. This is simply $v_R/v_c = 1/(1 + s/\omega_3)$, where the *modulation bandwidth* of the VCO is

$$\omega_3 = 1/R_c(C_T + C_s) \tag{5-7}$$

Since $\Delta\omega_o = K_o v_R$, the complete transfer function of the VCO is that given by Eq. (5-2):

$$\frac{\omega_o}{v_c} = \frac{K_o}{1 + s/\omega_3} \tag{5-8}$$

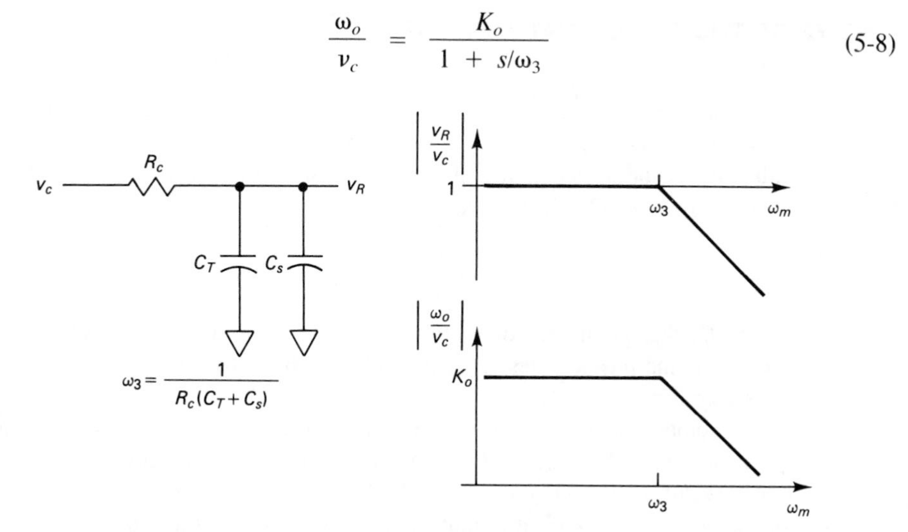

FIGURE 5–9 VCO modulation bandwidth

It is important to keep ω_3 high enough that it doesn't affect the loop response $H(s)$. A good rule is to keep $\omega_3 > 4K$, where K is the PLL bandwidth. Then for the worst case, we must look at the *maximum* C_T over the operating range of the VCO.

EXAMPLE 5–4

The VCO designed in Example 5–2 is to be used in a PLL with $K = 0.1$ Mrad/s. Choose R_c to maximize the Q of the tank.

In Example 5–2, the lower end of the range ($\omega_o = 67$ Mrad/s) corresponded to the maximum $C_T = 70$ pF. For $K = 0.1$ Mrad/s, we need at least $\omega_3 = 0.4$ Mrad/s. Since $C_s = 1000$ pF, Eq. (5-7) gives $R_c = \underline{2.38\ \text{k}\Omega}$. As we will see in the next section, this is a rather low value which spoils the Q of the tank circuit.

EXAMPLE 5–5

The VCO of Design #4 in Example 5–3 is to be used in a PLL with $K = 0.1$ Mrad/s. Choose R_c to maximize the Q of the tank.

In Design #4, the lower end of the range ($\omega_o = 95$ Mrad/s) corresponded to $C_T = 101$ pF. For $K = 0.1$ Mrad/s, we need at least $\omega_3 = 0.4$ Mrad/s. Since $C_s = 60$ pF, Eq. (5-7) gives $R_c = \underline{15\ \text{k}\Omega}$. This value will be high enough to preserve a good Q.

5–5 *Q* OF THE RESONANT CIRCUIT

The tank in a resonant VCO needs a high Q for low phase noise and low injection sensitivity for the VCO, as will be shown in sections 5–8 and 6–5. A high Q is associated with low lossiness, and much of this loss is in the inductor. The inductor loss can be modeled by a parallel resistance

$$r_L = Q_L \omega_o L \tag{5-9}$$

where Q_L, the quality factor of the inductor, is a weak function of ω_o. The function depends on the physical design of the inductor, but a rough relationship between ω_o and Q_L is given in Fig. 5–10a.

Another loss, one that we have introduced for the sake of control, is in the resistor R_c. In the model in Fig. 5–10b, the grounding of the left end of R_c assumes that the impedance of the v_c source is negligible. Since R_c is only across C_T, it doesn't see the full voltage across the resonant circuit, and it degrades the Q less than if it were across the

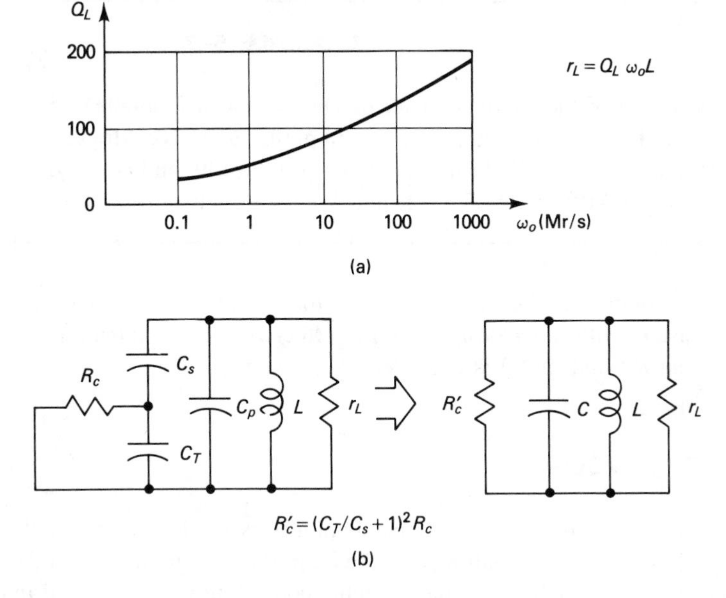

FIGURE 5–10 Q of resonant circuit

total C. Clarke and Hess [5] show that the loss due to an $R_{c'}$ across the total C (see the second model in Fig. 5–10b) is the same as that due to R_c if

$$R_{c'} = (C_T/C_s + 1)^2 R_c \tag{5-10}$$

Since small $R_{c'}$ degrades Q the most, the worst case must consider the *minimum* C_T.

In the final model in Fig. 5–10b, $R_{c'}$ and r_L are in parallel, constituting r_p. Then from Eq. (5-5), the Q of the resonant circuit is given by

$$Q = \frac{R_c' \parallel r_L}{\omega_o L} \tag{5-11}$$

EXAMPLE 5–6

Find the Q of the oscillator tank in Example 5–4.

The values in Example 5–4 were $C_{T\min} = 16$ pF for $\omega_o = 134$ Mrad/s, $C_s = 1000$ pF, $L = 3.7\ \mu$H, and $R_c = 2.38$ kΩ. From Fig. 5–10a, $Q_L \approx 135$, and $r_L = Q_L\omega_o L = 67$ kΩ. Then from Eq. (5-10), $R_c' = 2.45$ kΩ, and Eq. (5-11) gives $Q = \underline{4.75}$, which is quite low.

EXAMPLE 5–7

Find the Q of the oscillator tank of Design #4 in Example 5–5.

In Design #4 we had $C_{Tmin} = 28$ pF for $\omega_o = 105$ Mrad/s, $C_s = 60$ pF, $L = 1.08$ μH, and $R_c = 15$ kΩ. From Fig. 5–10a, $Q_L \approx 130$, and $r_L = Q_L\omega_o L = 14.7$ kΩ. Then from Eq. (5-10), $R_c' = 32$ kΩ, and Eq. (5-11) gives $Q = \underline{89}$, which is very good.

Comparing the two examples, the moral is to keep C_s as small as possible, consistent with other design criteria. Otherwise the modulation bandwidth requires too low an R_c, and the Q is degraded excessively.

5–6 CRYSTAL VCOs

When an extremely low PLL bandwidth K is needed, it is usually not satisfactory to simply make a very small K_h (see the loop filter design in Chapter 2). This would greatly attenuate v_c, which would have trouble competing with noise and injection signals. The better approach is to make K_o very small by using a crystal oscillator for the VCO. The result is a *voltage-controlled crystal oscillator*, called a "VCXO."

Figure 5–11a shows a circuit for a VCXO with a varactor used for tuning the

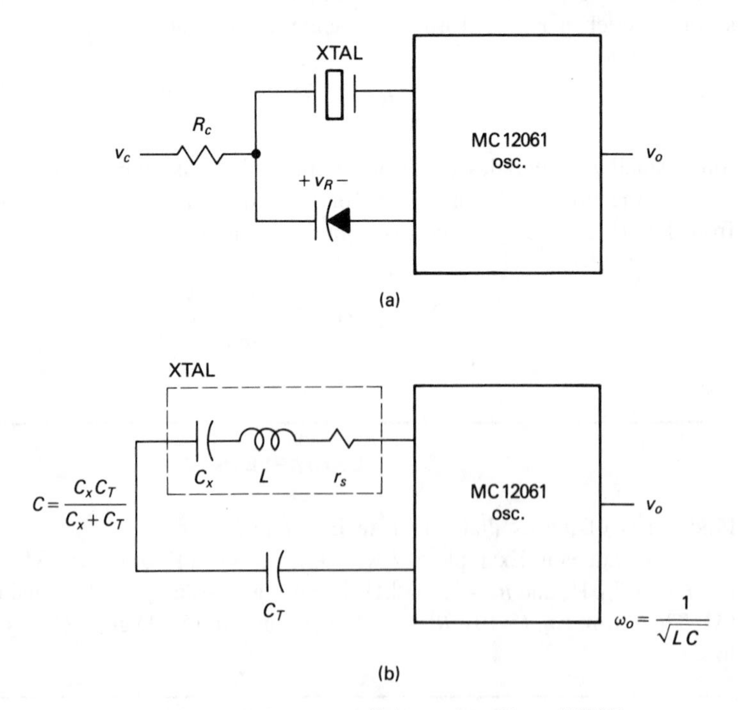

FIGURE 5–11 Voltage-controlled crystal oscillator (VCXO)

frequency. The amplifier portion of the oscillator is provided by a commercial circuit—the Motorola MC12061. In Fig. 5–11b the crystal has been replaced by its equivalent circuit—a series R-L-C circuit. The crystal manufacturer usually specifies the frequency ω_{oo} (the ω_o corresponding to $C_T = 30$ pF), the Q of the crystal (represented here by Q_x), and the equivalent series resistance r_s. From these the L and C_x for the model can be found through the relationships

$$Q_x = \omega_{oo}L/r_s \tag{5-12}$$

$$\omega_{oo} = 1/\sqrt{LC_{xo}} \tag{5-13}$$

$$1/C_{xo} \equiv 1/C_x + 1/(30 \text{ pF}) \tag{5-14}$$

(Sometimes the manufacturer specifies some capacitance other than 30 pF, such as 22 pF.)

The crystal oscillator is voltage-controlled by adjusting v_R across the varactor to vary C_T. As with a parallel resonant circuit, the oscillation frequency is given by

$$\omega_o = 1/\sqrt{LC} \tag{5-15}$$

where

$$\begin{aligned} 1/C &= 1/C_x + 1/C_T \\ &= 1/C_{xo} + 1/C_T - 1/(30 \text{ pF}) \end{aligned} \tag{5-16}$$

Because the effective capacitance C_x is extremely small—on the order of 0.01 pF—C_T must also be small to make any significant change in C. Figure 5–12 shows the characteristic of a low-capacitance varactor with a minimum C_T of 5 pF. (It is difficult to

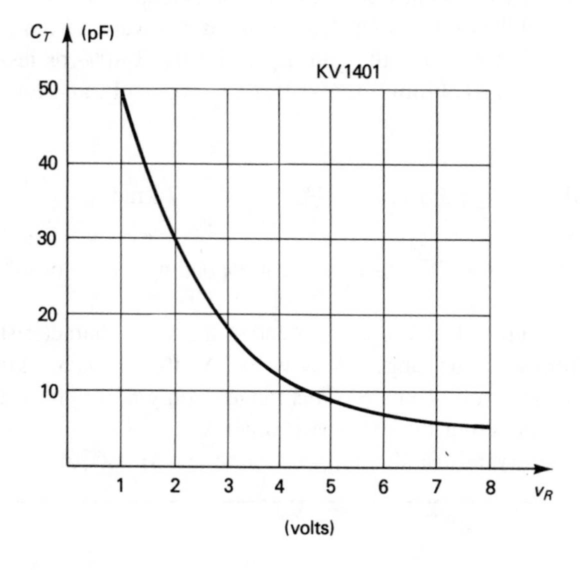

FIGURE 5–12 Low-capacitance varactor characteristic

go much lower than this because of case capacitance and stray wiring capacitance.) But even this small C_T is still about 500 times C_{xo}, and this allows us to make simplifying approximations in finding ω_o in terms of C_T.

From Eqs. (5-13) through (5-16), we have

$$\omega_o = \frac{\sqrt{1/C_{xo} + 1/C_T - 1/(30\text{ pF})}}{\sqrt{L}}$$

$$= \frac{\sqrt{1 + C_{xo}/C_T - C_{xo}/(30\text{ pF})}}{\sqrt{LC_{xo}}}$$

$$= \omega_{oo}\sqrt{1 + C_{xo}/C_T - C_{xo}/(30\text{ pF})}$$

$$\approx \omega_{oo}(1 + 0.5C_{xo}/C_T - 0.5C_{xo}/30\text{ pF})$$

or

$$\omega_o - \omega_{oo} \approx 0.5\omega_{oo}C_{xo}(1/C_T - 1/30\text{ pF}) \tag{5-17}$$

The approximation above uses $\sqrt{1 + x} \approx 1 + x/2$ for $x << 1$.

EXAMPLE 5–8

A crystal is specified to oscillate at $\omega_{oo} = 40$ Mrad/s for $C_T = 30$ pF. The Q_x is 50,000, and the effective series resistance is $r_s = 50\ \Omega$. Find L and C_{xo} for the crystal model. If a KV1401 varactor manufactured by Frequency Sources (see characteristics in Fig. 5–12) is put in series with the crystal, plot the oscillation frequency ω_o as v_R is varied from 1 V to 8 V. Find the range of the VCXO for which the gain K_o varies by $\pm 40\%$ or less.

From Eq. (5-12), $L = \underline{62.5\text{ mH}}$. From Eq. (5-13), $C_{xo} = \underline{0.01\text{ pF}}$, and Eq. (5-17) becomes

$$\omega_o - 40\text{ Mrad/s} = (200\text{ krad/s-pF})/C_T - 6.667\text{ krad}$$

Table 5–3 lists values of C_T taken from Fig. 5–12 and corresponding values of $\omega_o - 40$ Mrad/s.

These data are plotted in Fig. 5–13. At the upper end of the VCO characteristic, K_o degrades quite a bit. If we limit v_R to the range $1\text{ V} \le v_R \le 7\text{ V}$, then K_o is 5.7 krad/s/V $\pm 40\%$, and ω_o varies over a range of 34 krad/s. This range is only 850 ppm of the 40 Mrad/s oscillation frequency, reflecting the extremely small K_o.

For more information on crystal oscillators, see Clarke and Hess. [6]

TABLE 5–3

v_R	C_T	ω_o − 40 *Mrad/s*
1 V	50 pF	−2.667 krad/s
2	30	0.0
3	18	4.444
4	12	10.000
5	8.5	16.862
6	6.5	24.102
7	5.5	29.697
8	5.0	33.333

5–7 INJECTION IN MULTIVIBRATOR OSCILLATORS

Suppose that a periodic signal is injected into a multivibrator oscillator with a frequency near to that of the oscillator's frequency. It is possible for the frequency of the oscillator to be pulled to that of the injected signal and to be actually phase-locked to the signal. This is essentially the process by which the time base of an oscilloscope is synchronized to the viewed waveform. In the case of the VCO in a PLL, the injected signal of concern is the unintentional introduction of the input signal v_i directly into the VCO. This injection usually causes the PLL to behave in a way that it was not designed to behave.

The effect of injection in a multivibrator is illustrated in Fig. 5–14a. An injected

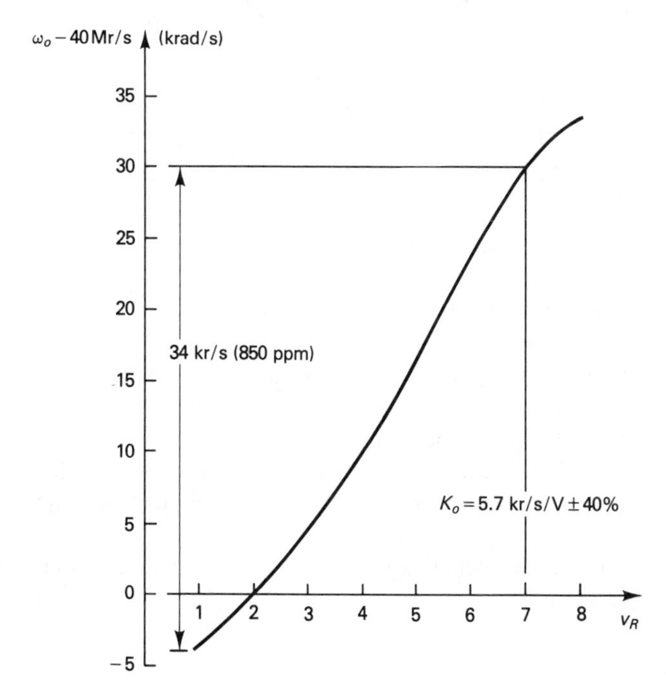

FIGURE 5–13 Voltage-controlled crystal oscillator characteristic

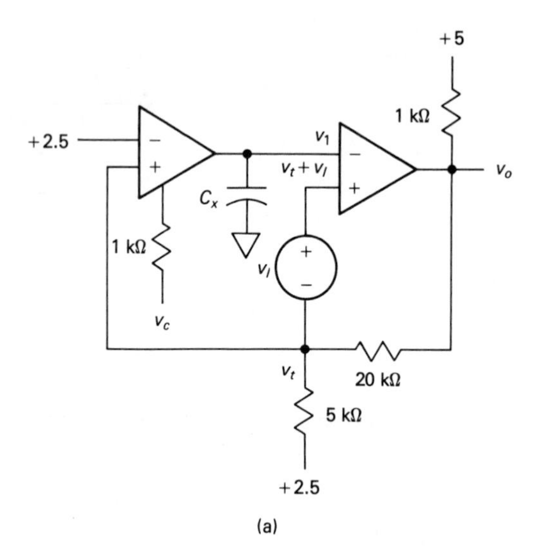

(a)

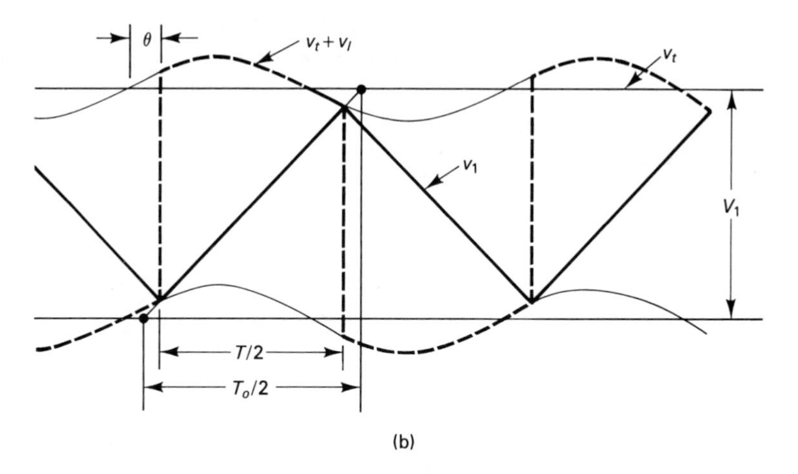

(b)

FIGURE 5–14 Frequency pulling by injection in a multivibrator

signal $v_I = V_I \sin(\omega_i t + \theta)$ is added to threshold voltage v_t (compare Fig. 5–2). The phase θ here uses the rising edge of v_t to define $t = 0$. In the steady state, the ramping voltage v_1 rises to meet the perturbed threshold at the same point in the cycle every time. In the example here, v_1 turns around before it would have if there were no injection. The result is that the oscillation frequency ω_o of the multivibrator is higher than normal—enough higher that $\omega_o = \omega_i$.

Let the frequency of the multivibrator without injection be ω_{oo}, and let the amount the frequency is pulled by injection be

$$\Delta\omega_o \equiv \omega_o - \omega_{oo} \tag{5-18}$$

The difference between the two (unperturbed) threshold voltages is V_1, the time v_1 would have risen without injection is $T_o/2$, and the time v_1 rises with injection is $T/2$ (see Fig. 5–14b). The slope of v_1 is a constant $\dot{v}_1$. Then

$$T_o/2 = V_1/\dot{v}_1 \tag{5-19}$$

and from the geometry in Fig. 5–14b it can be seen that

$$T/2 = (V_1 - 2V_I \sin\theta)/\dot{v}_1 \tag{5-20}$$

But $\omega_o = 2\pi/T$, and $\omega_{oo} = 2\pi/T_o$. Therefore

$$\begin{aligned}\omega_o/\omega_{oo} &= V_1/(V_1 - 2V_I \sin\theta)\\ &= 1/[1 - 2(V_I/V_1)\sin\theta]\\ &\approx 1 + 2(V_I/V_1)\sin\theta\end{aligned} \tag{5-21}$$

for $V_I << V_1$. But from Eq. (5-18)

$$\omega_o/\omega_{oo} = (\omega_{oo} + \Delta\omega_o)/\omega_{oo} = 1 + \Delta\omega_o/\omega_{oo} \tag{5-22}$$

Then from Eqs. (5-21) and (5-22),

$$\Delta\omega_o = 2\omega_{oo}(V_I/V_1)\sin\theta$$

This can be expressed more compactly as

$$\Delta\omega_o = K_I \sin\theta \tag{5-23}$$

where the *injection constant* K_I is given by

$$K_I \equiv 2\omega_{oo}(V_I/V_1) \tag{5-24}$$

The relation between $\Delta\omega_o$ and θ in Eq. (5-23) is actually part of a feedback loop since the phase θ_o of the oscillator is the integral of $\Delta\omega_o$, and θ is the difference between θ_o and the phase θ_I of the injected signal. This is summarized by the signal flow graph in Fig. 5–15. Note that this is identical to the flow graph for a first-order PLL with bandwidth K_I (compare Fig. 2–4). Therefore, it is possible to have phase locking with no phase detector—just an oscillator and an injected signal.

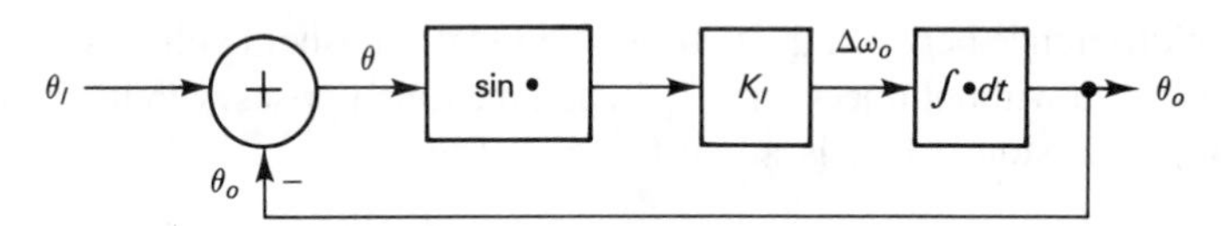

FIGURE 5–15 Model of oscillator with injection

5–8 INJECTION IN RESONANT OSCILLATORS

Resonant oscillators have a similar behavior in the presence of an injected signal (see Adler [7]). Figure 5–16a gives a model for a resonant oscillator with injection (compare Fig. 5–4). The voltage across the resonant circuit is $v_1 = V_1\sin(\omega_o t)$. The injected voltage is $v_I = V_I\sin(\omega_i t + \theta)$, where $\omega_o = \omega_i$ in lock, and θ is the phase of v_I relative to v_1. The effect of injection here is to produce $v_1' = v_1 + v_I$ with v_1' shifted in phase from v_1 by ϕ_2, where

$$\phi_2 \approx \tan^{-1}[(V_I\sin\,\theta)/V_1] \tag{5-25}$$

for $V_I << V_1$ (see phasor diagram in Fig. 5–16b). For oscillation to be sustained, the tank circuit must produce a compensating phase shift $-\phi_1$ between v_1' and v_1 (see Fig. 5–16c). Let v_1/v_1' be the transfer function of the transconductance amplifier and tank. This transfer function has a pair of poles a distance σ_o from the imaginary axis. The oscillation frequency for no injection is $\omega_{oo} = 1/\sqrt{LC}$. From the pole-zero plot in Fig. 5–16d, it is clear that

$$\begin{aligned}\phi_1 &= \tan^{-1}[(\omega_o - \omega_{oo})/\sigma_o]\\ &= \tan^{-1}[\Delta\omega_o/\sigma_o]\\ &= \tan^{-1}[2Q\Delta\omega_o/\omega_{oo}]\end{aligned} \tag{5-26}$$

where $\Delta\omega_o \equiv \omega - \omega_{oo}$ as in Eq. (5-18), and the Q of the tank is related to its bandwidth $2\sigma_o$ by

$$Q = \omega_{oo}/2\sigma_o \tag{5-27}$$

The phase of the transfer function v_1/v_1' function is $\text{ang}(v_1/v_1') = \tan^{-1}[(\omega - \omega_{oo})/\sigma_o]$; this is plotted in Fig. 5–16e.

For sustained oscillation, we must have $\phi_1 = \phi_2$. Then from Eqs. (5-25) and (5-26),

$$\Delta\omega_o = (\omega_{oo}/2Q)(V_I/V_1)\sin\,\theta \tag{5-28}$$

This can be put in the form we had for the multivibrator oscillator:

$$\Delta\omega_o = K_I\sin\,\theta \tag{5-29}$$

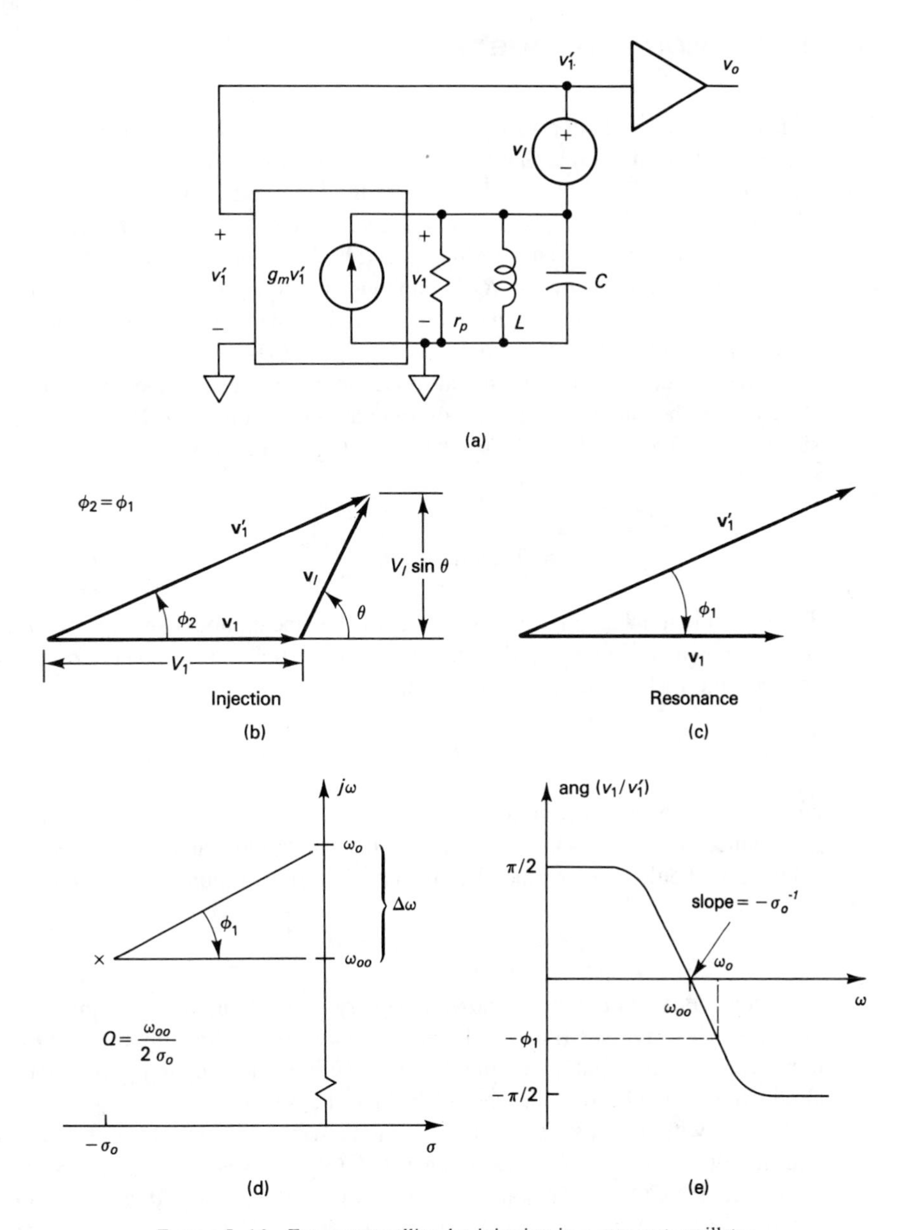

FIGURE 5–16 Frequency pulling by injection in a resonant oscillator

where for the resonant oscillator,

$$K_I \equiv (\omega_{oo}/2Q)(V_I/V_1) \tag{5-30}$$

Comparing Eq. (5-24), we see that Eq. (5-30) can include the case of a multivibrator if the "Q" of a multivibrator is taken as ¼.

5–9 PLL BEHAVIOR WITH INJECTION

When injection exists in a PLL, the VCO frequency $\Delta\omega_o$ is controlled in two ways: by a voltage v_c derived from a phase detector (as desired) and by injection [Eq. (5-29)]. This situation is modeled in Fig. 5–17a, where the loop filter is represented as in Fig. 3–1b. The injected signal is simply the input signal v_i attenuated to an amplitude V_I and shifted by a constant phase α. That is, the injection phase θ_I is given by $\theta_i + \alpha$. Using the relations $K = K_d K_h K_o$, $K_h = R_2/R_1$, and $\omega_2 = 1/R_2C$, we can simplify the model as in Fig. 5–17b. This is a large-signal model including the nonlinearities of the PD and of the injection. Note that $\theta \equiv \theta_I - \theta_o = \theta_I - (\theta_i - \theta_e) = \alpha + \theta_e$.

We can develop a small-signal model about the steady-state operating point. In the steady state, the integration in the loop filter assures that $\overline{\theta}_e = 0$. Therefore, θ_e is small, so $\sin\theta_e \approx \theta_e$, and $\cos\theta_e \approx 1$. We have the trigonometric identity

$$\sin\theta = \sin(\alpha + \theta_e) \equiv \sin\theta_e\cos\alpha + \cos\theta_e\sin\alpha$$
$$\approx \theta_e\cos\alpha + \sin\alpha$$

These small-signal approximations lead to the linear model in Fig. 5–17c. The term $(K + K_I\cos\alpha)\theta_e$ dominates the term $K\omega_2 \int \theta_e\,dt$ for $\omega > \omega_2$. Therefore, injection has changed the PLL bandwidth, from K to

$$K' = K + K_I\cos\alpha \qquad (5\text{-}31)$$

and a frequency offset $K_I\sin\alpha$ has been added.

Since we wish to be in control of the bandwidth, and since α is not known in general, K_I should be kept much less than K. As a rule of thumb, the designer should keep

$$K > 4K_I \qquad (5\text{-}32)$$

Then for a PLL with small bandwidth, every precaution should be made to minimize injection. This involves filtering the power to the VCO, shielding the VCO with a metal box, filtering the input v_c to the VCO, buffering the output v_o from the VCO, and eliminating ground loops that would include the VCO.

Even with the most care, some injection always exists. With careful circuit layout, a typical value of (V_I/V_1) is 1/1000. Then Eq. (5-24) gives a typical $K_I = \omega_{oo}/500$ for a multivibrator VCO. For a resonant VCO, Q is typically at least 20, and from Eq. (5-30) K_I is typically less than $\omega_{oo}/40{,}000$. The crystal in a VCXO typically has a Q of at least 20,000, and K_I is typically less than $\omega_{oo}/40{,}000{,}000$. Then Eq. (5-32) gives the following rules of thumb for a minimum practical bandwidth to avoid injection problems:

$$\begin{aligned} &K > \omega_{oo}/125; && \text{multivibrator VCO} \\ &K > \omega_{oo}/10{,}000; && \text{resonant VCO} \qquad (5\text{-}33) \\ &K > \omega_{oo}/10{,}000{,}000; && \text{crystal VCXO} \end{aligned}$$

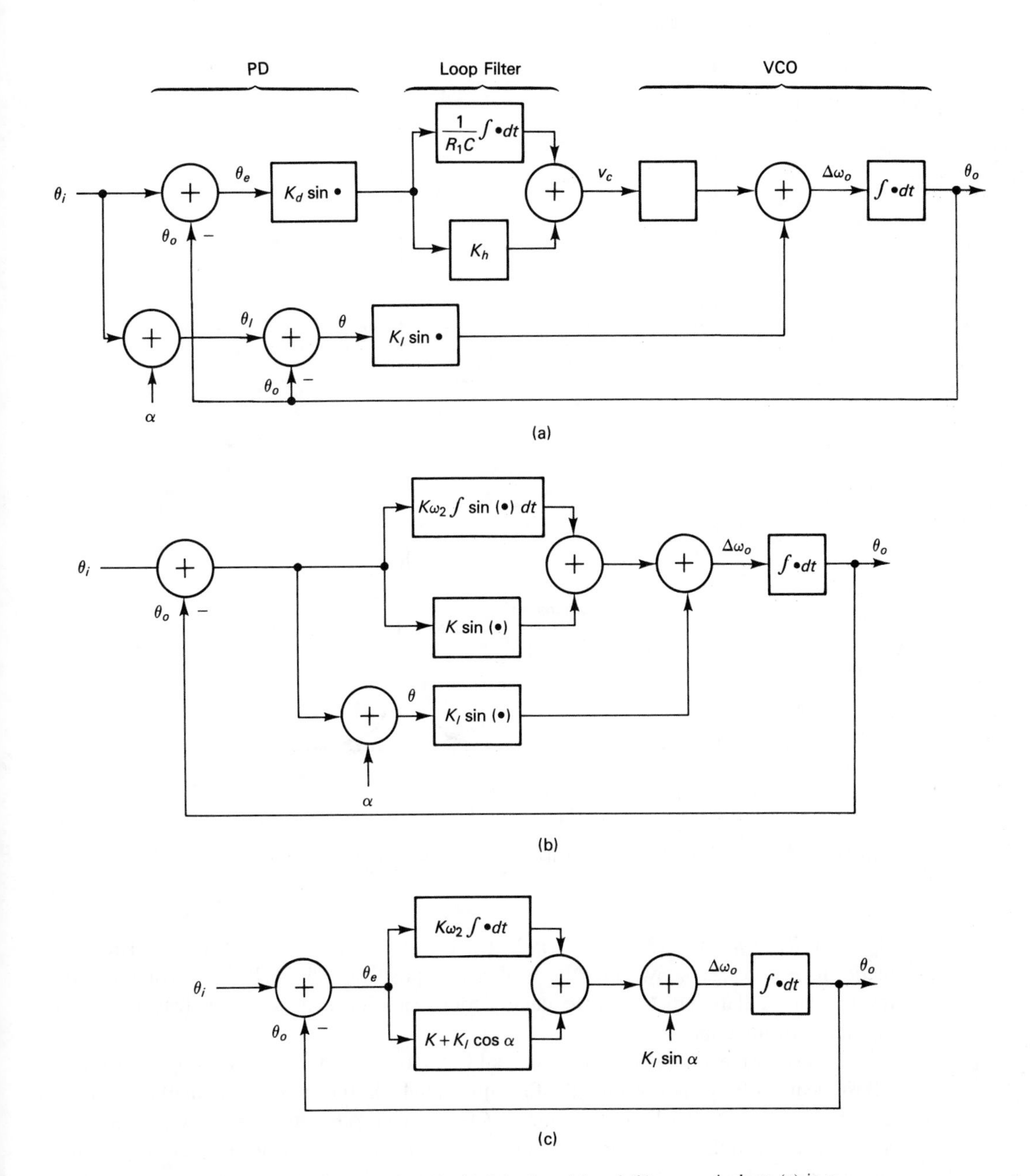

FIGURE 5–17 Models of PLL with injection. (a) and (b) are equivalent; (c) is a small-signal model

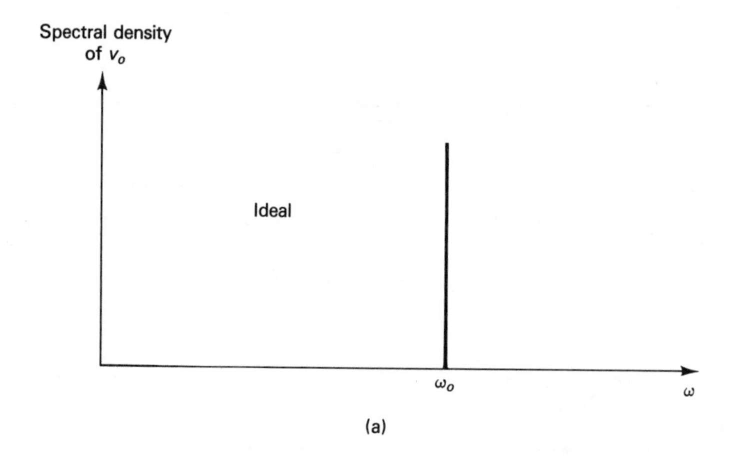

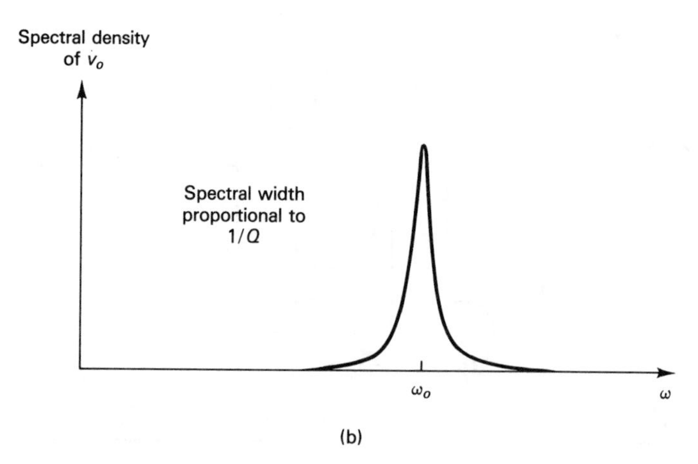

FIGURE 5–18 Oscillator spectral width due to internal noise

For $K_I > K$ and $\pi/2 < \alpha < 3\pi/2$, it is possible that $K + K_I\cos\alpha < 0$. Then the model in Fig. 5–17c shows that the feedback is positive and the PLL is unstable. The result is an oscillation of θ_o at a frequency around ω_2. This symptom is a clear indication of excessive injection.

A test for the amount of injection in a PLL is to ground the VCO input ($v_c = 0$), and adjust the input frequency ω_i until the PLL injection locks ($\omega_o = \omega_i$). For small K_I, ω_i may have to be very close to the free-running VCO frequency ω_{oo} (for $v_c = 0$) to obtain lock. Seek the center of the lock range so that $\theta \approx 0$. Then a small frequency change $\Delta\omega_i$ (with a corresponding $\Delta\omega_o = \Delta\omega_i$) will cause a $\Delta\theta$ in the phase between v_i and v_o. For $\theta \approx 0$, Eq. (5-29) gives

$$K_I = \Delta\omega_i/\Delta\theta \tag{5-34}$$

5–10 SPECTRAL PURITY

Ideally, an oscillator produces a single frequency ω_o; the output has a frequency spectrum consisting of a line of zero width (see Fig. 5–18a). In practice, the frequency is modulated by noise—thermal and shot noise originating within the oscillator itself. This causes the spectrum to have some width, as shown in Fig. 5–18b.

The mechanism causing the spectral width may be understood from the model for injection in an oscillator. Let the injected signal v_i in Fig. 5–16a be random noise rather than a sinusoid. It will be shown in Chapter 6 that noise can be represented as a sinusoid with random amplitude V_I and random phase θ. Then from Eq. (5-28), the noise causes frequency modulation of the VCO signal. The sidebands corresponding to the modulation give the spectral width. The modulation by the noise is inversely proportional to Q, so the spectral width decreases as Q increases. We will look at this more quantitatively in Chapter 6.

REFERENCES

[1] *Motorola MECL Data Book*, Motorola, Inc., Austin, Tex., 1986, Section 8.

[2] K. K. Clarke and D. T. Hess, *Communication Circuits: Analysis and Design*, Addison-Wesley: Reading, Mass., 1978, Section 2.3.

[3] Clarke and Hess, *Communication Circuits*, Chapter 6.

[4] U. L. Rohde, *Digital PLL Frequency Synthesizers*, Prentice-Hall: Englewood Cliffs, N.J., 1983, Section 4–1.

[5] Clarke and Hess, *Communication Circuits*, Section 2.4.

[6] Clarke and Hess, *Communication Circuits*, Section 6.7.

[7] R. Adler, "A Study of Locking Phenomena in Oscillators," *Proc. IRE and Waves and Electrons*, vol. 34 (June 1946), pp. 351–357.

CHAPTER

6

NOISE

We have considered the performance of phase-locked loops with perfect signals—perfect sinusoids or square waves with no noise. In fact, though, some noise is always present, and the noise may be significant, as in communications applications involving long distances. Noise appears at the input to the PLL along with the input voltage v_i, and there is a small amount of noise actually generated within the VCO. Both of these sources of noise contribute to phase noise at the PLL output. This chapter reviews the characterization of noise in the frequency domain and resolves the noise into an amplitude component and a phase component. Once we can view noise as phase, we can bring to bear on it the phase responses that we have developed for a PLL.

We begin with a review of some noise and random signal concepts. See Davenport and Root [1] and Papoulis [2] for more on these basics.

6–1 POWER SPECTRAL DENSITY

An example of a random noise waveform $n(t)$ is shown in Fig. 6–1a. A measure of its strength is the mean-square value $\overline{n^2}$, also called its "power." This is defined by

$$\overline{n^2} \equiv \lim_{T\to\infty} \int_0^T n^2(t)\, dt \qquad (6\text{-}1)$$

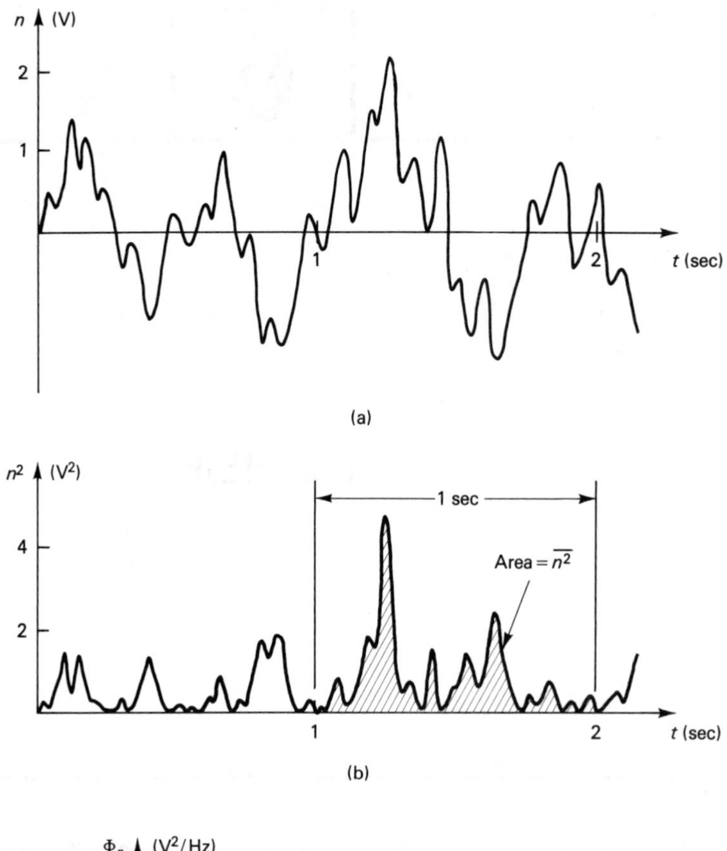

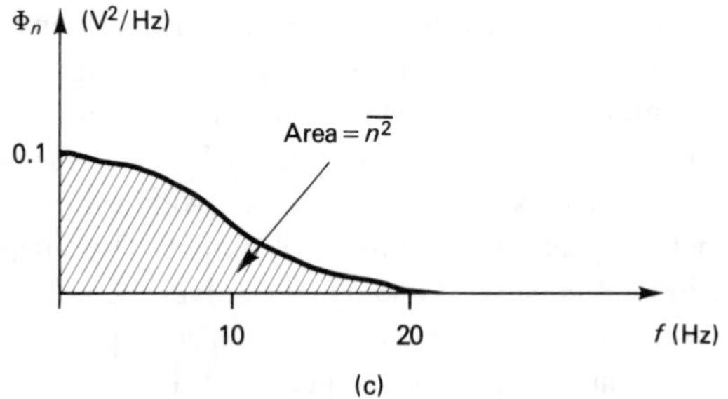

FIGURE 6–1 (a) Noise as a function of time; (b) mean-square noise in the time domain; (c) power spectral density of the noise

The integral in Eq. (6-1) can't strictly be shown as an area because it involves a limit. However, it can be thought of as the area under $n^2(t)$ for one unit of time, if $n(t)$ is "typical" during that time. Figure 6–1b shows an example. The root-mean-square (rms) value is defined as

$$n_{\text{rms}} \equiv \sqrt{\overline{n^2}} \tag{6-2}$$

The distribution of the power in the frequency domain is given by the one-sided *power spectral density* $\Phi_n(f)$ for the variable $n(t)$. [The term *one-sided* means that $\Phi_n(f)$ is zero for $f < 0$.] A typical power spectral density is shown in Fig. 6–1c; it gives for each frequency the mean-square value within a 1-Hz band centered on that frequency. Therefore, the total mean-square value is the area under the curve:

$$\overline{n^2} = \int_0^\infty \Phi_n(f)\, df \tag{6-3}$$

$\Phi(f)$ is traditionally called the *power* spectral density rather than the *mean-square* spectral density; it is simpler, and the two concepts are related by some constant impedance. It is also traditional to define $\Phi(f)$ so that the mean-square is given by the integral over f (in cycles per second) rather than by the integral over ω (in radians per second).

The equivalence of the two areas in Fig. 6–1 is a convenient tool for finding time averages [Eq. (6-1)] from frequency-domain information [Eq. (6-3)].

6–2 NOISE BANDWIDTH

Most sources of noise, such as shot noise and thermal noise, are white; they have a spectral density that is essentially flat for all frequencies, as in Fig. 6–2b. In accord with standard notation, N_o is the constant value of the white spectral density. Now, Eq. (6-3) says that this white noise n' has infinite power. In practice, though, there is always some filtering function limiting the bandwidth and therefore the power of the noise. The *noise bandwidth* of a filter is a value related to the 3-dB bandwidth that makes it easy to calculate this finite power.

Suppose n' is filtered by a transfer function $H_L(s)$ to produce an output signal n, as shown in Fig. 6–2a. The spectral density Φ_n of n is given in terms of the spectral density Φ_n' of n' by

$$\Phi_n(f) = \Phi_n'(f)\ |H_L(j2\pi f)|^2 \tag{6-4}$$

Since the noise is white, we can replace Φ_n' by a constant:

$$\Phi_n(f) = N_o\ |H_L(j2\pi f)|^2 \tag{6-5}$$

We will be concerned first with $H_L(s)$ a low-pass filter:

$$H_L(s) = \omega_{3dB}/(s + \omega_{3dB}) \tag{6-6}$$

This function has unity gain at dc and a first-order cutoff at ω_{3dB}. The corresponding spectral density Φ_n given by Eq. (6-6) is plotted in Fig. 6–2c.

Now we use Eq. (6-3) to get the mean-square noise at the output:

$$\overline{n^2} = \int_0^\infty N_o\ |H_L(j2\pi f)|^2\, df \tag{6-7}$$

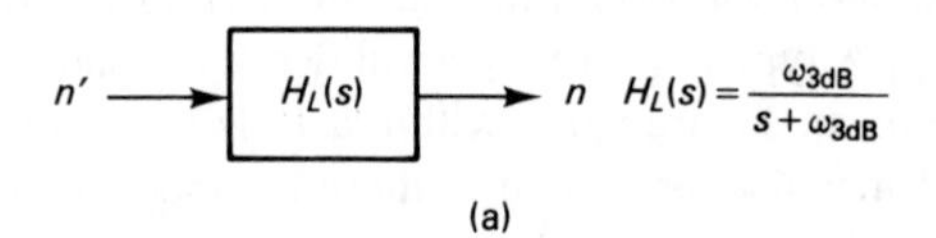

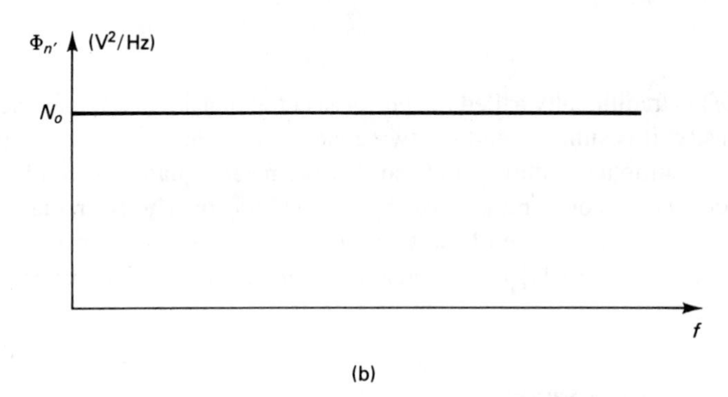

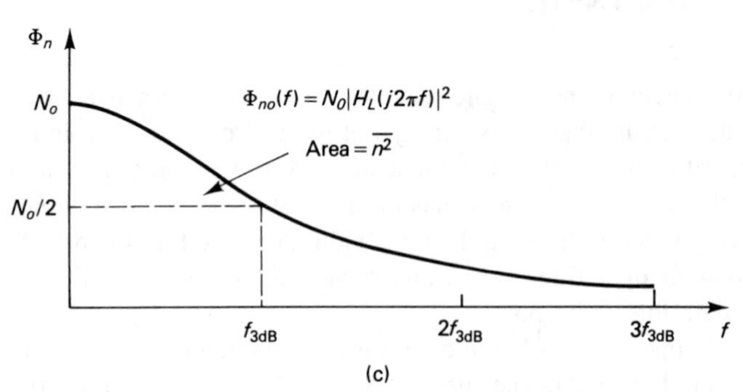

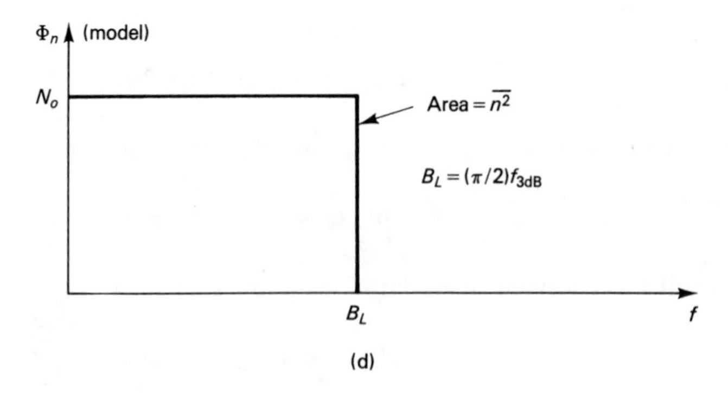

FIGURE 6–2 Noise through a low-pass function $H_L(s)$, and equivalent noise bandwidth B_L

It is convenient to define a *noise bandwidth* B_L for a low-pass transfer function $H_L(s)$ such that

$$\overline{n^2} = N_o B_L \tag{6-8}$$

[This is for the case $H_L(0) = 1$.] From Eqs. (6-7) and (6-8) it follows that the noise bandwidth is given by

$$B_L \equiv \int_0^\infty |H_L(j2\pi f)|^2 \, df \tag{6-9}$$

For the case of the low-pass filter in Eq. (6-6), carrying out the integration in Eq. (6-9) gives

$$B_L = \omega_{3dB}/4 = (\pi/2) f_{3dB} \tag{6-10}$$

where $f_{3dB} \equiv \omega_{3dB}/2\pi$. For the purpose of calculating the mean-square noise, Φ_n may be modeled by a rectangular spectrum with bandwidth B_L as shown in Fig. 6–2d. The area under this curve is the same as that in Fig. 6–2c.

We similarly define a noise bandwidth B_i for a bandpass filter

$$H_i(s) = \frac{\omega_{3dB} s}{s^2 + \omega_{3dB} s + \omega_i^2} \tag{6-11}$$

with unity gain at ω_i and a 3-dB bandwidth of ω_{3dB}. This filter acts on noise n' to produce an output n (see Fig. 6–3a). Again, n' has a flat spectral density N_o, as shown in Fig. 6–3b. Then $\Phi_n(f) = N_o |H_i(j2\pi f)|^2$, as shown in Fig. 6–3c. The noise bandwidth is defined such that

$$\overline{n^2} = N_o B_i \tag{6-12}$$

It follows that B_i is given by

$$\begin{aligned} B_i &\equiv \int_0^\infty |H_i(j2\pi f)|^2 df \\ &= (\pi/2) f_{3dB} \end{aligned} \tag{6-13}$$

where $f_{3dB} \equiv \omega_{3dB}/2\pi$. For the purpose of calculating mean-square noise, Φ_n may be modeled by a rectangular spectrum with bandwidth B_i centered on f_i, where $f_i = \omega_i/2\pi$ (see Fig. 6–3d).

6–3 NOISE-INDUCED PHASE

In applications where noise is significant, the PLL performance is improved by using a *prefilter* before the PLL to reduce the noise n' as much as possible without materially

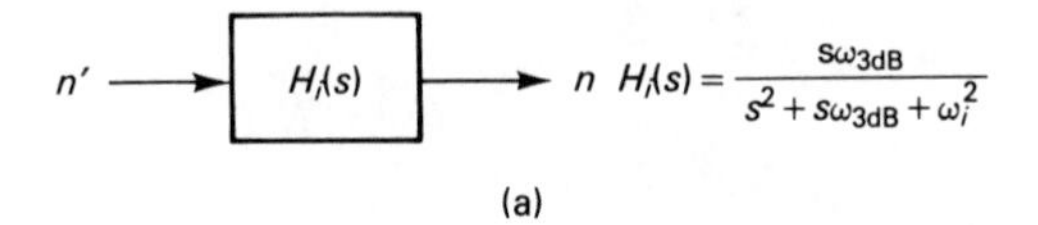

(a)

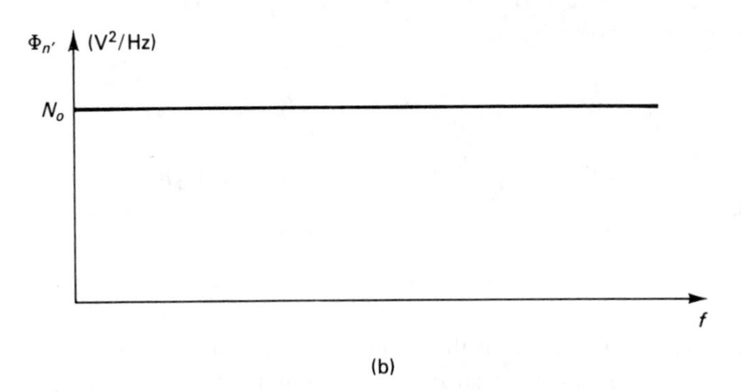

(b)

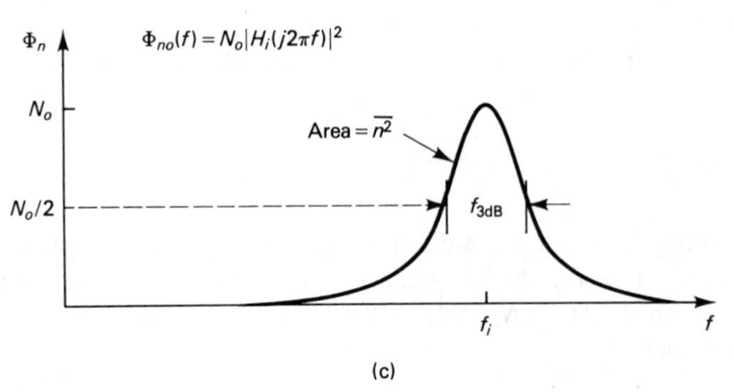

(c)

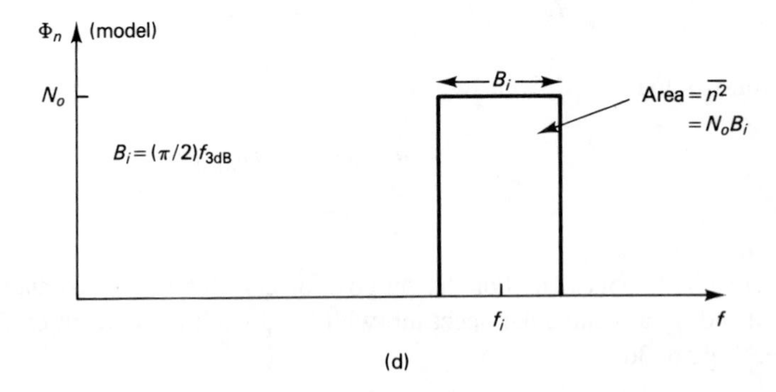

(d)

FIGURE 6–3 Noise through a bandpass function $H_i(s)$, and equivalent noise bandwidth B_i

affecting the signal v_i (see Fig. 6–4a). Let the input signal be a sinusoid without any modulation:

$$v_i = V_i \sin(\omega_i t)$$

We will see that the filtered noise n induces both amplitude modulation x and phase modulation θ_i:

$$v_i + n = (V_i + x) \sin(\omega_i t + \theta_i)$$

The effect of the amplitude modulation x is only to vary the PLL bandwidth K slightly by changing the PD gain [see Eq. (4-10)]. The PLL responds mostly to the phase θ_i, acting as a transfer function $H(s)$ to pass phase θ_o to the output:

$$v_o = V_o \cos(\omega_i t + \theta_o)$$

(See Fig. 6–4b.)

Consider the case of a bandpass prefilter with noise bandwidth B_i centered on $\omega_i/2\pi$. If the unfiltered noise n' is white with spectral density N_o, then the noise n after the filter has the spectral density model Φ_n shown in Fig. 6–5. (The spectrum is not actually rectangular, but the area under the curve is correct.) The mean-square noise appearing at the input to the PLL is $\overline{n^2} = N_o B_i$, as in Eq. (6-12).

The noise n can be divided into two components:

$$n = n_x + n_y \tag{6-14}$$

where

$$n_x = x \sin \omega_i t, \; n_y = y \cos \omega_i t \tag{6-15}$$

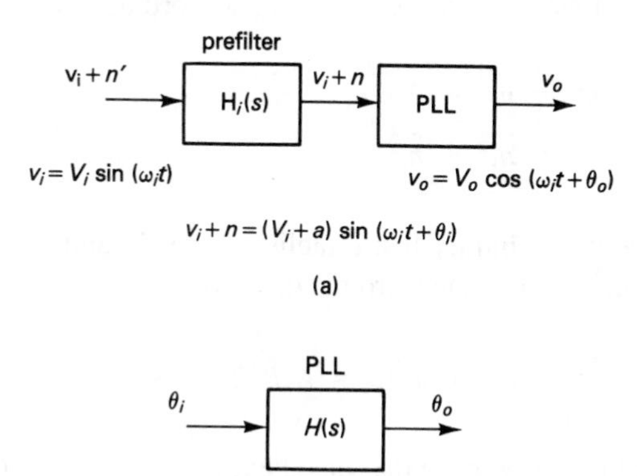

FIGURE 6–4 Prefiltered noise n inducing phase θ_i at PLL input

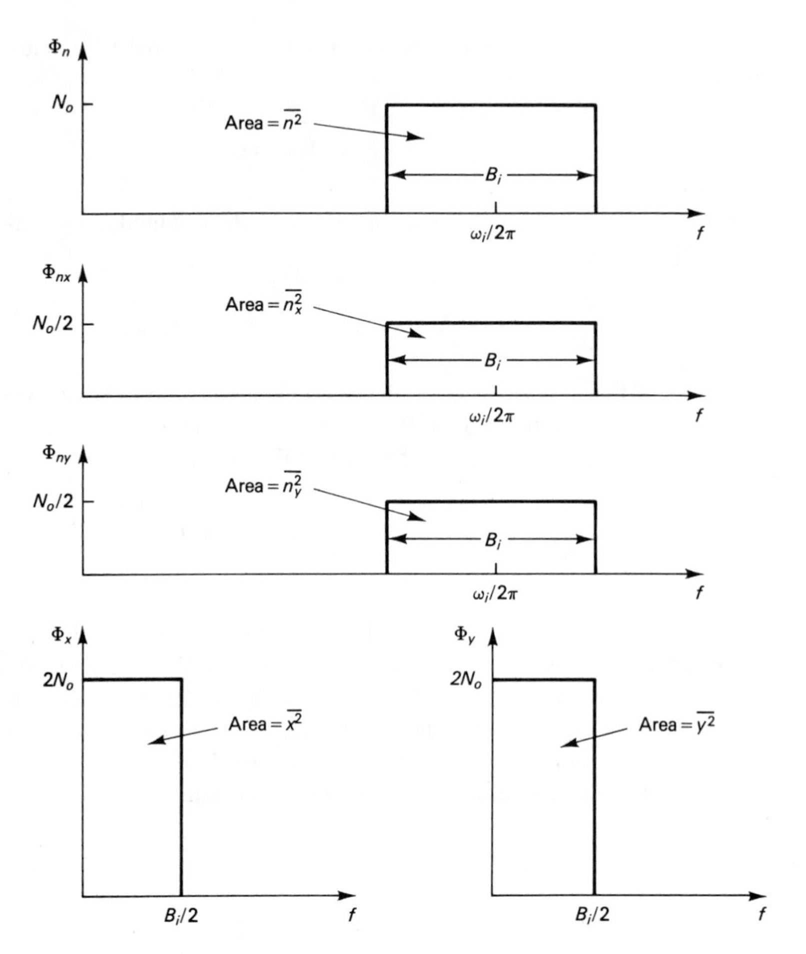

FIGURE 6–5 Power spectral densities of noise components

and x and y are random time functions. Since n_x and n_y are orthogonal

$$\begin{aligned}\overline{n^2} &= \overline{n_x^2} + 2\,\overline{n_x n_y} + \overline{n_y^2} \\ &= \overline{n_x^2} + \overline{n_y^2}\end{aligned} \tag{6-16}$$

But since the phase reference was arbitrary that established the sin and cos functions in Eq. (6-15), it must be that $\overline{n_x^2} = \overline{n_y^2}$. Then from Eq. (6-16)

$$\overline{n_x^2} = \overline{n_y^2} = \overline{n^2}/2 = N_o B_i/2 \tag{6-17}$$

Since the spectral densities of n_x and n_y have the same bandwidth B_i as n, it follows from Eq. (6-17) that the power spectral density of n_x and n_y is $N_o/2$ (see Φ_{nx} and Φ_{ny} in Fig. 6–5). We will see that n_x, comprising half the noise power, induces amplitude modulation of v_i, and n_y, comprising the other half of the noise power, induces phase modulation of v_i.

The expression for n_x in Eq. (6-15) has the form of suppressed-carrier amplitude modulation, where x is the baseband signal occupying a bandwidth of $B_i/2$ (see Fig. 6–5). What is the spectral density of x? Since x and $\sin \omega_i t$ are independent variables

$$\overline{n_x^{\,2}} = \overline{(x \sin \omega_i t)^2} = \overline{x^2}\ \overline{\sin^2 \omega_i t} = \overline{x^2}/2$$

But Eq. (6-17) has $\overline{n_x^{\,2}} = \overline{n^2}/2 = N_o B_i/2$. Therefore

$$\overline{x^2} = \overline{n^2} = N_o B_i \tag{6-18}$$

Because the bandwidth of Φ_x is $B_i/2$, its height must be $2N_o$ to give the area in Eq. (6-18). A similar development leads to the spectral density Φ_y shown in Fig. 6–5 and to

$$\overline{y^2} = \overline{n^2} = N_o B_i \tag{6-19}$$

Now let the noise n be added to the input signal $v_i = V_i \sin \omega_i t$. Using the expressions in Eqs. (6-14) and (6-15)

$$\begin{aligned} v_i + n &= V_i \sin \omega_i t + x \sin \omega_i t + y \cos \omega_i t \\ &= (V_i + x) \sin \omega_i t + y \cos \omega_i t \end{aligned}$$

For $x << V_i$ and $y << V_i$, the phasor diagram in Fig. 6–6 shows that

$$v_i + n \approx (V_i + x) \sin (\omega_i t + \theta_i) \tag{6-20}$$

where the random time function θ_i is given by

$$\theta_i = \tan^{-1}[y/(V_i + x)] \approx \tan^{-1}(y/V_i) \approx y/V_i$$

for $y << V_i$. This together with Eq. (6-19) gives

$$\overline{\theta_i^2} = \overline{y^2}/V_i^2 = N_o B_i/V_i^2 \tag{6-21}$$

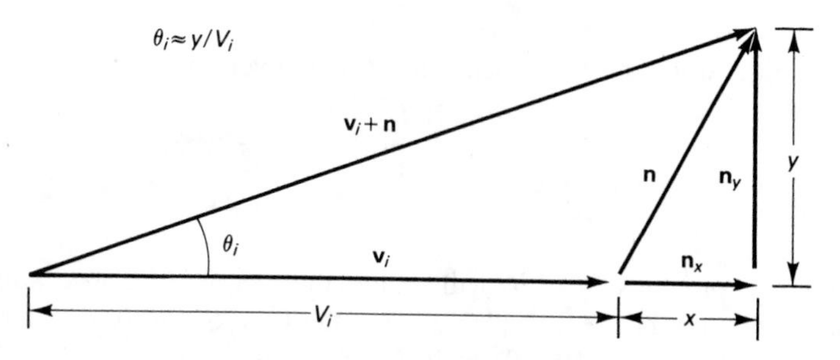

FIGURE 6–6 Noise-induced phase

The spectral density $\Phi_{\theta i}(f)$ for θ_i is shown in Fig. 6–7, where Θ_o represents $\Phi_{\theta i}(0)$. The area under the curve is

$$\overline{\theta_i^2} = \Theta_o B_i/2 \tag{6-22}$$

which, together with Eq. (6-21), gives the phase spectral density at low frequencies:

$$\Theta_o = 2\, N_o/V_i^2 \tag{6-23}$$

By comparing Φ_n in Fig. 6–5 with $\Phi_{\theta i}$ in Fig. 6–7, it can be seen that

$$\Phi_{\theta i}(f) = (2/V_i^2)\Phi_n(f_i + f); \qquad f \geq 0 \tag{6-24}$$

Sometimes the noise is specified in terms of its spectral density N_o, as in Eq. (6-23). Sometimes the relative noise power is specified as a signal-to-noise ratio at the PLL input:

$$SNR_i \equiv \overline{v_i^2}/\overline{n^2} = V_i^2/2N_oB_i \tag{6-25}$$

where the signal power is $\overline{v_i^2}$, and the noise power is $\overline{n^2}$. Then from Eqs. (6-23) and (6-25), the phase spectral density is

$$\Theta_o = 1/B_iSNR_i \tag{6-26}$$

See Blanchard [3] for a more rigorous development of the results in this section.

6–4 OUTPUT PHASE NOISE DUE TO INPUT NOISE

Part of the noise-induced θ_i passes through the PLL to produce phase noise θ_o at the output. In Chapter 3 we saw that a PLL acts as a phase low-pass filter. The transfer function of a second-order PLL with an active loop filter is

$$\frac{\theta_o}{\theta_i} \equiv H(s) = \frac{Ks + K\omega_2}{s^2 + Ks + K\omega_2} \tag{3-13}$$

As in Eq. (6-9), the noise bandwidth B_L is given by

$$B_L \equiv \int_0^\infty |H(j2\pi f)|^2\, df$$

$$B_L = (K + \omega_2)/4 \tag{6-27}$$

From now on, the symbol B_L will be used only for the *PLL noise bandwidth* as defined in Eq. (6-27). For $\omega_2 \rightarrow 0$, Eq. (3-13) has the form of Eq. (6-6) with $K = \omega_{3dB}$, and Eq. (6-27) reduces to Eq. (6-10) with $B_L = (\pi/2)f_{3dB}$, where $f_{3dB} = K/2\pi$. Note that Eq.

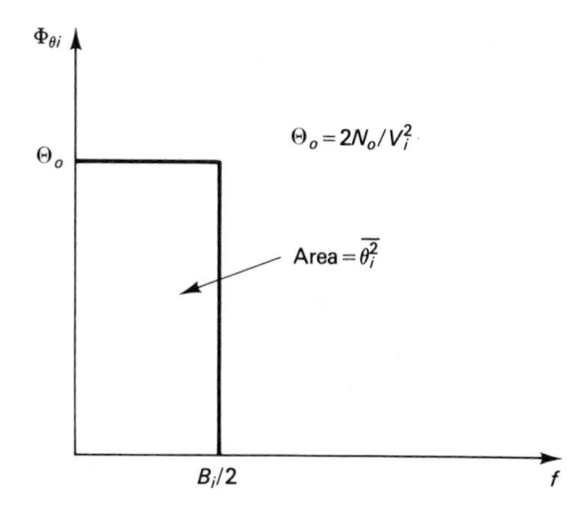

FIGURE 6–7 Spectral density of noise-induced phase

(6-27) has the dimensions Hz on the left and rad/sec on the right; the factor of 2π is included already. For the purpose of phase noise calculations, the frequency response $|H|$ of the PLL can be modeled as in Fig. 6–8a; it has unity gain up to $f = B_L$ and zero gain beyond that. Then as in Eq. (6-4), the spectral density of the output phase θ_o is given by

$$\Phi_{\theta o} = \Phi_{\theta i}|H(j2\pi f)|^2 \tag{6-28}$$

These spectral densities are shown in Figs. 6–8b and 6–8c for the case $B_L < B_i/2$, which is usually true in practice. The mean-square of θ_o is the area under $\Phi_{\theta o}$:

$$\overline{\theta_o^{\,2}} = \Theta_o B_L = 2B_L N_o/V_i^2 \tag{6-29}$$

or in terms of the signal-to-noise ratio

$$\overline{\theta_o^{\,2}} = B_L/B_i SNR_i \tag{6-30}$$

It is also of interest to know how much of the phase noise θ_i appears as phase error θ_e at the phase detector. For example, Eq. (4-46) shows the effect on K_d of noise-induced θ_e. The transfer function from θ_i to θ_e is

$$\frac{\theta_e}{\theta_i} \equiv H_e(s) = \frac{s^2}{s^2 + Ks + K\omega_2} \tag{3-28}$$

From Eqs. (6-3) and (6-4),

$$\begin{aligned}\overline{\theta_e^{\,2}} &= \int_0^\infty \Phi_{\theta i}(f)\,|H_e(2\pi ft)|^2\,df \\ &= \Theta_o \int_0^{B_i/2} |H_e(2\pi ft)|^2\,df \\ &= \Theta_o \int_0^{B_i/2} [1 - (1 - |H_e|^2)]\,df\end{aligned}$$

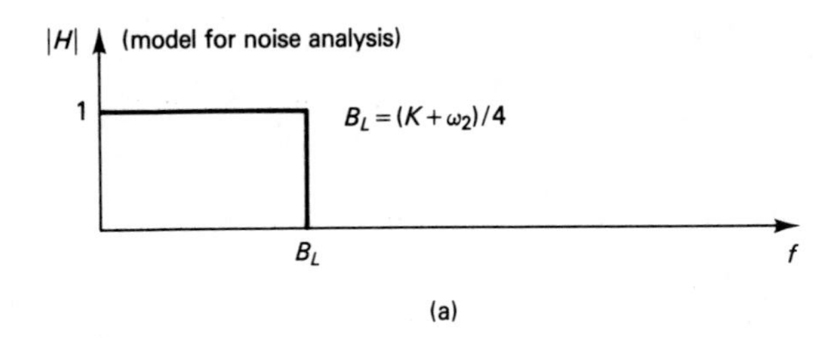

(a)

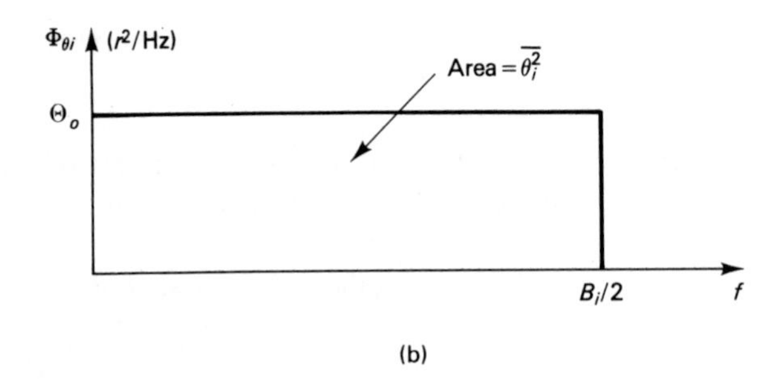

(b)

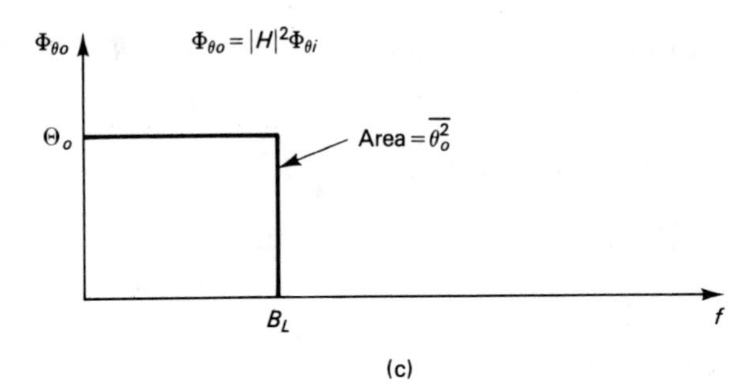

(c)

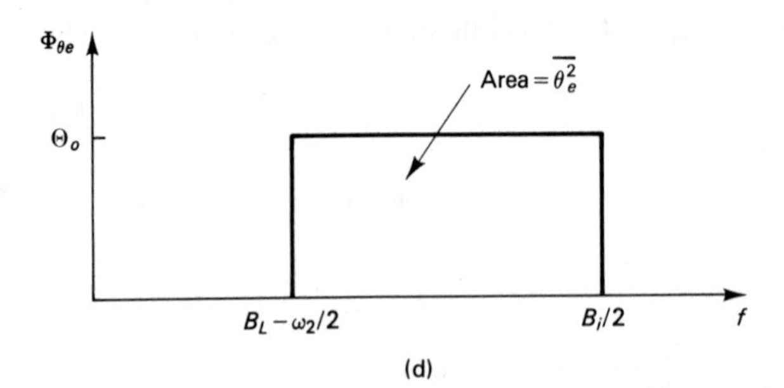

(d)

FIGURE 6–8 Phase noise in a PLL

$$= \Theta_o B_i/2 - \Theta_o \int_0^{B_i/2} (1 - |H_e|^2)\, df$$

$$\approx \Theta_o B_i/2 - \Theta_o \int_0^{\infty} (1 - |H_e|^2)\, df \tag{6-31}$$

for $B_L < B_i/2$. (The second term is the ''missing area'' between $f = 0$ and $f = B_L - \omega_2/2$ in Fig. 6–8d.) Evaluating the integral in Eq. (6-31) for H_e given in Eq. (3-28), we find

$$\int_0^{\infty} (1 - |H_e(2\pi f)|^2) df = (K - \omega_2)/4 \tag{6-32}$$

Therefore

$$\overline{\theta_e^2} = \Theta_o[B_i/2 - (K - \omega_2)/4] \tag{6-33}$$

Using Eq. (6-27), this can be put in terms of B_L:

$$\begin{aligned} \overline{\theta_e^2} &= \Theta_o[B_i/2 - (B_L - \omega_2/2)] \\ &= (2\, N_o/V_i^2)\, [B_i/2 - (B_L - \omega_2/2)] \end{aligned} \tag{6-34}$$

On this basis, a ''noise low-frequency cutoff'' for H_e can be thought of as $(K - \omega_2)/4$ or $B_L - \omega_2/2$, as indicated in Fig. 6–8d. Note that while it is true that $\theta_e = \theta_i - \theta_o$, according to Eqs. (6-22), (6-29), and (6-34), it is *not* true in general that $\overline{\theta_e^2} = \overline{\theta_i^2} - \overline{\theta_o^2}$. However, it is approximately true as ω_2 becomes much less than K.

EXAMPLE 6–1

Given: $f_o = 10$ MHz, the Q of the input bandpass filter is 20, $SNR_i = 10$ at the PLL input, $K = 400$ krad/s, $\omega_2 = 100$ krad/s, and the PD has a sinusoidal characteristic. Find $\theta_{i\ \mathrm{rms}}$, $\theta_{o\ \mathrm{rms}}$, and $\theta_{e\ \mathrm{rms}}$. Find the effect of $\theta_{e\ \mathrm{rms}}$ on the gain K_d of the PD.

The 3dB-bandwidth of the input filter is $f_o/Q = 500$ kHz. From Eq. (6-13), the noise bandwidth is $B_i = (\pi/2)f_{3\mathrm{dB}} = 785$ kHz. From Eq. (6-27), the noise bandwidth of the PLL is $B_L = (1/4)(400 + 100)$ krad/s $= 125$ kHz. From Eq. (6-26), the spectral density is $\Theta_o = 1/B_i SNR_i = 1.27 \times 10^{-7}$ rad^2/Hz. From Eq. (6-22), $\overline{\theta_i^2} = \Theta_o B_i/2 = 0.05$ rad^2, and $\theta_{i\ \mathrm{rms}} = \sqrt{0.05} = \underline{0.224 \text{ rad}}$. From Eq. (6-29), $\overline{\theta_o^2} = \Theta_o B_L = 0.016$ rad^2, and $\theta_{o\ \mathrm{rms}} = \sqrt{0.016} = \underline{0.126 \text{ rad}}$. From Eq. (6-34), $\overline{\theta_e^2} \approx \Theta_o(B_i/2 - B_L + \omega_2/2) = 0.040$ rad^2, and $\theta_{e\ \mathrm{rms}} = \sqrt{0.040} = \underline{0.201 \text{ rad}}$. From Eq. (4-46), $K_d' = K_d \exp(-\overline{\theta_e^2}/2) = 0.98\ K_d$, only a 2% reduction.

6-5 VCO PHASE NOISE

Noise within an oscillator causes phase noise θ_n at its output: $v_o = V_o \sin(2\pi f_o + \theta_n)$. In the next section we will look at how θ_n contributes to θ_o when the oscillator is in a PLL. First we will characterize the spectral density of θ_n when the oscillator is not in a PLL.

A model of a resonant oscillator is shown in Fig. 6–9a. This is the same model used in Fig. 5–16a to analyze injection in an oscillator. Here a noise voltage $n(t)$ rather than v_I is added to the oscillation voltage v_1. The amplitude of v_1 is V_1, and the oscillation frequency is $f_o = 1/2\pi\sqrt{LC}$. The $n(t)$ is thermal noise, shot noise, and flicker noise generated within the oscillator and represented here as all originating at the amplifier input. The thermal and shot noise contribute a flat spectral density

$$\Phi_n(f_o + f_m) = N_o; \qquad f_m > f_a \tag{6-35a}$$

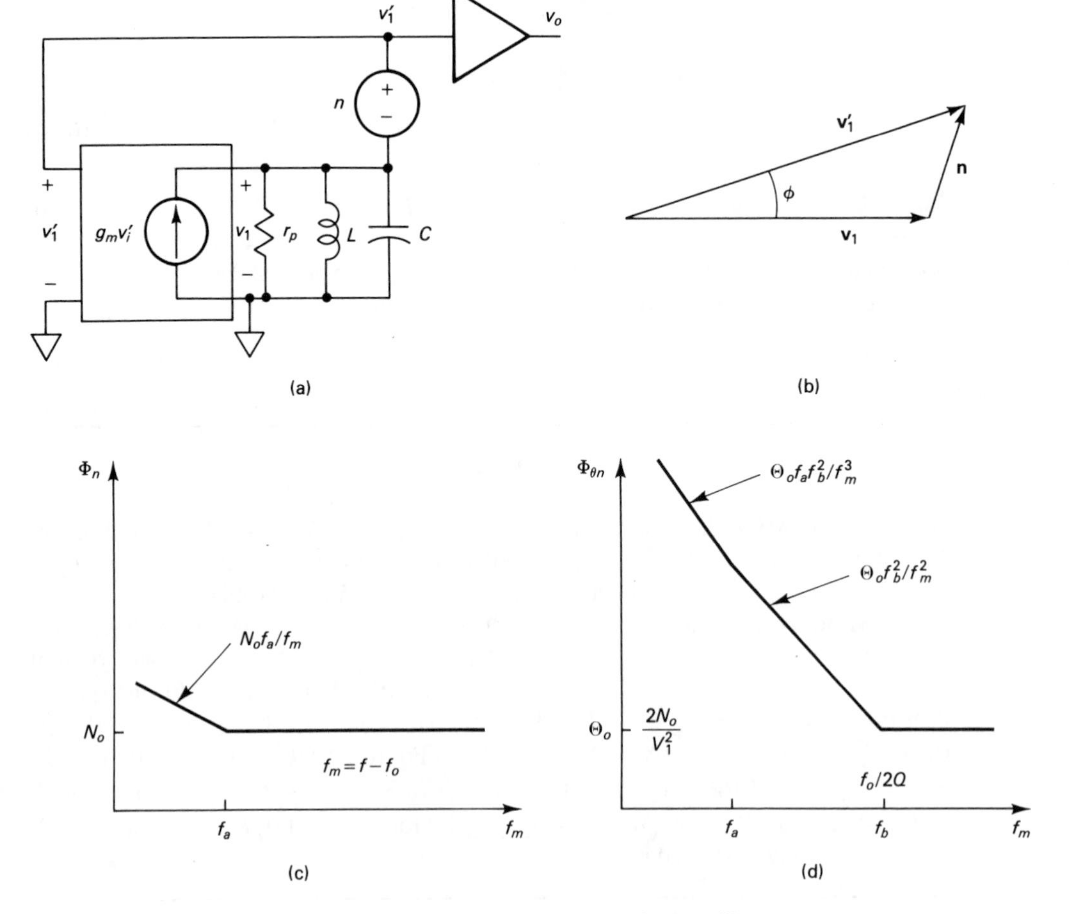

FIGURE 6–9 Phase noise θ_n in an oscillator

where

$$f_m \equiv f - f_o$$

is called the *offset frequency* from f_o. At frequencies near f_o (small f_m), the flicker noise dominates with a spectral density proportional to $1/f_m$:

$$\Phi_n\,(f_o + f_m) = N_o\,f_a/f_m; \qquad f_m < f_a \tag{6-35b}$$

The spectral density Φ_n is shown as a function of the offset frequency f_m in Fig. 6–9c for $f_m > 0$. (The portion of Φ_n for $f_m < 0$, not shown, is symmetrical about f_o.) The frequency f_a below which flicker noise dominates cannot be calculated; it must be measured. The value of f_a depends on the construction, materials, and environment of the oscillator, but it is typically around $10^{-5} \times f_o$.

The addition of $n(t)$ in Fig. 6–9a is represented as a phasor sum $\mathbf{v}_1' = \mathbf{v}_1 + \mathbf{n}$ in Fig. 6–9b. As in the phasor diagram in Fig. 5–16b, the effect is that $\mathbf{n}$ produces a phase difference ϕ between $\mathbf{v}_1$ and $\mathbf{v}_1'$. But since $\mathbf{n}$ is a noise phasor, we need the analysis in section 6–3 to find the relationship between $\mathbf{n}$ and ϕ. By a development similar to that of $\Phi_{\theta i}$ in Eq. (6-24),

$$\Phi_\phi(f_m) = (2/V_1^2)\ \Phi_n(f_o + f_m)$$

where Φ_ϕ is the spectral density of ϕ, and V_1 is the amplitude of the oscillation voltage $v_1 = V_1 \sin 2\pi f_o t$. Then with Eq. (6-35),

$$\Phi_\phi = \Theta_o\,f_a/f_m; \qquad f_m < f_a \tag{6-36a}$$

$$\Phi_\phi = \Theta_o; \qquad f_m > f_a \tag{6-36b}$$

where

$$\Theta_o \equiv 2N_o/V_1^2 \tag{6-37}$$

Because the phase ϕ is a baseband effect, f_m is no longer offset frequency but the entire frequency for which the spectral density Φ_ϕ is defined. At this point, convention supports either f or f_m for notation; we will retain the f_m through this section.

The bandwidth of the oscillator tank is f_o/Q, where the Q of the tank is discussed in section 5–5. Let the half-bandwidth be represented by f_b:

$$f_b = f_o/2Q \tag{6-38}$$

Spectral components of $n(t)$ that fall within the tank's bandwidth ($f_m < f_b$) cause frequency modulation of the oscillator as v_I did in the case of injection. Let the frequency deviation $\Delta\omega_o$ due to $n(t)$ be represented by ω_n. Then Eq. (5-26) for injection becomes

$$\phi = \tan^{-1}(2Q\omega_n/2\pi f_o) \approx 2Q\omega_n/2\pi f_o = \omega_n/2\pi f_b$$

$$\omega_n = (2\pi f_b)\phi$$

for $\phi << 1$. (The unperturbed oscillator frequency ω_{oo} is represented by $2\pi f_o$ in the analysis here.) Then the corresponding spectral densities $\Phi_{\omega n}$ and Φ_ϕ are related by

$$\Phi_{\omega n} = (2\pi f_b)^2 \Phi\phi; \qquad f_m < f_b$$

Then with Eq. (6-36)

$$\Phi_{\omega n} = \Theta_o\, (2\pi f_b)^2 f_a / f_m; \qquad f_m < f_a \tag{6-39a}$$

$$\Phi_{\omega n} = \Theta_o (2\pi f_b)^2; \qquad f_a < f_m < f_b \tag{6-39b}$$

Since the phase modulation θ_n is the integral of the frequency modulation ω_n, then $\theta_n(s) = (1/s)\ \omega_n(s)$ and

$$\theta_n(\, j2\pi f_m) = (1/j2\pi f_m)\ \omega_n(\, j2\pi f_m)$$

Therefore

$$\Phi_{\theta n} = (1/2\pi f_m)^2 \Phi_{\omega n}; \qquad f_m < f_b \tag{6-40a}$$

For $f_m > f_b$, the feedback in the oscillator is effectively broken, and $\theta_n = \phi$. Therefore

$$\Phi_{\theta n} = \Phi_\phi = \Theta_o; \qquad f_b < f_m \tag{6-40b}$$

Then from Eqs. (6-37) through (6-40),

$$\Phi_{\theta n}(\, f_m) = \Theta_o \frac{f_a f_b^2}{f_m^3} = \frac{N_o f_a f_o^2}{2V_1^2 Q^2} \cdot \frac{1}{f_m^3}; \qquad f_m < f_a \tag{6-41a}$$

$$\Phi_{\theta n}(\, f_m) = \Theta_o \frac{f_b^2}{f_m^2} = \frac{N_o f_o^2}{2V_1^2 Q^2} \cdot \frac{1}{f_m^2}; \qquad f_a < f_m < f_b \tag{6-41b}$$

$$\Phi_{\theta n}(\, f_m) = \Theta_o = \frac{2N_o}{V_1^2}; \qquad f_b < f_m \tag{6-41c}$$

This piecewise approximation to the phase spectral density is shown in Fig. 6–9d. Leeson [4] has confirmed experimentally the phase noise model expressed by Eq. (6-41).

In communication applications, it is desirable to keep the phase noise θ_n as small as possible. Equation (6-41) shows that θ_n is minimized by maximizing Q, the oscillation amplitude V_1, and the noise figure of the active components. Rohde [5] and Manassewitsch [6] give more detailed suggestions for the reduction of oscillator phase noise.

The flat portion of $\Phi_{\theta n}$ does not extend forever; otherwise the phase noise would have an infinite mean-square. In practice, the curve breaks at some cutoff frequency f_c, as shown in Fig. 6–10. The full expression for the phase spectral density of an oscillator not in a PLL is

$$\Phi_{\theta n}(f_m) = \Theta_o \frac{(f_m + f_a)(f_m^2 + f_b^2)\, f_c^2}{f_m^3\,(f_m^2 + f_c^2)} \tag{6-42}$$

where $\Theta_o = 2N_o/V_1^2$, and $f_b = f_o/2Q$.

EXAMPLE 6–2

A 50-MHz oscillator has a tank with a Q of 25. The signal amplitude at the tank is $V_1 = 1$ V, and noise with spectral density $N_o = 5 \times 10^{-14}$ V²/Hz adds internally to this signal. Below the frequency $f_a = 3$ kHz, flicker noise dominates. The output of the oscillator is filtered by a bandpass filter with half-bandwidth $f_c = 10$ MHz. Find the spectral density of the oscillator phase noise.

From Eq. (6-37), the spectral density for $f_b < f_m < f_c$ is $\Theta_o = 10^{-13}$ rad²/Hz, where Eq. (6-38) gives $f_b = 1.0$ MHz. Therefore, the spectral density $\Phi_{\theta n}(f_m)$ is that shown in Fig. 6–10.

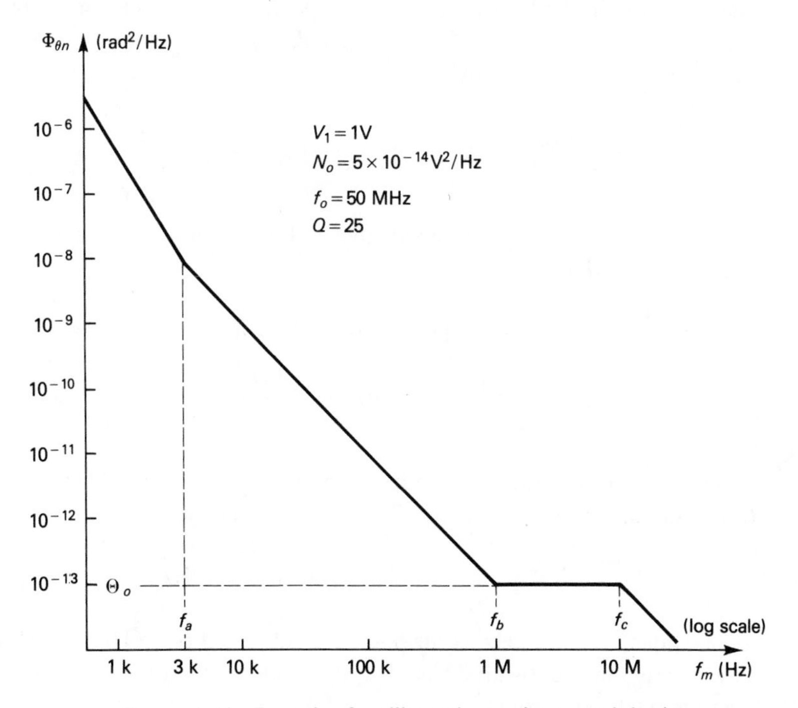

FIGURE 6–10 Example of oscillator phase noise spectral density

Since it is impractical to fully characterize the noise source $n(t)$, $\Phi_{\theta n}$ must be determined by measurement. There are several techniques for making this measurement. [7, 8] A simple method is to use a spectrum analyzer to measure the spectral density Φ_{vo} of the oscillator output v_o. [9] The oscillator output is

$$v_o = V_o \sin(2\pi f_o t + \theta_n) \tag{6-43}$$

Just as Eq. (6-24) gives $\Phi_{\theta i}$ for θ_i in Eq. (6-20), in a similar manner $\Phi_{\theta n}$ for θ_n in Eq. (6-43) is given by

$$\Phi_{\theta n}(f_m) = (2/V_o^2)\, \Phi_{vo}(f_o + f_m); \qquad f_m \geq 0 \tag{6-44}$$

where f_o is the oscillation frequency (or ''carrier'') and f_m is the offset frequency: $f_m = f - f_o$.

The dimension of Φ_{vo} is V^2/Hz, but a spectrum analyzer reads $\Phi_{vo}(f)$ in dBm/Hz, as shown in Fig. 6–11a. The two dimensions are related by

$$\text{dBm/Hz} = 10 \log \frac{\text{V}^2/\text{Hz}}{50\ \Omega \times 1\ \text{mW}} \tag{6-45}$$

The spectrum is converted to dBc/Hz (dB relative to the carrier per Hz) by dividing the spectral density in V^2/Hz by the ''carrier'' power $V_o^2/2$, as shown in Fig. 6–11b. This is part of the conversion from Φ_{vo} to $\Phi_{\theta n}$ indicated by Eq. (6-44). The carrier power is usually expressed in dBm by

$$\text{dBm (carrier)} = 10 \log \frac{V_o^2/2}{50\ \Omega \times 1\ \text{mW}} \tag{6-46}$$

For example, $V_o = 1$ V corresponds to 10 dBm. Then from Eqs. (6-45) and (6-46), we have the dimension

$$\begin{aligned} \text{dBc/Hz} &= \text{dBm/Hz} - \text{dBm (carrier)} \\ &= 10 \log \frac{\text{V}^2/\text{Hz}}{V_o^2/2} \end{aligned} \tag{6-47}$$

When the spectrum is translated down to baseband, as indicated by the conversion in Eq. (6-44), the dimension in Eq. (6-47) has the interpretation

$$\text{dBc/Hz} = 10 \log \text{rad}^2/\text{Hz} \tag{6-48}$$

For example, -30dBc/Hz corresponds to 10^{-3} rad^2/Hz. This translated spectrum is shown as $\Phi_{\theta n}$ in Fig. 6–11c. Note that this plot is with a linear frequency axis, as most spectrum analyzers provide. When converted to a log frequency axis, the plot of $\Phi_{\theta n}$ appears as in Fig. 6–10.

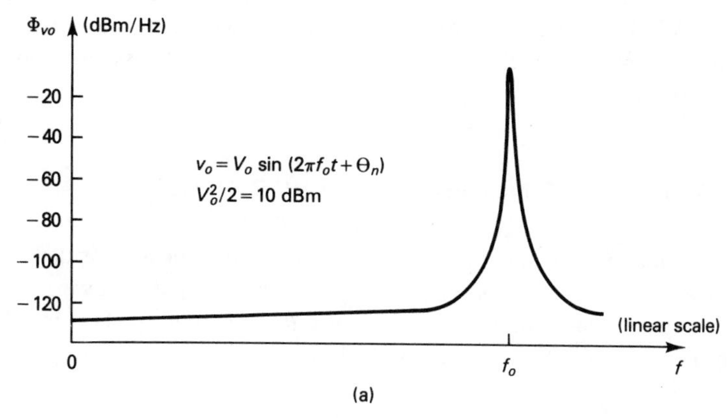

(a)

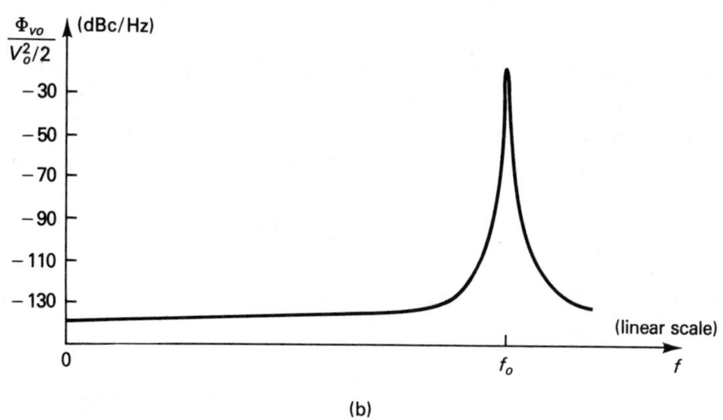

(b)

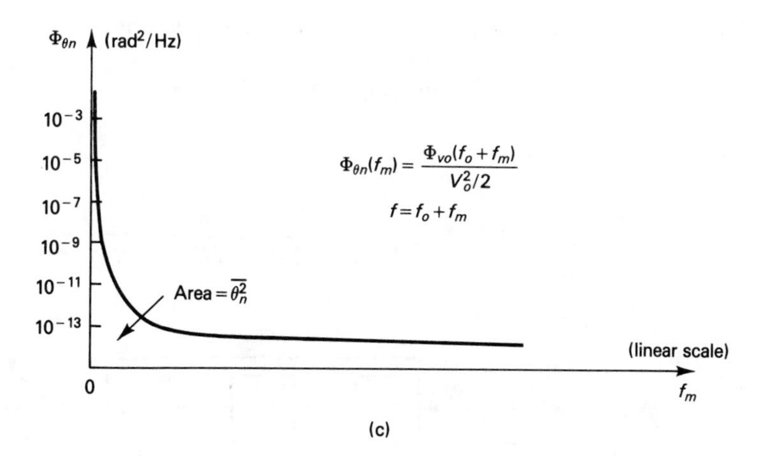

(c)

FIGURE 6–11 Phase spectral density from voltage spectral density

The characterization here of oscillator phase noise was for the case of an oscillator (or a VCO with v_c grounded) not in a PLL. We will see that a PLL can improve the phase noise of a VCO by locking it to a frequency source with less phase noise, such as a crystal oscillator.

6–6 OUTPUT PHASE NOISE DUE TO VCO NOISE

In Chapter 5 we ignored phase noise, and the VCO phase was determined by the control voltage: $\theta_o = v_c(K_o/s)$ (see Fig. 2–4). In the previous section we assumed $v_c = 0$, and the VCO phase was entirely the phase noise θ_n. When the VCO is in a PLL, both v_c and θ_n influence the VCO, and the VCO phase is

$$\theta_o = v_c(K_o/s) + \theta_n \tag{6-49}$$

This is represented by the signal flow graph in Fig. 6–12a. Here the input phase θ_i is assumed to be zero so we can see the effect of θ_n alone. Within the PLL bandwidth, θ_n is mostly canceled by the first term in Eq. (6-49) as the PLL tries to lock θ_o to θ_i.

From control theory, Fig. 6–12 leads to the transfer function

$$\theta_o/\theta_n = 1/[1 + G(s)] \tag{6-50}$$

But this is $H_e(s)$. Then from Eq. (3-28),

$$\frac{\theta_o}{\theta_n} = H_e(s) = \frac{s^2}{s^2 + Ks + K\omega_2} \tag{6-51}$$

From Eq. (6-4), the spectral density of the output phase is

$$\Phi_{\theta o}(f) = \Phi_{\theta n}(f)\ |H_e(j2\pi f)|^2 \tag{6-52}$$

where from Eq. (6-51)

$$|H_e(j2\pi f)|^2 = \frac{(2\pi f)^4}{(2\pi f)^4 + (K^2 - 2K\omega_2)(2\pi f)^2 + K^2\omega_2{}^2} \tag{6-53}$$

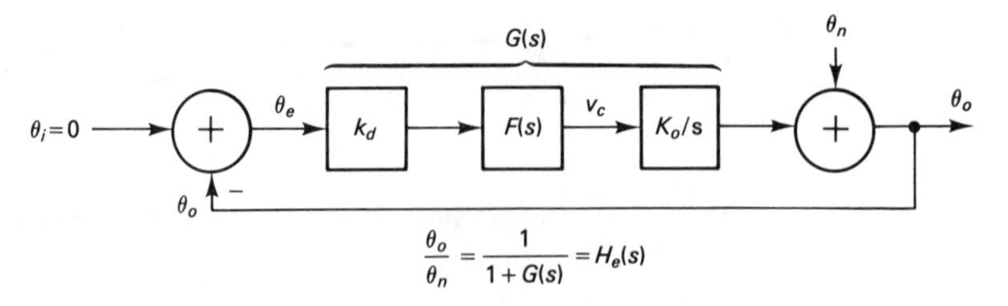

FIGURE 6–12 Transfer function from θ_n to θ_o

(We return to the simpler symbol f here rather than f_m.) The mean-square phase noise $\overline{\theta_o^2}$ is the area under the curve given by Eq. (6-52):

$$\overline{\theta_o^2} = \int_0^{\infty} \Phi_{\theta o}(f)df \tag{6-54}$$

Substituting Eqs. (6-42), (6-44), (6-52), and (6-53) into this equation and evaluating the integral yields

$$\overline{\theta_o^2} \approx \Phi_{\theta n}(K/2\pi)\left[\frac{K}{4} + \frac{f_a}{y}\,\ell\text{n}\frac{1+y}{1-y}\right] + \frac{\pi}{2}(f_c - f_b)\,\Theta_o \tag{6-55}$$

where

$$y \equiv \sqrt{1 - 4\omega_2/K}$$

This approximation holds for $f_a < K/2\pi < f_b/4$ and $f_b \leq f_c$. For the case $\omega_2 = K/4$, Eq. (6-55) becomes

$$\overline{\theta_o^2} \approx \Phi_{\theta n}(K/2\pi)[K/4 + 2f_a] + (\pi/2)(f_c - f_b)\,\Theta_o; \quad \omega_2 = K/4 \tag{6-56}$$

For the case $\omega_2 << K/4$, Eq. (6-55) can be approximated by

$$\overline{\theta_o^2} \approx \Phi_{\theta n}(K/2\pi)[K/4 + f_a\,\ell\text{n}(K/\omega_2)] + (\pi/2)(f_c - f_b)\,\Theta_o; \quad \omega_2 << K/4 \tag{6-57}$$

Note that for $f_a << K/4$, $\overline{\theta_o^2}$ is not a function of ω_2.

EXAMPLE 6–3

The phase noise spectral density of a VCO is that shown in Fig. 6–10. A bandpass filter with a 10-MHz half-bandwidth reduces the phase noise above $f = 10$ MHz (see $\Phi_{\theta n}$ in Fig. 6–13). The PLL has $K = 2\pi(100\text{ kHz})$ and $\omega_2 = 2\pi(10\text{ kHz})$. Find $\theta_{o\text{ rms}}$ for the VCO in the PLL.

A Bode plot of $|H_e|^2$ is shown in Fig. 6–13. It is unity for $f > K/2\pi$ and is proportional to f^2 between $\omega_2/2\pi$ and $K/2\pi$. Therefore, the product $\Phi_{\theta o} = \Phi_{\theta n}|H_e|^2$ follows $\Phi_{\theta n}$ for $f > 100$ kHz and is flat for 10 kHz $< f <$ 100 kHz. The value where it is flat is $\Phi_{\theta n}(K/2\pi) \approx \Theta_o\,(2\pi f_b/K)^2 = 10^{-11}$ rad^2/Hz [see Eq. (6-41b)]. Then from Eq. (6-55), $y = 0.775$, and

$$\begin{aligned}\overline{\theta_o^2} &= (10^{-11}\text{rad}^2/\text{Hz})[(\pi/2)(100\text{ kHz}) + 30.5\text{ kHz}] + (10^{-13}\text{rad}^2/\text{Hz})(\pi/2)(9\text{ MHz}) \\ &= 1.87 \times 10^{-6} + 1.41 \times 10^{-6} = 3.28 \times 10^{-6}\text{ rad}^2\end{aligned}$$

Then the square root gives $\theta_{o\text{ rms}} = \underline{0.0018\text{ rad}}$.

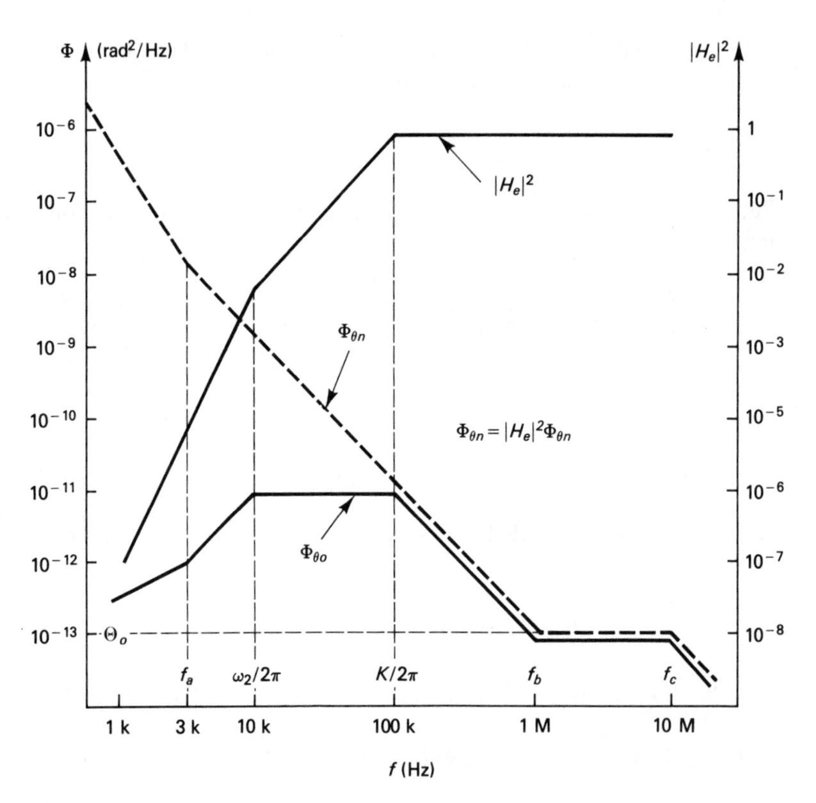

FIGURE 6–13 PLL phase noise θ_o due to VCO phase voise θ_n

6–7 OUTPUT PHASE NOISE DUE TO BOTH NOISE SOURCES

Let $\Phi_{\theta on}$ be the portion of $\Phi_{\theta o}$ due to $\Phi_{\theta n}$, as analyzed in section 6–6. Let $\Phi_{\theta oi}$ be the portion of $\Phi_{\theta o}$ due to $\Phi_{\theta i}$, as analyzed in section 6–4. Then, as in Eq. (6-52),

$$\Phi_{\theta on}(f) = \Phi_{\theta n}(f)\ |H_e(j2\pi f)|^2 \tag{6-58}$$

and as in Eq. (6-28),

$$\Phi_{\theta oi}(f) = \Phi_{\theta i}(f)\ |H(j2\pi f)|^2 \tag{6-59}$$

Since these phase noise components are from different sources, they are independent, and the powers sum to give the spectral density at the output:

$$\Phi_{\theta o}(f) = \Phi_{\theta on}(f) + \Phi_{\theta oi}(f) \tag{6-60}$$

Then $\overline{\theta_o^{\,2}}$ is the area under this curve.

EXAMPLE 6–4

$\Phi_{\theta i}$ at the input to a PLL is flat with spectral density 10^{-9} rad^2/Hz. The VCO phase noise spectral density $\Phi_{\theta n}$ falls as $1/f^2$ and is 10^{-9} rad^2/Hz for $f = 10$ kHz (see Fig. 6–13). (Here we are making the simplifying assumptions $f_a \approx 0$ and $f_b \approx f_c$.) Also, for simplicity, we assume $\omega_2 << K$ so that $B_L \approx K/4$. Find $\theta_{o\ \mathrm{rms}}$ for (a) $K = 2\pi(3.2$ kHz), (b) $K = 2\pi(10$ kHz), and (c) $K = 2\pi(32$ kHz).

The five spectra are plotted in Fig. 6–14a for the case $K = 2\pi(3.2$ kHz). The flat portion has the value $\Phi_{\theta n}(K/2\pi) = 10^{-8}$ rad^2/Hz. $\Phi_{\theta oi}$ follows $\Phi_{\theta i} = 10^{-9}$ for $f < K/2\pi$ and falls as $1/f^2$ for $f > K/2\pi$. As can be seen from the plots, $\Phi_{\theta oi}$ is always a factor of 10 below $\Phi_{\theta on}$. Therefore, $\Phi_{\theta oi}$ is negligible, and $\Phi_{\theta o} \approx \Phi_{\theta on}$ for this case. For $f_b = f_c$, the second term in Eq. (6-55) disappears, and $\overline{\theta_o^2} = \Phi_{\theta n}(K/2\pi)\ K/4 = 10^{-8} \times (\pi/2)$ (3.2 kHz) $= 5.0 \times 10^{-5}$ rad^2, and $\theta_{o\ \mathrm{rms}} =$ 0.00707 radians.

For the case $K = 2\pi(10$ kHz), the spectra are as shown in Fig. 6–14b. The plots for $\Phi_{\theta on}$ and $\Phi_{\theta oi}$ coincide here because $K/2\pi$ happens to be at the intersection of $\Phi_{\theta n}$ and $\Phi_{\theta i}$. The flat density of each is 10^{-9} rad^2/Hz for $f < 10$ kHz. The total $\Phi_{\theta o}$ is their sum, with a flat density of 2×10^{-9} for $f < 10$ kHz. Then the area under the curve is $\overline{\theta_o^2} = (2 \times 10^{-9}$ rad^2/Hz)$K/4 = (2 \times 10^{-9}$ rad^2/Hz) 15.7 kHz $= 3.14 \times 10^{-5}$ rad^2, and $\theta_{o\ \mathrm{rms}} =$ 0.0056 radians.

For the case $K = 2\pi(32$ kHz), the spectra are as shown in Fig. 6–14c. As can be seen from the plots, $\Phi_{\theta on}$ is always a factor of 10 below Φ_{oi}. Therefore, $\Phi_o \approx \Phi_{\theta oi}$ for this case. The area under the curve $\Phi_{\theta o}$ is $\overline{\theta_o^2} = (10^{-9}$ rad^2/Hz)$K/4 = (10^{-9}$ rad^2/Hz) 50 kHz $= 5.0 \times 10^{-5}$ rad^2, and $\theta_{\mathrm{rms}} =$ 0.00707 radians.

Of the three cases, note that the minimum $\theta_{o\ \mathrm{rms}}$ is realized when $K/2\pi$ is at the intersection of $\Phi_{\theta n}$ and $\Phi_{\theta i}$. Let this value of K be represented by K':

$$\Phi_{\theta n}(K'/2\pi) = \Phi_{\theta i} \tag{6-61}$$

This value of K minimizes $\overline{\theta_o^2}$ when $\Phi_{\theta i}(f)$ is flat and $\Phi_{\theta n}$ is dominated by Eq. (6-41b), falling as $1/f^2$. Then the minimized mean-square phase noise is

$$\overline{\theta_o^2} = 2\ \Phi_{\theta i}K'/4 = \Phi_{\theta i}K'/2 \tag{6-62}$$

Because K affects other performance parameters, the best choice of K may not be K'. In Chapter 9 we will see that reducing K reduces spurious modulation of θ_o. Therefore, the best tradeoff is sometimes achieved by some K less than K'. In that case, $\Phi_{\theta on}$ dominates as in Fig. 6–14a, and

$$\overline{\theta_o^2} \approx (K/4)\ \Phi_{\theta n}(K/2\pi)$$

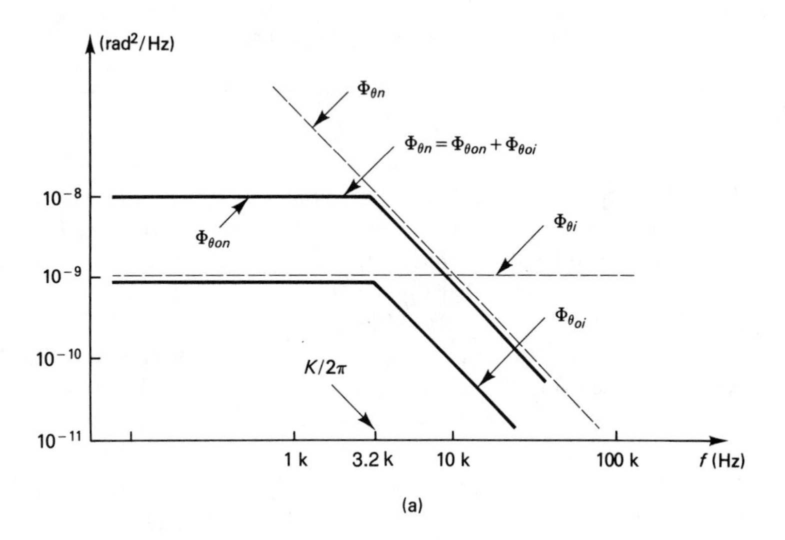

(a)

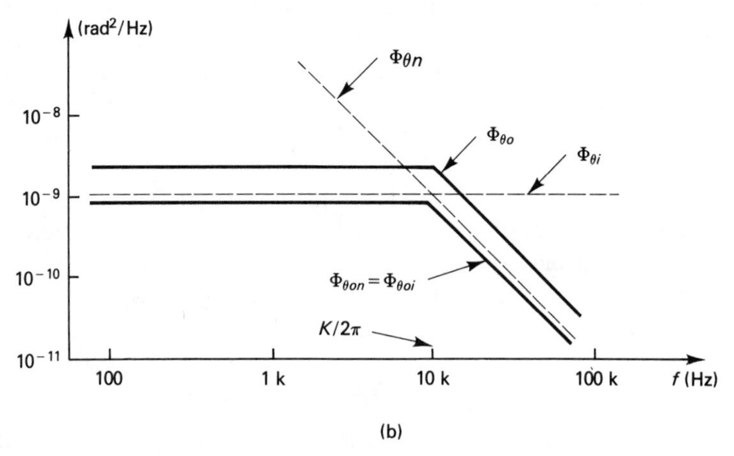

(b)

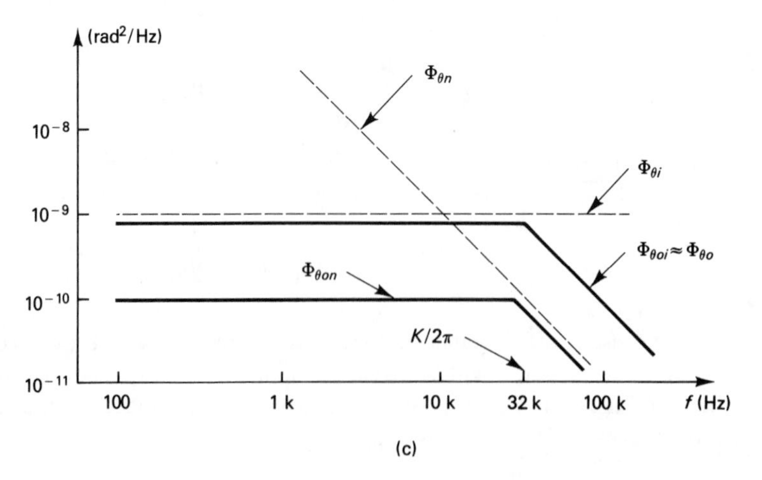

(c)

FIGURE 6–14 Phase noise both at input and in VCO

Here we are assuming that $\Phi_{\theta n}(f)$ falls as $1/f^2$ [see Eq. (6-41b)]. Then $\Phi_{\theta n}$ $(K/2\pi)$ is proportional to $1/K^2$, and

$$\overline{\theta_o^2} = a/K; \qquad K < K' \tag{6-63}$$

where a is a constant. Equation (6-63) indicates that, for the conditions assumed here, increasing K reduces phase noise. But increasing K increases spurious modulation of θ_o (see section 9–1–2 and section 11–4). Therefore, a tradeoff is involved.

6–8 CYCLE SLIPS

In the noise analysis so far we have assumed that the noise n was small enough compared with the signal v_i that the resulting phase θ_i was small—small enough that tan $\theta_i \approx \theta_i$ (see Fig. 6–6). But Gaussian noise has occasional large excursions, even for small $\theta_{i\ \mathrm{rms}}$. These can cause cycle slips (jumps of 2π) between the phases of v_i and v_o. The energy in these cycle slips is much greater than the amount of phase noise predicted by Eq. (6-30). This effect becomes significant at low SNR_i and causes the ''clicks'' in FM demodulation.

Consider the phasors for $\mathbf{v}_i$ and for $\mathbf{v}_i + \mathbf{n}$ shown in Fig. 6–15a. Since $v_i(t)$ is unmodulated, $\mathbf{v}_i$ stays horizontal. The noise $\mathbf{n}$ causes $\mathbf{v}_i + \mathbf{n}$ to deviate from $\mathbf{v}_i$, but usually not too far, as shown by the locus. Occasionally $\mathbf{n}$ becomes larger than V_i (the length of $\mathbf{v}_i$) and of the opposite phase from $\mathbf{v}_i$ so that $\mathbf{v}_i + \mathbf{n}$ goes around the origin (see the counterclockwise loop in Fig. 6–15a). This is a noise-induced phase change of 2π in θ_i [see the plot of $\theta_i(t)$ in Fig. 6–15c].

If the PLL bandwidth B_L is large enough to pass all the frequency components of θ_i, then $\mathbf{v}_o$ will follow the phase of $\mathbf{v}_i + \mathbf{n}$, as shown by the locus in Fig. 6–15b. Unlike the magnitude of $\mathbf{v}_\mathrm{i} + \mathbf{n}$, the magnitude of $\mathbf{v}_o$ is constant. (The loops in the locus are separated only so they are distinct.) Figure 6–15c shows $\theta_o \approx \theta_i$ for large B_L. As θ_i and θ_o settle down around 2π, there is one cycle difference in phase v_o and the noiseless input v_i, which never varied in phase. This is called a *cycle slip*.

Let B_L be reduced to the point that θ_o can't follow θ_i as it rises rapidly toward 2π radians. This is shown by the dashed curve of $\theta_o(t)$ in Fig. 6–15c. Once the difference between θ_i and θ_o exceeds π, $\mathbf{v}_o$ finds it closer to approach $\mathbf{v}_i + \mathbf{n}$ in a clockwise direction, and θ_o starts to decrease. Eventually, θ_i settles down around 2π, and θ_o settles down around zero; θ_o has avoided a cycle slip.

The probability of a cycle slip depends on the power of n that falls within the bandwidth B_L compared with the power $V_i^2/2$ of v_i. But from Eq. (6-29), this is proportional to $\overline{\theta_o^2}$. The exact analysis of cycle slips is difficult because of the nonlinear nature of the PLL for large noise. Viterbi [10] has derived an approximate expression for the mean time T_s between cycle slips for a first-order PLL:

$$T_s \approx (\pi/4B_L) \exp(2/\overline{\theta_o^2}); \qquad \omega_2 << K \tag{6-64}$$

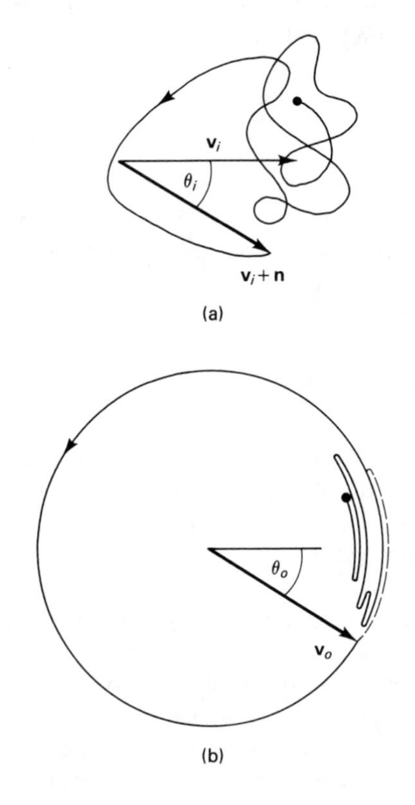

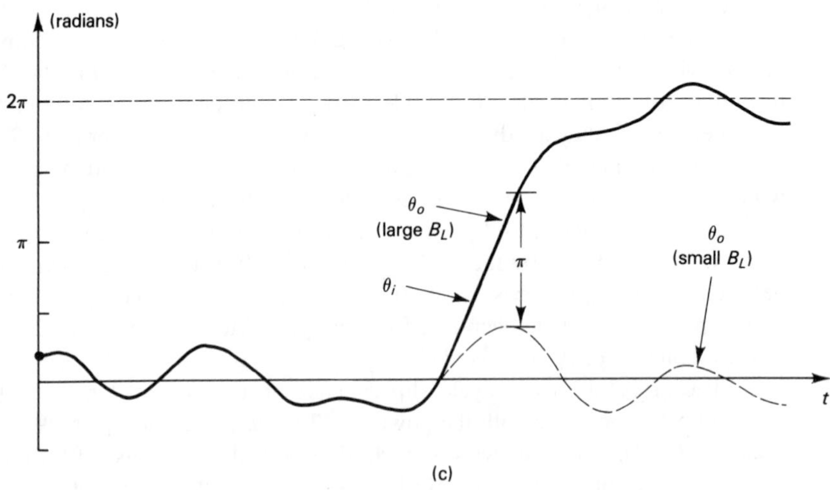

FIGURE 6–15 Cycle slip due to noise

We see the expected dependence on $\overline{\theta_o^2}$. The exponential function is about inversely proportional to the probability that θ_o exceeds 2 radians.

Equation (6-64) holds for highly damped cases with ω_2 much less than K. Experimental data by Ascheid and Meyr [11] show that for the critically damped case,

$$T_s \approx (\pi/5.6B_L)\exp(1.66/\overline{\theta_o^2}); \qquad \omega_2 = K/4 \tag{6-65}$$

Almost all applications fall between these two situations. Therefore, Eqs. (6-64) and (6-65) effectively serve as upper and lower bounds on T_s. See Blanchard [12] for a mathematical treatment of cycle slips.

EXAMPLE 6–5

A PLL has a noise bandwidth $B_L = 125$ krad/s and $\omega_2 = K/4$. For $\overline{\theta_o^2} = 0.016\ \text{rad}^2$ (see Example 6–1), find the mean time between cycle slips.

Equation (6-65) gives $T_s = 5.1 \times 10^{39}$ sec $= 1.6 \times 10^{32}$ yr. If the output phase noise is increased to $\overline{\theta_o^2} = 0.09\ \text{rad}^2$, then $T_s = 7.6$ minutes. If the damping were increased so that $\omega_2 << K$, then from Eq. (6-64), the 7.6 minutes would be increased to 7.8 hours.

This example indicates that cycle slips are generally negligible for $\overline{\theta_o^2} \leq 0.09\ \text{rad}^2$, or

$$\theta_{o\ \text{rms}} \leq 0.3\ \text{rad} \tag{6-66}$$

REFERENCES

[1] W. B. Davenport and W. L. Root, *Random Signals and Noise*, McGraw-Hill: New York, 1958.

[2] A. Papoulis, *Probability, Random Variables, and Stochastic Processes*, McGraw-Hill: New York, 1965.

[3] A. Blanchard, *Phase-Locked Loops: Application to Coherent Receiver Design*, Wiley: New York, 1976, Chapter 7.

[4] D. B. Leeson, ''A Simple Model of Feedback Oscillator Noise Spectrum,'' *Proc. IEEE*, February 1966, pp. 329–330.

[5] U. L. Rohde, *Digital PLL Frequency Synthesizers*, Prentice-Hall: Englewood Cliffs, NJ, 1983, section 4–1–2.

[6] V. Manassewitsch, *Frequency Synthesizers*, Wiley: New York, 1987, section 2–2.

[7] Manassewitsch, *Frequency Synthesizers.*

[8] Rohde, *Digital PLL Frequency Synthesizers*, section 2–8.

[9] Hewlett Packard Associates, ''Phase Noise Characterization of Microwave Oscillators,'' *Product Note 11729C-2* Palo Alto, Calif., September, 1985.

[10] A. J. Viterbi, *Principles of Coherent Communications*, McGraw-Hill: New York, 1966, Chapter 4.

[11] G. Ascheid and H. Meyr, ''Cycle Slips in Phase-Locked Loops: A Tutorial Survey,'' *IEEE Trans. on Com.*, Vol. COM-30, No. 10 (October 1982), pp. 2228–41, Fig. 18.

[12] Blanchard, *Phase-Locked Loops*, Chapter 12.

CHAPTER

7

MAINTAINING LOCK

Our analysis of phase-locked loops in Chapters 2 and 3 used linear models of the PD and the VCO. If the input frequency deviation is too large, those models no longer hold. When the PD reaches the limit of the voltage it can put out or the VCO reaches the limit of the frequency it can generate, the PLL loses lock. In this chapter, we look at limitations on input frequency so the PLL will maintain lock.

7–1 HOLD-IN RANGE

Consider a PLL that is in lock ($\omega_o = \omega_i$), and let the input frequency ω_i be changed very slowly. The range of frequencies over which the PLL can stay in lock is the *hold-in range*. This is essentially the range of frequencies the VCO can reach given the limitations of its VCO characteristic and the limitations on its control voltage v_c. This rather simple concept of hold-in range is best illustrated by an example rather than by developing general formulas.

EXAMPLE 7–1

A PLL has a PD with the characteristic shown in Fig. 7–1a, a VCO with the characteristic shown in Fig. 7–1b, and one of the loop filters shown in Fig. 7–1c, d, or e. (Note that $K_h = 0.4$ for each of the loop filters.) Find the hold-in range of the PLL with each of the loop filters.

The hold-in range can't exceed the range from 7 Mrad/s to 23 Mrad/s since the VCO characteristic doesn't extend beyond this range. Any further restriction of the range is due to restriction of v_c. The maximum steady-state control voltage is

$$v_{c\ \max} = F(0)\ V_{dm} \tag{7-1}$$

where $F(0)$ is the dc gain of the loop filter and V_{dm} is the maximum v_d the PD can put out. By the symmetry of the PD characteristic in this example, $v_{c\ \min} = -\ v_{c\ \max}$.

For the voltage divider "loop filter" in Fig. 7–1c, $F(0) = 0.4$. Since $V_{dm} = 5$ V, Eq. (7-1) gives $v_{c\ \max} = 2$ V. From the VCO characteristic in Fig. 7–1b, the corresponding upper limit of the hold-in range is $\omega_{o\ \max} = \underline{9 \text{ Mrad/s}}$. The minimum v_c is -2 V, which exceeds the lower end of the VCO characteristic. Therefore, the lower limit of the hold-in range is $\omega_{o\ \min} = \underline{7 \text{ Mrad/s}}$.

For the passive loop filter in Fig. 7–1d, $F(0) = 1.0$. Then Eq. (7-1) gives $v_{c\ \max} = 5$ V. From the VCO characteristic, the corresponding upper limit of the hold-in range is $\omega_{o\ \max} = \underline{15 \text{ Mrad/s}}$. The minimum v_c is -5 V, which exceed the lower end of the VCO characteristic. Therefore, the lower limit of the hold-in range is still $\omega_{o\ \min} = \underline{7 \text{ Mrad/s}}$.

For the active loop filter in Fig. 7–1e, $F(0) = \infty$. Then v_c is unlimited, or in practice as high a voltage as the op amp can provide. We will assume that this is greater than 9 V, and the upper limit of the hold-in range is 23 Mrad/s. As before, the lower limit is 7 Mrad/s.

7–2 INPUT FREQUENCY DEVIATION $\Delta\omega_i$

When the input frequency changes rapidly, the PLL may lose lock before the limits of the hold-in range. This is because $|F(j\omega)| \leq F(0)$ which, together with Eq. (7-1), says v_c is more restricted for high modulation frequency ω. The physical reason is that the capacitor in the loop filter doesn't have time to charge in following rapid changes. The rest of this chapter will find limits on input frequency deviation so the PLL will be able to maintain lock.

We need to define input frequency deviation. In Chapter 2, ω_i was defined as a constant in the expression for the input signal: $v_i = \sin(\omega_i t + \theta_i)$, where θ_i is a function of time. Therefore, any deviation of the input frequency must come through θ_i. For zero-mean θ_i, ω_i is the *average* input frequency, but the *complete* frequency is the time

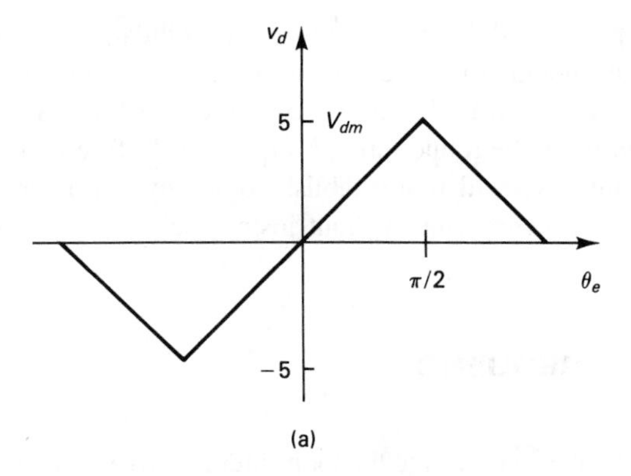

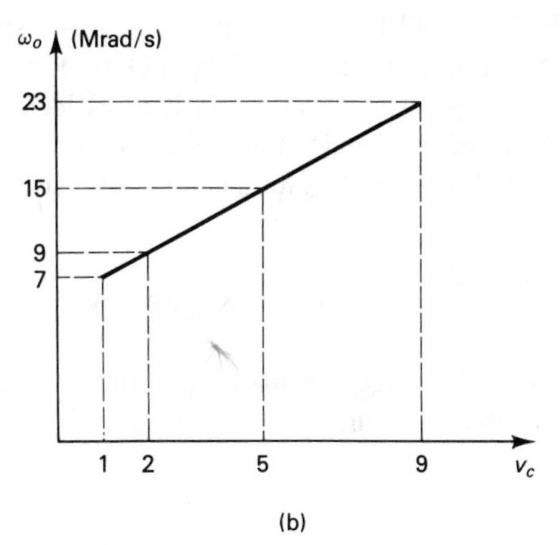

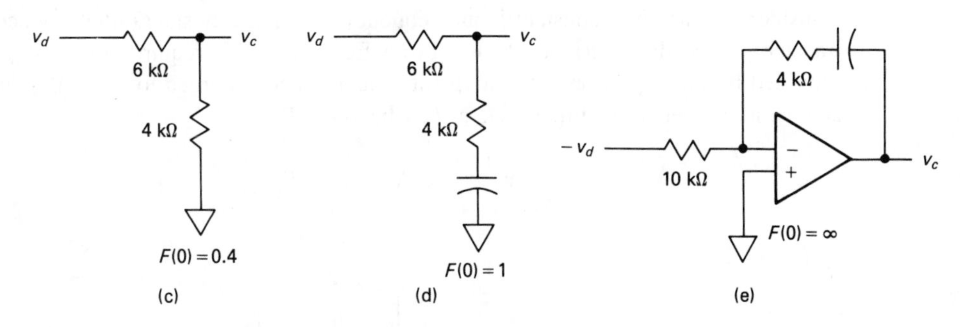

FIGURE 7–1 Example of hold-in range

derivative of the argument $\omega_i t + \theta_i$. Let this complete input frequency be represented by $\omega_i + \Delta\omega_i$, where $\Delta\omega_i$, is the frequency deviation. Then

$$\omega_i + \Delta\omega_i \equiv d/dt\,(\omega_i t + \theta_i) = \omega_i + d\theta_i/dt$$

$$\Delta\omega_i \equiv d\theta_i/dt \tag{7-2}$$

In fact, phase modulation and frequency modulation are indistinguishable. Deviations that are large enough to cause loss of lock are usually thought of as $\Delta\omega_i$ rather than θ_i. Therefore, we will find limits on $\Delta\omega_i$ so the lock is maintained. If desired, corresponding limits on θ_i can be found through Eq. (7-2). Before looking at a more exact derivation of these limits, we will first establish some physical insight as to what allows the PLL to handle a frequency step without losing lock.

7–3 LOCK-IN FREQUENCY ω_L

In this chapter, we assume the loop filter is active, which provides the best performance in maintaining lock. The PLL circuit in Fig. 7–2 uses a simple version of such a loop filter (see Fig. 3–5 for other versions). The PD in Fig. 7–2 is modeled by one of the characteristics in Fig. 7–3. Usually a PLL loses lock when θ_e exceeds the PD range—when the PD is asked to provide more voltage than V_{dm}. Figure 7–3 shows the phase error θ_{em} that corresponds to $v_d = V_{dm}$ for a number of phase detectors. The VCO is modeled as in Chapter 2 by

$$\omega_o = K_o (v_c - V_{co}) + \omega_i \tag{7-3}$$

Note that the control voltage v_c is made up of the voltage across R_2 plus the voltage across the capacitor in the loop filter:

$$v_c = v_2 + v_3 \tag{7-4}$$

Consider the case of a constant input frequency ω_i. In steady state (after any acquisition transients have died out), $\omega_o = \omega_i$. From Eq. (7-3), this requires $v_c = V_{co}$. This is provided by the capacitor in the loop filter having been charged to $v_3 = V_{co}$ during the acquisition. Then from Eq. (7-4), v_2 can be zero. But

$$v_2 = (R_2/R_1)\, v_d = K_h\, v_d \tag{7-5}$$

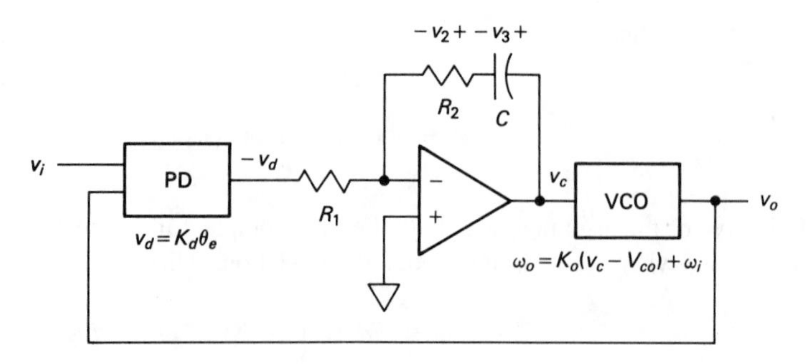

FIGURE 7–2 Example of second-order PLL circuit

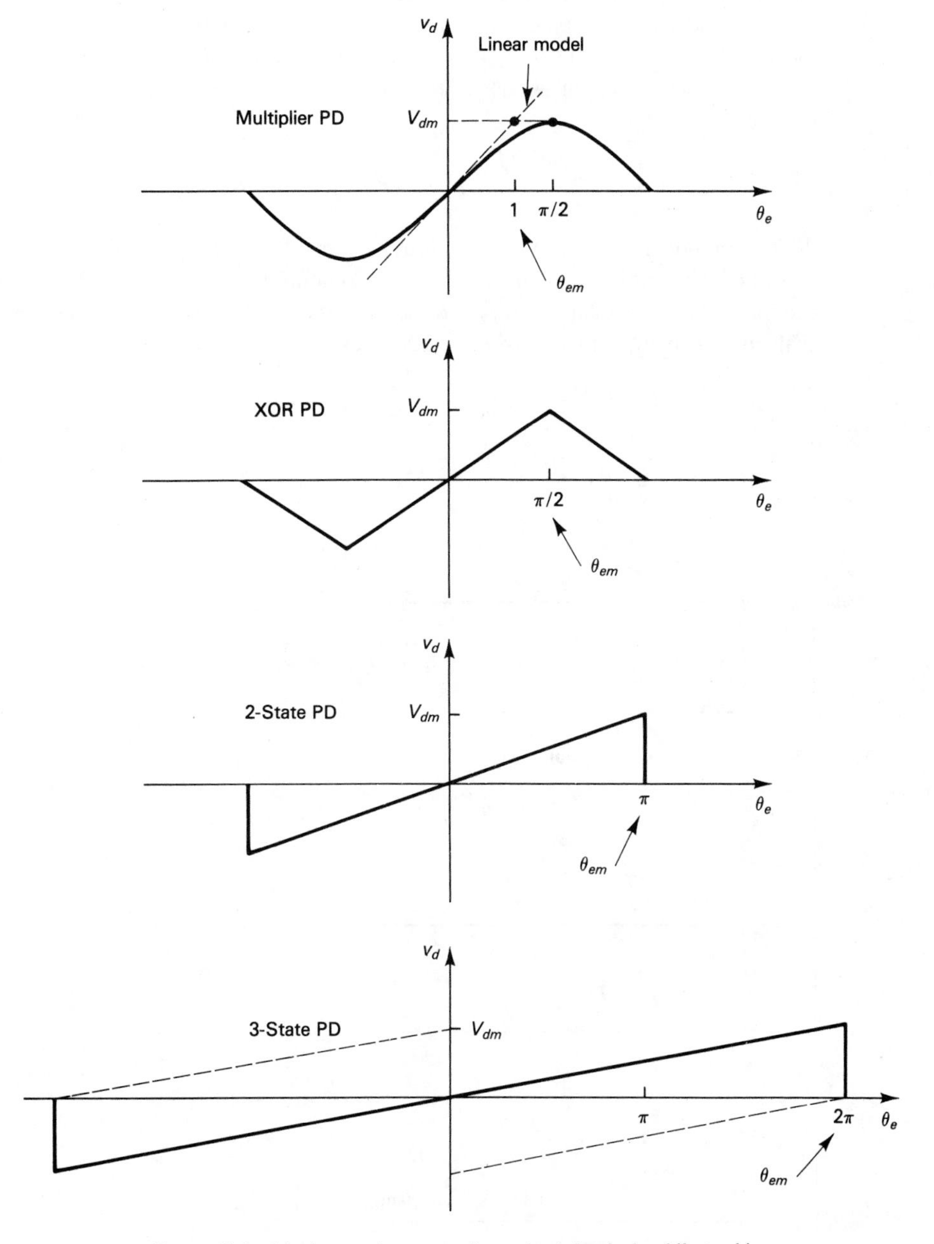

FIGURE 7–3 Maximum phase error for maintaining lock while tracking

Therefore, v_d is also zero, and from the PD characteristics in Fig. 7–3, $\theta_e = 0$. (This assumes the PD offset voltage V_{do} is zero.)

Consider now a step change of height $\Delta\omega$ for the input frequency deviation $\Delta\omega_i$, as shown in Figure 7–4a. Then if ω_o is to maintain lock, the VCO frequency ω_o must follow with a step $\Delta\omega_o$ of height $\Delta\omega$ as in Fig. 7–4b. This is achieved by a step Δv_c in the control voltage:

$$\Delta v_c = \Delta\omega_o/K_o = \Delta\omega/K_o$$

Which component of v_c in Eq. (7-4) provides this Δv_c? Since v_3 can't change instantly, it plays no part in the early part of the step. To simplify the study in this section, we will assume that C is very large ($\omega_2 << K$) so v_3 is essentially a constant for the duration of the analysis. Then v_2, which was zero, becomes

$$v_2 = \Delta v_c = \Delta\omega/K_o$$

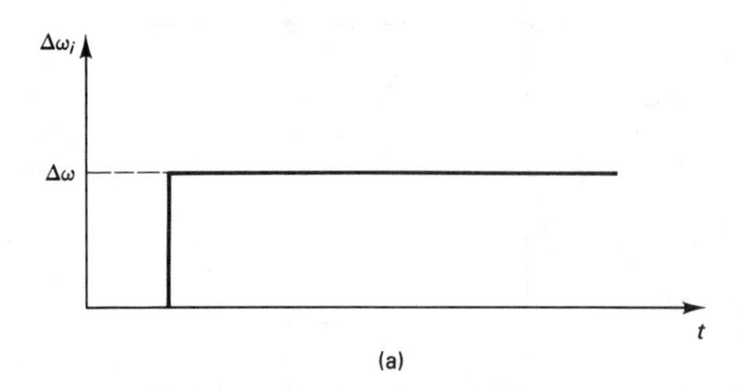

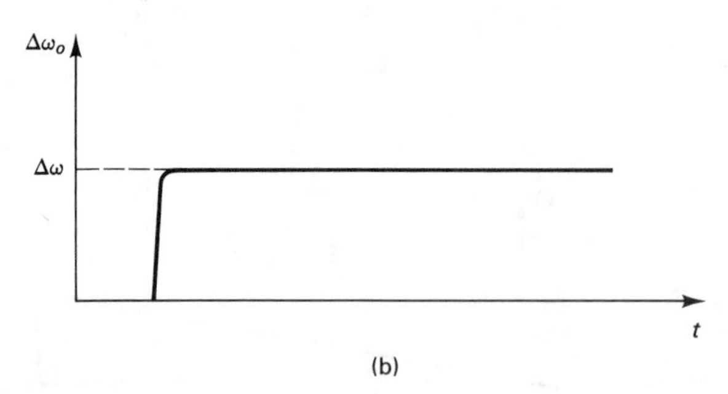

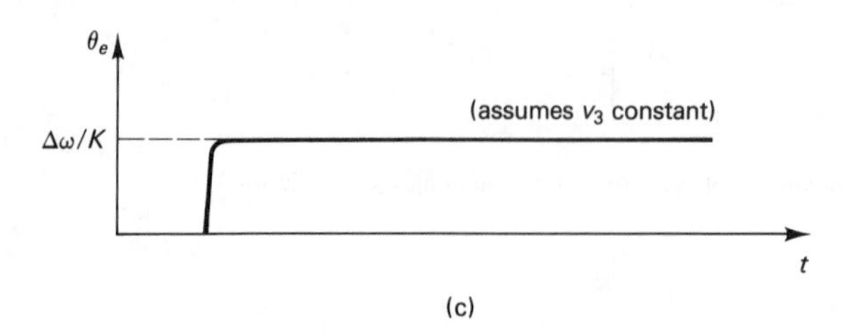

FIGURE 7–4 Simplified responses to a step change in input frequency

and from Eq. (7-5),

$$v_d = v_2/K_h = \Delta\omega/K_oK_h \qquad (7\text{-}6)$$

If $\Delta\omega$ is not too large, then v_d doesn't try to exceed the V_{dm} that the PD can provide, and the PLL stays in lock. The maximum $\Delta\omega$ for which this holds is called the *lock frequency* ω_L:

$$V_{dm} = \omega_L/K_oK_h$$

$$\omega_L = K_oK_hV_{dm} \qquad (7\text{-}7)$$

(In Chapter 8 we will see this ω_L has another interpretation in terms of acquisition behavior.)

For a piecewise-linear PD characteristic, $\theta_e = v_d/K_d$. Then from Eq. (7-6),

$$\theta_e = \Delta\omega/K_oK_hK_d = \Delta\omega/K$$

This response is plotted in Fig. 7–4c. Let θ_{em} be the maximum θ_e for which the characteristic is still linear. Figure 7-3 shows that

$$\theta_{em} = \pi/2 \text{ radians}, \quad \text{XOR PD}$$

$$\theta_{em} = \pi \text{ radians}; \quad \text{2-State PD}$$

$$\theta_{em} = 2\pi \text{ radians}; \quad \text{3-State PD} \qquad (7\text{-}8)$$

For a piecewise-linear PD characteristic, there is a simple relationship between the V_{dm} in Eq. (7-7) and θ_{em}:

$$V_{dm} = K_d\theta_{em} \qquad (7\text{-}9)$$

Then Eq. (7-7) can be represented in the simpler form

$$\omega_L = K_oK_hK_d\theta_{em} = K\theta_{em} \qquad (7\text{-}10)$$

where K is the PLL bandwidth.

For a multiplier PD with a sinusoidal characteristic, $V_{dm} = K_d$, as shown in section 4–1. Then from Eq. (7-7), $\omega_L = K_oK_hK_d = K$. This case can be included in Eq. (7-10) if we use

$$\theta_{em} = 1 \text{ radian}; \text{ Multiplier PD}$$

θ_e will actually go to $\pi/2$ radians before losing lock, but using the value $\theta_{em} = 1$ radian will give the correct results for a linear analysis. (The dashed line in Fig. 7–3a is the linear model for the sinusoidal characteristic.) Then for the various PD characteristics, Eqs. (7-8) and (7-10) give

$$\omega_L = K; \qquad \text{Multiplier PD}$$

$$\omega_L = (\pi/2)K; \qquad \text{XOR PD}$$

$$\omega_L = \pi K; \qquad \text{2-State PD}$$

$$\omega_L = 2\pi K; \qquad \text{3-State PD}$$

The PLL will maintain lock for a step change $\Delta\omega$ if $\Delta\omega < \omega_L$.

The analysis here was greatly simplified. It assumed the phase error θ_e could change instantly to provide the needed voltage for the VCO to follow $\Delta\omega_i$. In fact, though, θ_e is the integral of the frequency error (which equals $\Delta\omega$ immediately after the step), and it takes θ_e a short time to get to its new value. We also assumed v_3 across the capacitor was constant. Actually v_3 changes slowly, gradually taking over the voltage that v_2 had to provide initially. These details are treated more formally in section 7–5.

7–4 TRANSFER FUNCTION FROM $\Delta\omega_i$ TO θ_e

We will find the peak θ_e for different waveforms of $\Delta\omega_i$ and then establish bounds on $\Delta\omega_i$ so that $\theta_e < \theta_{em}$—so that the PLL maintains lock. The waveforms of $\Delta\omega_i$ examined in the following sections are a step, a ramp, a sinusoid, and a random signal.

An ac model for a PLL is shown in Fig. 7–5. In particular, the PD is modeled by the gain K_d, although this holds only near $\theta_e = 0$ for a multiplier PD. From Eq. (7-2), θ_i is the integral of $\Delta\omega_i$. Therefore, θ_i is shown in Fig. 7–5 as $(1/s)\Delta\omega_i$, where $1/s$ represents integration, and $\Delta\omega_i$ represents the change in the input frequency. For an active loop filter, we determined in section 3–8 that the transfer function from θ_i to θ_e is

$$\frac{\theta_e(s)}{\theta_i(s)} \equiv H_e(s) = \frac{s^2}{s^2 + Ks + K\omega_2} \tag{7-11}$$

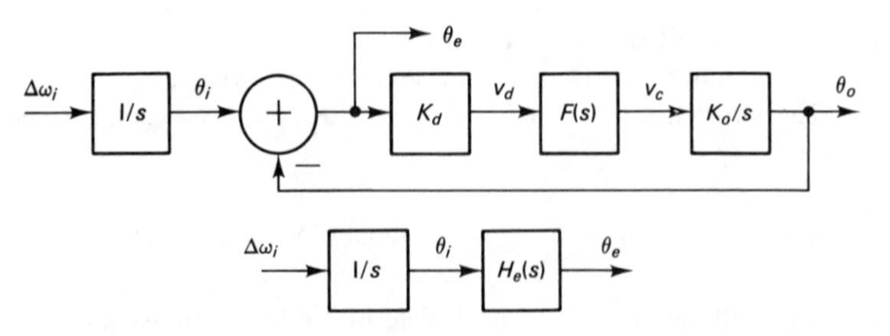

FIGURE 7–5 Transfer function from ω_i to θ_e—a linear model

Then the transfer function from $\Delta\omega_i$ to θ_e is given by

$$\frac{\theta_e(s)}{\Delta\omega_i(s)} = \frac{1}{s} H_e(s) = \frac{s}{s^2 + Ks + K\omega_2} \tag{7-12}$$

For $\omega_2 << K$, the transfer function in Eq. (7-12) can be approximated by

$$\frac{\theta_e(s)}{\Delta\omega_i(s)} \approx \frac{s}{(s + \omega_2)(s + K)} \tag{7-13}$$

with a corresponding frequency response plotted in Fig. 7–6. The response has high- and low-frequency cutoffs at ω_2 and at K and a "midband gain" of $1/K$.

7–5 HANDLING A FREQUENCY STEP

Suppose the change $\Delta\omega_i$ in the input frequency is a step of height $\Delta\omega$, as shown in Fig. 7–7a. This is essentially what happens when a user changes the frequency of a frequency synthesizer, for instance. Since $\theta_e/\Delta\omega_i \approx 1/K$ for the band of frequencies from ω_2 to K,

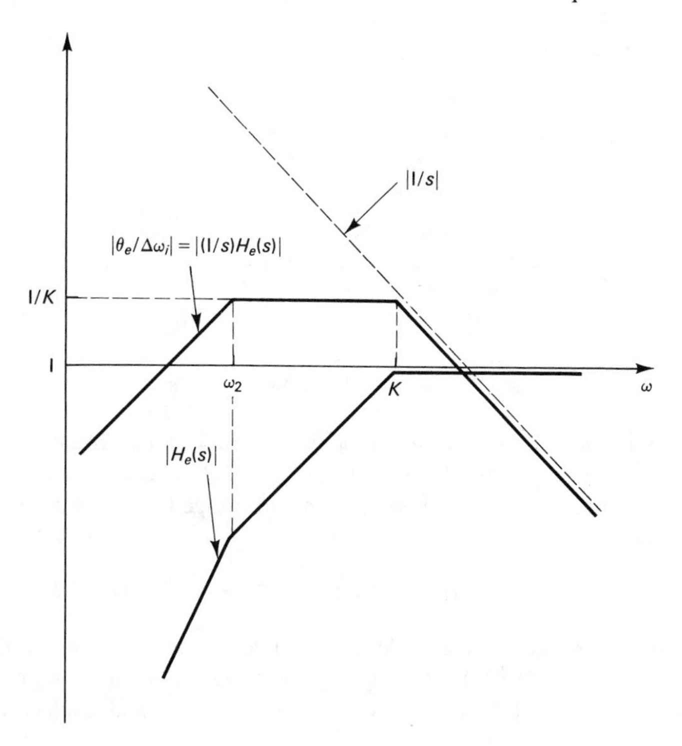

FIGURE 7–6 Frequency response of $\theta_e/\Delta\omega_i$ transfer function

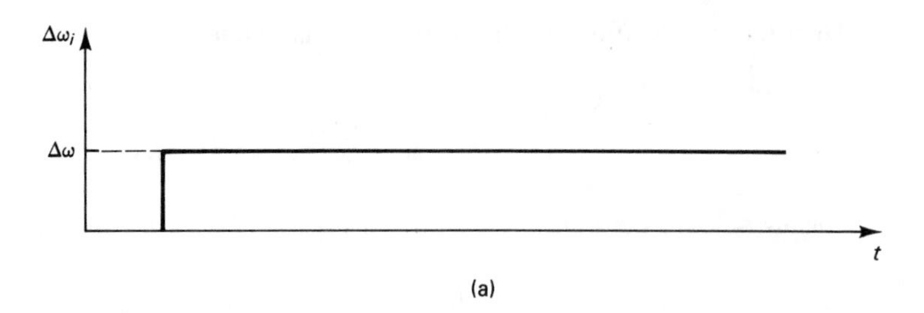

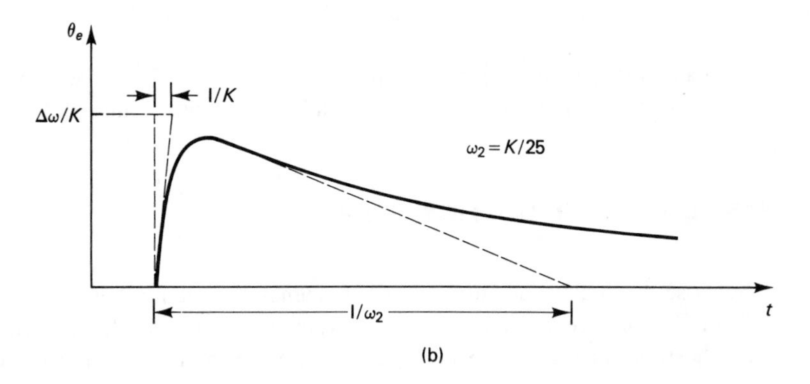

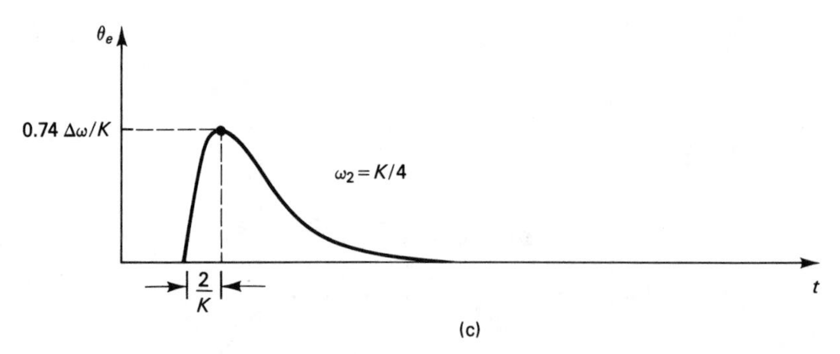

FIGURE 7–7 Response to a frequency step

then θ_e is approximately a step of height $\Delta\omega/K$. The high-frequency cutoff at K causes a finite rise time, and the low-frequency cutoff at ω_2 causes sag in the θ_e waveform, as shown in Fig. 7–7b. Using Laplace transforms together with Eq. (7-13), we can solve for the transient

$$\theta_e(t) \approx (\Delta\omega/K)\,(e^{-\omega_2 t} - e^{-Kt});\ \omega_2 < K/4 \tag{7-14}$$

For $\omega_2 << K$, the θ_e waveform reaches a peak value of almost $\Delta\omega/K$. (For $\omega_2 = K/25$, the peak value is 0.87 $\Delta\omega/K$.) The highest design value of ω_2 in practice is $\omega_2 = K/4$. In that case, Eq. (7-12) gives $\theta_e/\Delta\omega_1 = s/(s + K/2)^2$, and Laplace transforms give

$$\theta_e(t) = \Delta\omega\ t\ e^{-Kt/2}; \qquad \omega_2 = K/4$$

which has a peak value of 0.74 $\Delta\omega/K$ (see Fig. 7–7c). In any case,

$$\theta_e(t) < \Delta\omega/K$$

holds for all ω_2. In order to maintain lock, we need to ensure that $\theta_e(t) < \theta_{em}$ for all t. This will be satisfied for all ω_2 if

$$\Delta\omega/K < \theta_{em} \tag{7-15}$$

This bound for a frequency step relies on a parameter within the PLL design—θ_{em} of the PD characteristic. By contrast, ω_L is a parameter that can be measured externally to the PLL. Therefore, it is often useful to use $\theta_{em} = \omega_L/K$ from Eq. (7-10) to present the bound for a frequency step in the form

$$\Delta\omega < \omega_L \tag{7-16}$$

This is, in fact, our definition of ω_L from the previous section, but now it has been established on a more formal basis.

The bound in Eqs. (7-15) and (7-16) assumes that the PLL has an active filter and that the initial phase error is $\theta_e = 0$. If there is a static phase error θ_{eo}, or if the PLL had not reached steady-state before the step change in $\Delta\omega_i$, then the bound may not be adequate to ensure lock is maintained. Suppose $\omega_2 = K/25$ and a second step occurs at a time $25/K$ after the first step, as in Fig. 7–8a. Then from Eq. (7-14), $\theta_e = 0.37\ \Delta\omega/K$ at the time of the second step, and the peak value of θ_e due to the second step will be about $\theta_{e\ \max} = (0.37 + 0.87)\Delta\omega/K = 1.24\ \Delta\omega/K$ (see Fig. 7–8b). Then to satisfy $\theta_e(t) < \theta_{em} = \omega_L/K$, we would need to restrict $1.24\ \Delta\omega < \omega_L$, or $\Delta\omega < 0.8\ \omega_L$. This is a tighter bound than Eq. (7-16).

7–6 HANDLING A FREQUENCY RAMP

The input frequency change $\Delta\omega_i$ is said to be a ramp if its time derivative is a constant $\Delta\dot{\omega}$ (see Figure 7–9a). Since a ramp primarily consists of lower frequencies, we can no longer characterize the transfer function $\theta_e/\Delta\omega_i = (1/s)\,H_e$ by its midband gain $1/K$ shown in Fig. 7–9b. (For convenience, Fig. 7–9b is drawn for the case $K < 1$.)

For low frequencies, Eq. (7-12) reduces to the approximation

$$\theta_e'/\Delta\omega_i = s/K\omega_2; \qquad \text{for } \omega < \omega_2 \tag{7-17}$$

The frequency response of this approximate transfer function is plotted in Fig. 7–9c. This is just a differentiator, which is basically due to having the loop filter (an integrator) in the feedback path. The corresponding phase response θ_e' is shown in Fig. 7–9d. It is a constant of height $\Delta\dot{\omega}/K\omega_2$.

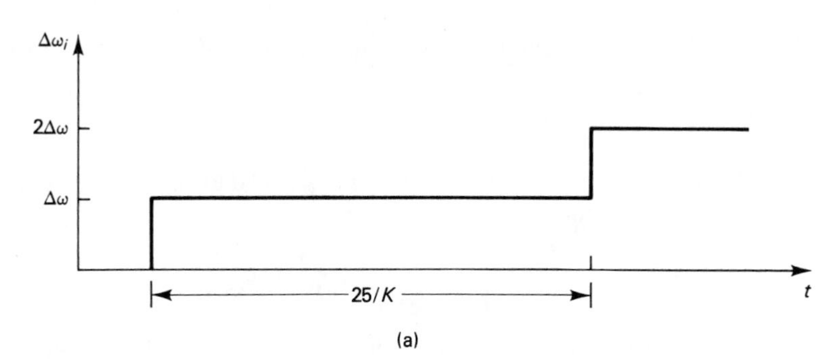

(a)

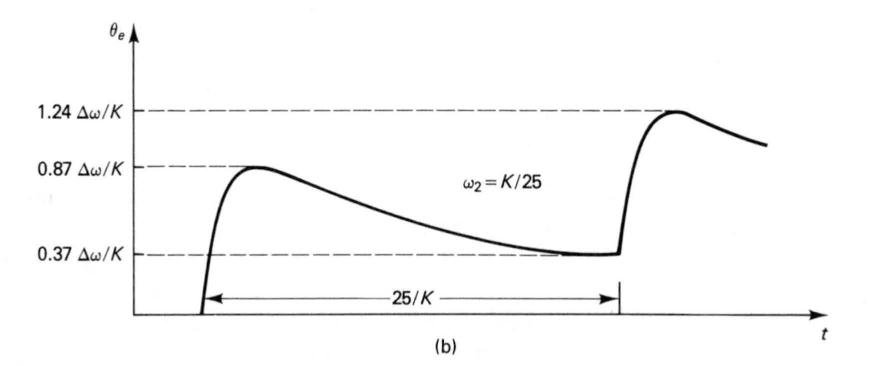

(b)

FIGURE 7–8 Response to multiple steps

We can get a better approximation to the transfer function by including the second term in the denominator of Eq. (7-12);

$$\theta_e''/\Delta\omega_i = \frac{s}{Ks + K\omega_2} = \frac{s}{K\omega_2} \cdot \frac{1}{s/\omega_2 + 1} \tag{7-18}$$

The frequency response now includes the break at ω_2, as shown in Fig. 7–9e. The effect of the break is to low-pass the response θ_e' to give the better approximation

$$\theta_e'' = (\Delta\dot{\omega}\,/K\omega_2)\,(1 - e^{-\omega_2 t})$$

which is graphed in Fig. 7–9f. Note that the response still doesn't exceed $\Delta\dot{\omega}\,/K\omega_2$.

We could go on to include the effect of the break at K, but for $K \geq 4\,\omega_2$ the effect is small, and the phase error still doesn't exceed $\Delta\dot{\omega}\,/K\omega_2$. To maintain lock, we must have $\theta_e'(t) < \theta_{em}$, which is satisfied if

$$\Delta\dot{\omega}\,/K\omega_2 < \theta_{em} \tag{7-19}$$

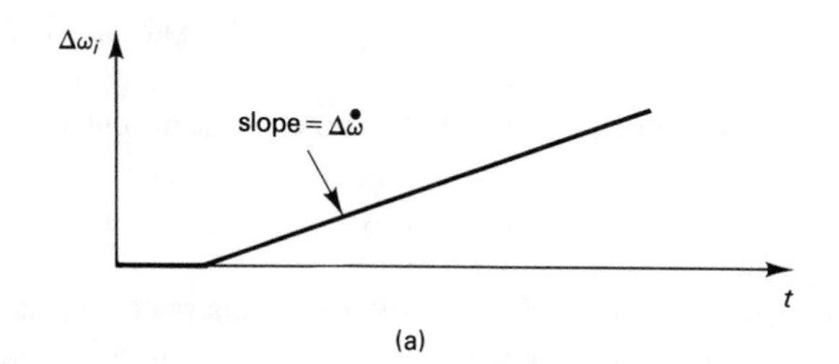

(a)

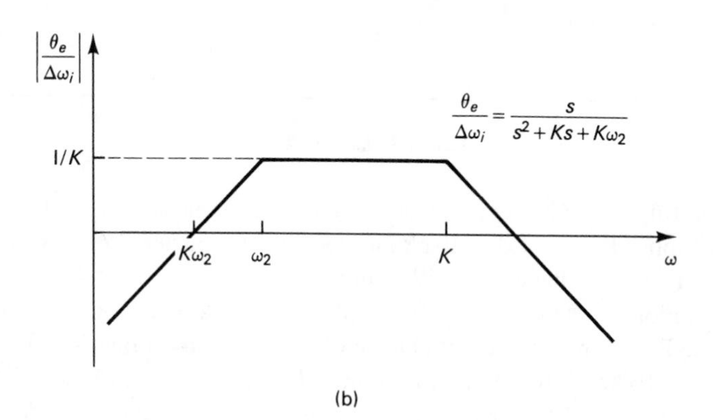

(b)

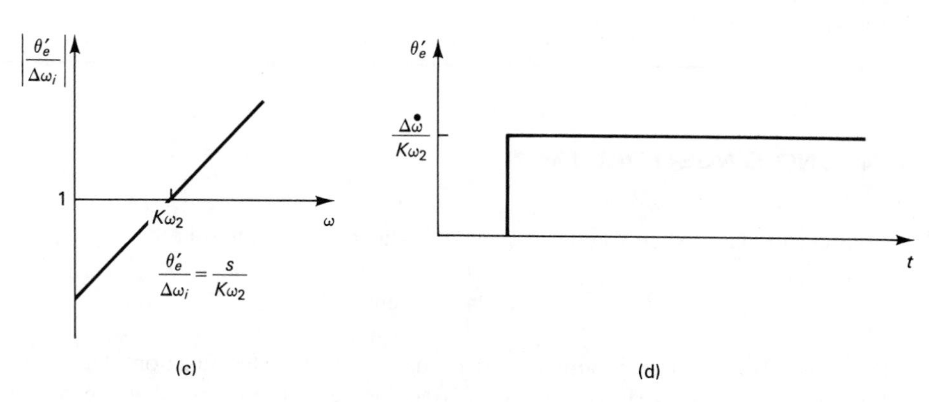

(c)

(d)

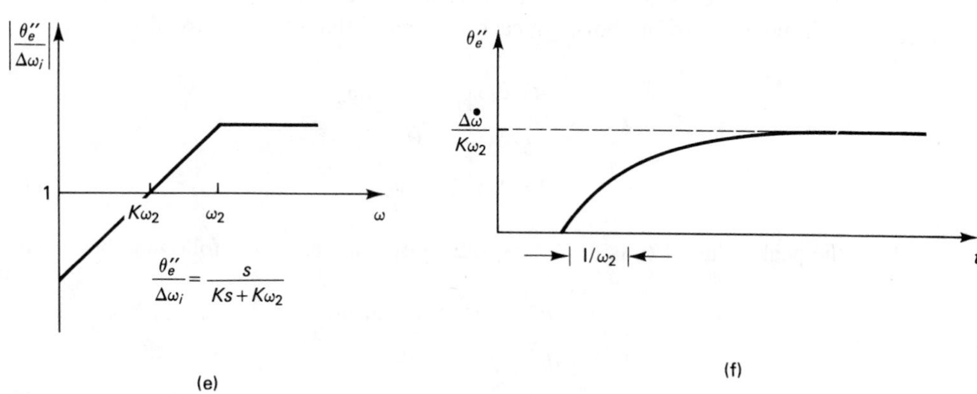

(e)

(f)

FIGURE 7–9 Response to a frequency ramp

Together with $\theta_{em} = \omega_L/K$ from Eq. (7-10), the bound can be put in the form

$$\Delta\dot{\omega} < \omega_L\omega_2 \tag{7-20}$$

This corresponds to the maximum rate the PD voltage can charge the capacitor in the loop filter. The PLL will eventually lose lock when it reaches the limit of its hold-in range, and Eq. (7-20) no longer holds.

EXAMPLE 7–1

A rocket accelerating at 30 Gs receives a communication signal with a 1.6-GHz carrier. A PLL with a multiplier PD is used to track and recover the carrier. What is the minimum bandwith K the PLL may have and still maintain lock?

An acceleration of 30 Gs is $a = 30 \times 9.8\ \text{m/s}^2 = 294\ \text{m/s}^2$. The speed of light is $c = 3 \times 10^8$ m/s. Therefore, the rate of change of $\Delta\omega_i$ is $\Delta\dot{\omega} = (a/c)\omega_i = (0.98 \times 10^{-6})\ 2\pi \times 1.6\ \text{GHz} = 10000\ \text{rad/s}^2$. For a multiplier PD, $\omega_L = K$. Then Eq. (7-20) becomes $\Delta\dot{\omega} < K\omega_2$. K is minimized here by maximizing ω_2. Let $\omega_2 = K/4$. Then $K^2/4 > \Delta\dot{\omega} = 10000\ \text{rad/s}^2$, or $K >$ 200 rad/s.

7–7 HANDLING SINUSOIDAL FM

Consider the case when $\Delta\omega_i$ is frequency modulated by a sinusoid:

$$\Delta\omega_i = \Delta\omega \sin(\omega_m t) \tag{7-21}$$

(See Fig. 7–10c.) The resulting θ_e depends on the transfer function $\theta_e/\Delta\omega_i$, whose frequency response is shown in Fig. 7–10a. Figure 7–10b shows that the magnitude $|\theta_e/\Delta\omega_i|$ is bounded from above by each of three different functions of ω_m:

$$|\theta_e/\Delta\omega_i| \leq \omega_m/K\omega_2$$
$$|\theta_e/\Delta\omega_i| \leq 1/K$$
$$|\theta_e/\Delta\omega_i| \leq 1/\omega_m \tag{7-22}$$

Then the peak value of θ_e is $\theta_{e\ \max} = |\theta_e/\Delta\omega_i|\ \Delta\omega$, and all of the following bounds hold:

$$\theta_{e\ \max} \leq \Delta\omega\ \omega_m/K\omega_2$$
$$\theta_{e\ \max} \leq \Delta\omega/K$$
$$\theta_{e\ \max} \leq \Delta\omega/\omega_m \tag{7-23}$$

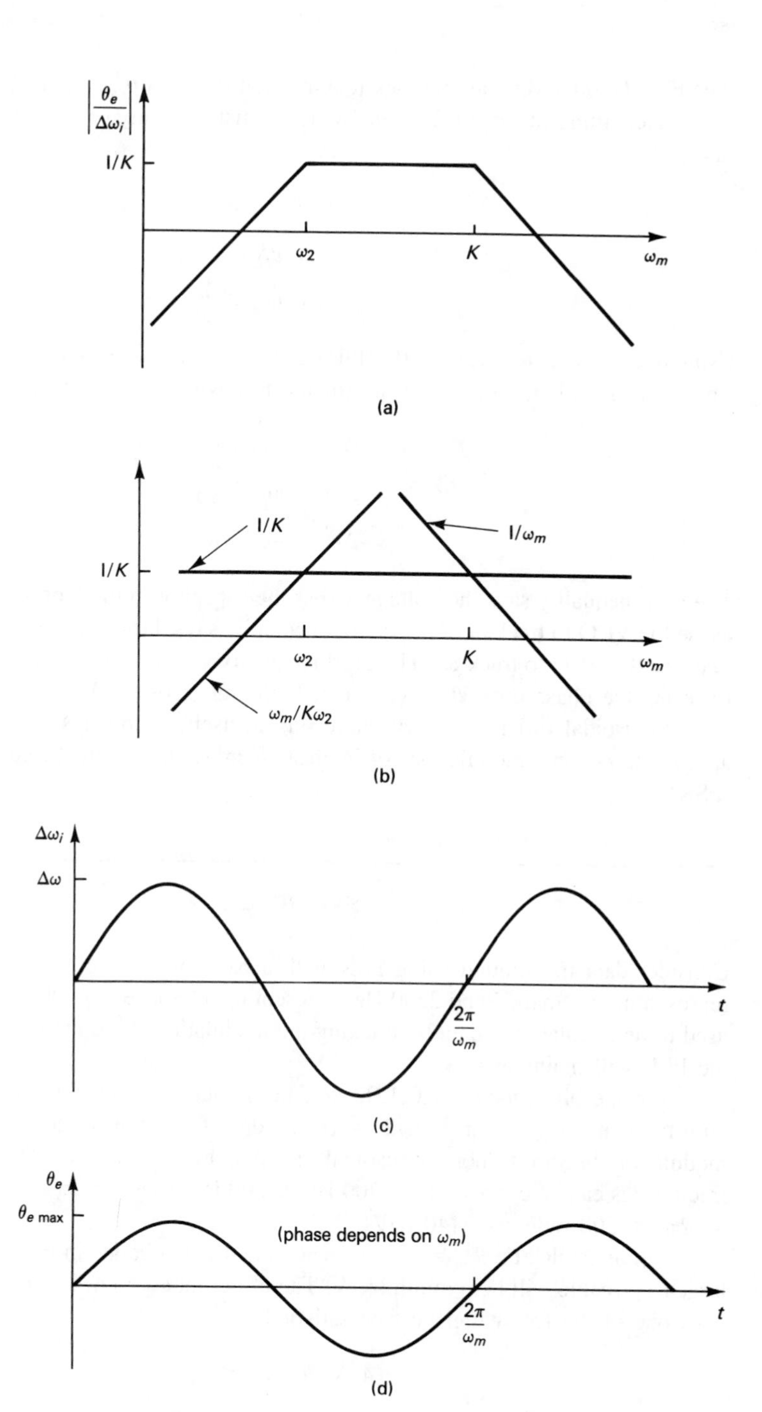

FIGURE 7–10 Response to sinusoidal FM

(See Fig. 7–10d.) Maintaining lock requires that $\theta_{e\ \max} \leq \theta_{em}$, where θ_{em} is the PD phase range. According to Eq. (7-23), this will be satisfied if any of the following bounds are satisfied:

$$\Delta\omega\ \omega_m/K\omega_2 \leq \theta_{em}$$

$$\Delta\omega/K \leq \theta_{em}$$

$$\Delta\omega/\omega_m \leq \theta_{em} \tag{7-24}$$

Using $\theta_{em} = \omega_L/K$ from Eq. (7-10), this can be stated in terms of ω_L. The PLL will handle sinusoidal modulation of $\Delta\omega_i$ if any of the following are satisfied:

$$\Delta\omega \leq \omega_L\omega_2/\omega_m$$

$$\Delta\omega \leq \omega_L$$

$$\Delta\omega \leq \omega_m\omega_L/K \tag{7-25}$$

The first inequality says the voltage across the capacitor in the loop filter is alone able to cause the VCO to track ω_i. The second inequality says the voltage across R_2 is alone able to cause the VCO to track ω_i. The third inequality says it is not necessary for the VCO to track ω_i; the phase error θ_e never exceeds the range of the PD.

Sinusoidal FM is not very interesting in itself; it conveys no information. But it approximates some modulations of interest, such as band-limited frequency shift keying (FSK).

EXAMPLE 7–2

Consider data transmitted using FSK with a baud of $f_B = 600$ bits/sec with 1200 Hz representing a "mark" and 2200 Hz representing a "space." A PLL with an XOR PD is used to demodulate the data by tracking the modulation. Find the minimum K for which the PLL will maintain lock.

For the bit sequence 1,0,1,0, etc., the frequency deviation $\Delta\omega_i$ is a square wave with frequency $\omega_m = 2\pi \times f_B/2 = 1880$ rad/s. If the transmission is band limited, the modulation becomes more sinusoidal, as $\Delta\omega_i$ in Fig. 7–10c. The mark and space frequencies can be expressed as 1700 Hz $\pm$ 500 Hz, so $\omega_i = 2\pi \times 1700$ rad/s, and $\Delta\omega = 2\pi \times 500$ rad/s $= 3140$ rad/s.

For an XOR PD, $\theta_{em} = \pi/2$. Considering the first bound in Eq. (7-24), choosing ω_2 large as possible will help minimize K. Therefore, let $\omega_2 = K/4$. Then Eq. (7-24) says at least one of the following must be satisfied:

$$4\ \Delta\omega\ \omega_m/K^2 \leq \pi/2$$

$$\Delta\omega/K \leq \pi/2$$

$$\Delta\omega/\omega_m \leq \pi/2$$

Then we must choose K to satisfy at least one of the following:

$$K \geq \sqrt{(8/\pi)\Delta\omega\ \omega_m} = 3880 \text{ rad/s}$$

$$K \geq (2/\pi)\Delta\omega = 2000 \text{ rad/s}$$

$$\Delta\omega/\omega_m = 1.67 \leq \pi/2 = 1.57$$

The last bound cannot be satisfied. Then to minimize K, we choose to just satisfy the second bound: $K = \underline{2000 \text{ rad/s}}$, and $\omega_2 = K/4 = 500$ rad/s.

7–8 HANDLING RANDOM FM

When a carrier is frequency modulated by an analog signal, the modulation usually has a broad spectrum and is best characterized as a random signal. For simplicity, we assume that the power spectral density $\Phi_{\omega i}$ is flat out to a bandwidth of B_m, as shown in Fig. 7–11a. (See Chapter 6 for a discussion of power spectral density.) The area under the curve is $\overline{\Delta\omega_i^2}$, where $\Delta\omega_i$ has a zero mean. Then the spectral density must be

$$\Phi_{\omega i} = \begin{cases} \overline{\Delta\omega_i^2}/B_m; & f < B_m \\ 0; & f > B_m \end{cases} \tag{7-26}$$

The power spectral density for θ_e is given by

$$\Phi_{\theta e} = |\theta_e/\Delta\omega_i|^2\Phi_{\omega i} \tag{7-27}$$

where $|\theta_e/\Delta\omega_i|$ is plotted in Figure 7–11b. This is the same response as the plot in Fig. 7–10a, except the function is in terms of f, where $f = \omega_m/2\pi$. As in the previous section, $|\theta_e/\Delta\omega_i|$ is bounded by either of the functions shown in Fig. 7–11b:

$$|\theta_e/\Delta\omega_i| \leq 2\pi f/K\omega_2 \tag{7-28a}$$

$$|\theta_e/\Delta\omega_i| \leq 1/K \tag{7-28b}$$

(A third bound $|\theta_e/\Delta\omega_i| \leq 1/2\pi f$ is not useful here.) From Eqs. (7-26), (7-27), and (7-28), we have the two bounds

$$\Phi_{\theta e} \leq \begin{cases} \dfrac{(2\pi f)^2\overline{\Delta\omega_i^2}}{B_m K^2 \omega_2^2}; & f < B_m \\ 0; & f > B_m \end{cases} \tag{7-29a}$$

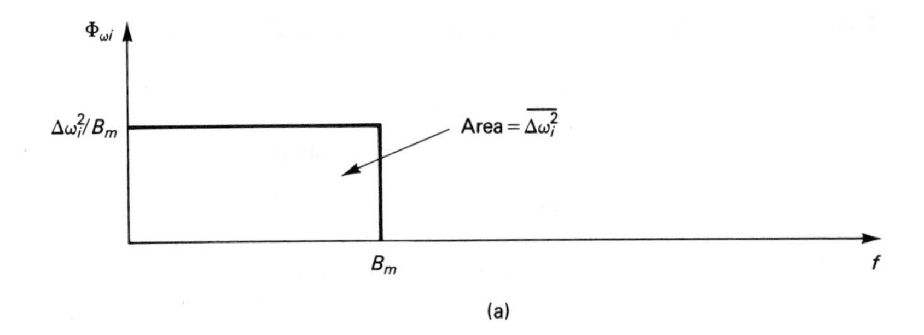

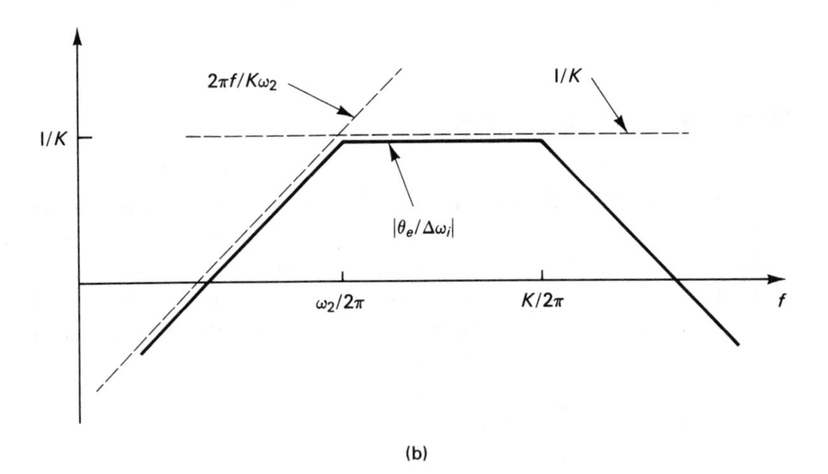

FIGURE 7–11 Response to random FM

$$\Phi_{\theta e} \leq \begin{cases} \overline{\omega_i^2}/B_m K^2; & f < B_m \\ 0; & f > B_m \end{cases} \tag{7-29b}$$

Since the mean square of θ_e is given by $\overline{\theta_e^2} = \int_0^\infty \Phi_{\theta e} df$, carrying out the integration of Eq. (7-29) gives the two bounds

$$\overline{\theta_e^2} \leq \frac{(2\pi B_m)^2 \overline{\Delta\omega_i^2}}{3K^2\omega_2^2} \tag{7-30a}$$

$$\overline{\theta_e^2} \leq \overline{\Delta\omega_i^2}/K^2 \tag{7-30b}$$

Taking the square root of both sides of Eq. (7-30) gives bounds on the root–mean–square of θ_e:

$$\theta_{e\ \text{rms}} \leq \frac{2\pi B_m \Delta\omega_{i\ \text{rms}}}{\sqrt{3}\ K\ \omega_2} \tag{7-31a}$$

$$\theta_{e\ \mathrm{rms}} \le \Delta\omega_{i\ rms}/K \tag{7-31b}$$

We would like to be in a position now to use Eq. (7-31) to get bounds of θ_e in terms of bounds on $\Delta\omega_i$. But for Gaussian random $\Delta\omega_i$ there is no bound on either $\Delta\omega_i$ or θ_e. As a practical matter, we define the maximum θ_e as that value that is exceeded only 0.05% of the time. From a normal distribution table, this corresponds to

$$\theta_{e\ \max} = 3.5\ \theta_{e\ \mathrm{rms}} \tag{7-32}$$

Maintaining lock requires $\theta_{e\ \max} \le \theta_{em}$. Then Eqs. (7-31) and (7-32) give two bounds, either of which being satisfied will assure that lock is maintained:

$$\frac{7\pi B_m \Delta\omega_{i\ \mathrm{rms}}}{\sqrt{3}\ K\ \omega_2} \le \theta_{em} \tag{7-33a}$$

$$3.5\ \Delta\omega_{i\ \mathrm{rms}}/K \le \theta_{em} \tag{7-33b}$$

Using $\theta_{em} = \omega_L/K$ from Eq. (7-10), this can be stated in terms of ω_L. The PLL will handle random FM of $\Delta\omega_i$ if either of the following is satisfied:

$$\Delta\omega_{i\ \mathrm{rms}} \le \frac{\sqrt{3}\ \omega_L\ \omega_2}{7\pi B_m} \tag{7-34a}$$

$$\Delta\omega_{i\ \mathrm{rms}} \le \omega_L/3.5 \tag{7-34b}$$

EXAMPLE 7–3

An FM radio station transmits music with a uniform spectrum out to a bandwidth $B_m = 15$ kHz. The frequency deviation $\Delta\omega_i$ corresponds to the amplitude of the music, where the Gaussian distribution of the amplitude causes the frequency deviation to exceed 75 kHz only 0.05% of the time. A PLL with a multiplier PD is used to demodulate the signal. Find the minimum bandwidth K for which the PLL will maintain lock all but 0.05% of the time.

The maximum frequency deviation in rad/sec is $3.5\ \Delta\omega_{i\ \mathrm{rms}} = 2\pi \times 75$ kHz $= 471$ krad/s, or $\Delta\omega_{i\ \mathrm{rms}} = 135$ krad/s. For a multiplier PD, $\omega_L = K$. Let $\omega_2 = K/4$. Then Eq. (7-34a) gives

$$K^2 \ge 4 \times 7\pi B_m \omega_{i\ \mathrm{rms}}/\sqrt{3} = 102500\ (\mathrm{krad/s})^2$$
$$K \ge 320\ \mathrm{krad/s}$$

and Eq. (7-34b) gives

$$K \geq 3.5\ \Delta\omega_{i\ \mathrm{rms}} = 471\ \text{krad/s}$$

To minimize K, we choose to satisfy the first bound with $K = \underline{320\ \text{krad/s}}$. Then $\omega_2 = K/4 = 80$ krad/s.

CHAPTER

8

Lock Acquisition

We have been assuming in our analyses of PLLs that they are in lock—that the average VCO frequency $\overline{\omega}_o$ equals the average input frequency ω_i. However, the two frequencies are not the same when power is first applied to the circuit or when v_i is first applied to the PLL input. The process of bringing ω_o to equal ω_i is called *frequency acquisition*. After the frequency error has been brought to zero, there is a transient process during which the phase error is also brought to zero. There is no clear definition of when the PLL is actually in lock, but it is reasonable to define lock acquisition as synonymous with frequency acquisition. Any remaining phase transient is similar to the θ_e transient studies in section 7–5; it can be considered as taking place in lock.

In a second-order PLL, most of the voltage to bring the VCO to the proper frequency is provided by the capacitor in the loop filter. For example, the active filter in the PLL shown in Fig. 8–1a provides the control voltage $v_c = v_2 + v_3$, where v_2 is across R_2 and v_3 is across the capacitor. The frequency acquisition process amounts to charging (or discharging) the capacitor until v_3 provides the correct control voltage for $\overline{\omega}_o = \omega_i$. After lock, v_2 provides the small changes in frequency necessary to provide phase error correction.

In some cases, a frequency detector is added to a PLL to help charge the capacitor. But we will see that the unaided PLL is able to acquire lock for a limited initial frequency error.

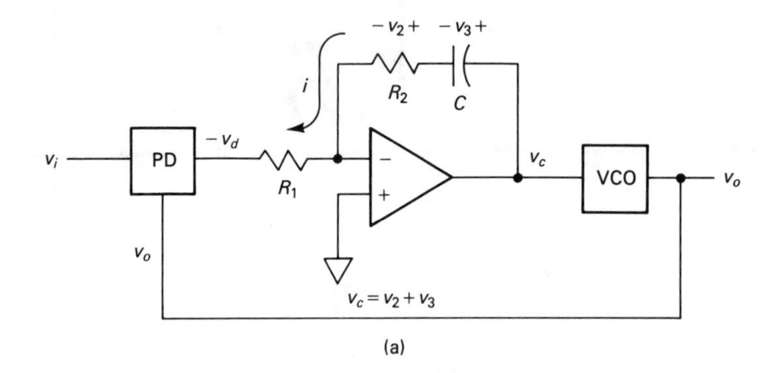

(a)

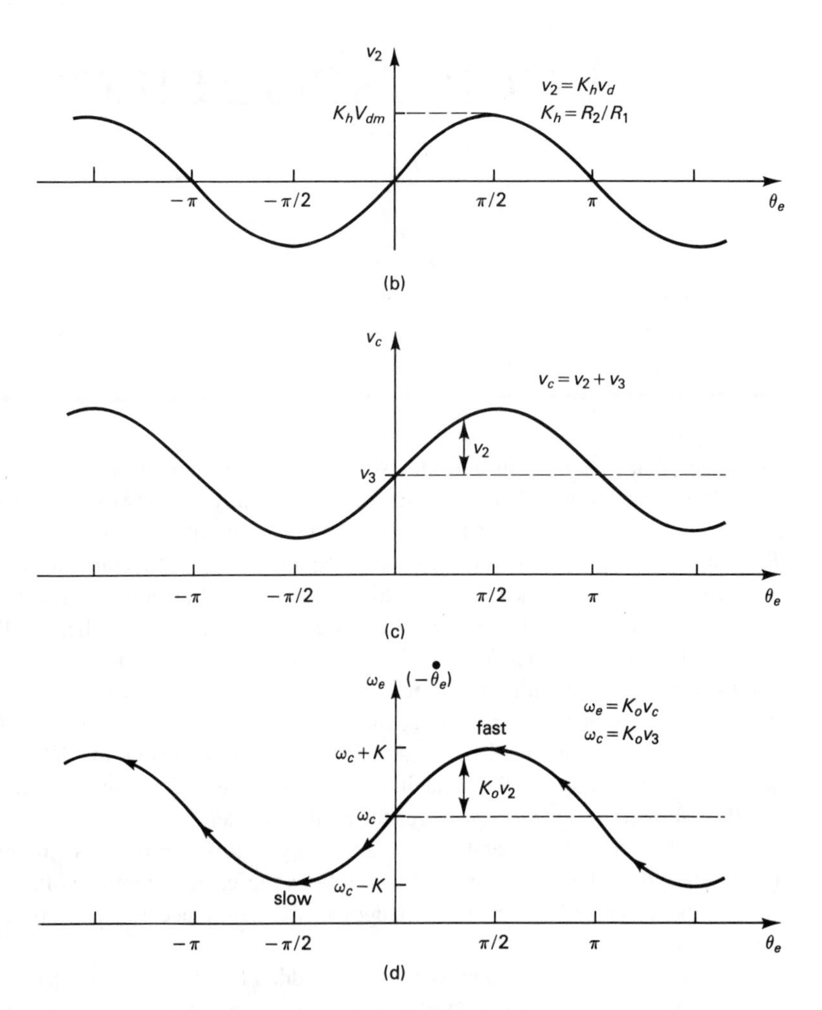

FIGURE 8–1 PLL out of lock

To simplify the analysis of lock acquisition, the input frequency and phase will be assumed constant ($\theta_i = 0$). The effects of $\theta_i \neq 0$ can be taken into account through a modified PD characteristic (see section 4–13). For a constant input frequency, any deviation of ω_o from ω_i is considered a frequency error. Therefore, in this chapter we relabel $\Delta\omega_o$ as the *frequency error* ω_e:

$$\omega_e \equiv \omega_o - \omega_i \tag{8-1}$$

where ω_i is a constant and $\theta_i = 0$. [Compare Eq. (2-5).] The maximum ω_e for which the PLL will attain lock is called the *pull-in range*. In this chapter we will find the pull-in range and pull-in time for a PLL to acquire lock with different phase detectors and different acquisition aids.

8–1 SELF ACQUISITION: ACTIVE LOOP FILTER

The process of a PLL acquiring lock without the aid of a frequency detector is called *self acquisition*. In this case, the PD is responsible for charging the capacitor in the active loop filter. There must be at least a small dc component to the PD output v_d, and the polarity must be such that the VCO frequency ω_o is brought closer to ω_i. When a PLL is out of lock, the dc component is small, but it can be enough to pull the PLL into lock.

8–1–1 Pull-In Voltage v_p

The dc component (or average) of v_d during acquisition is called the *pull-in voltage* v_p. To understand the origin of v_p, consider a PLL that has no modulation of the input signal: $\theta_i = 0$. Then the phase error is $\theta_e \equiv \theta_i - \theta_o = -\theta_o$. When the PLL is out of lock, $\omega_o \neq \omega_i$ on average, and the frequency error $\omega_e \equiv \omega_o - \omega_i$ is not zero on average. But

$$\omega_e = \dot{\theta}_o = -\dot{\theta}_e \tag{8-2}$$

Then θ_e is either constantly increasing or constantly decreasing.

What is the control voltage v_c as a function of θ_e? Consider a PD with a sinusoidal characteristic:

$$v_d = V_{dm} \sin \theta_e \tag{8-3}$$

The voltage v_2 across R_2 is proportional to v_d:

$$v_2 = K_h v_d = K_h V_{dm} \sin \theta_e$$

as shown in Fig. 8–1b. To simplify the analysis, let the capacitor be large enough (ω_2 is small enough; $\omega_2 \leq K/4$) that its voltage v_3 is essentially independent of θ_e. Then the total control voltage is

$$v_c = v_2 + v_3 = K_h V_{dm} \sin \theta_e + v_3$$

as shown in Fig. 8–1c. If there is some θ_e for which v_c causes $\omega_e = 0$, then the PLL will lock at that phase. Without loss of generality, let us assume that $v_c = 0$ causes $\omega_e = 0$. Then the VCO characteristic is simply

$$\begin{aligned} \omega_e &= K_o v_c \\ &= K_o K_h V_{dm} \sin \theta_e + K_o v_3 \end{aligned} \tag{8-4}$$

But $V_{dm} = K_d$ for a sinusoidal PD, and $K = K_o K_h K_d$. Then

$$\omega_e = K \sin \theta_e + \omega_c \tag{8-5}$$

where

$$\omega_c \equiv K_o v_3 \tag{8-6}$$

is the portion of the frequency error due to the incorrect capacitor voltage v_3. A plot of the function ω_e in Eq. (8-5) is shown in Figure 8–1d. For the case of $\theta_i = 0$ during acquisition, Eq. (8-2) gave $\omega_e = -\dot{\theta}_e$. Then Eq. (8-5) can also be represented as

$$-\dot{\theta}_e = K \sin \theta_e + \omega_c \tag{8-7}$$

This gives the rate of change of θ_e in terms of θ_e. As shown in Fig. 8–1d, when $\theta_e \approx \pi/2$, $-\dot{\theta}_e$ is large, meaning that θ_e decreases rapidly (the motion to the left along the curve is fast). For $\theta_e \approx -\pi/2$, $-\dot{\theta}_e$ is small, and the motion to the left is slow. Since ω_e never goes to zero, the PLL never locks for any θ_e.

In order for lock to occur, v_3 must decrease (with a corresponding decrease in ω_c) until the curve in Fig. 8–1d touches the $\omega_e = 0$ axis. Can v_d supply the necessary average current to discharge the capacitor in Fig. 8–1a? Consider the PD characteristic in Fig. 8–2a. If θ_e moves smoothly to the left when the PLL is out of lock, then v_d is sinusoidal in time, and its average is zero. There is no average current, and the capacitor neither charges nor discharges on average. But as we have seen, θ_e does not move smoothly; it lurches along, spending more time in the vicinity of 1.5π, -0.5π, etc. (see Fig. 8–2b). This means that v_d spends more time with negative values, as shown in Fig. 8–2c. What is important is that the asymmetry of the waveform about the t axis has produced a nonzero average component $\bar{v}_d$, which in this case is negative.

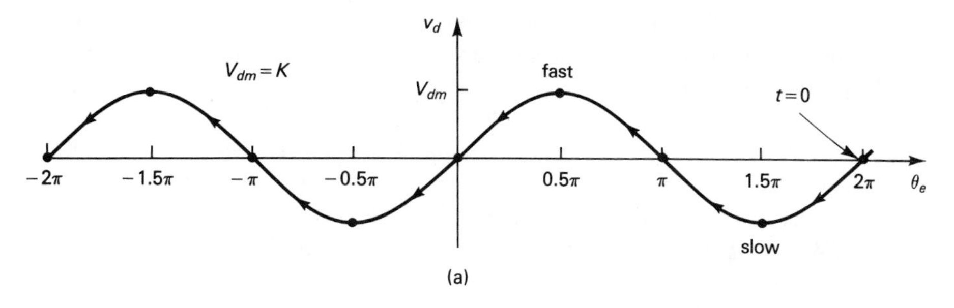

(a)

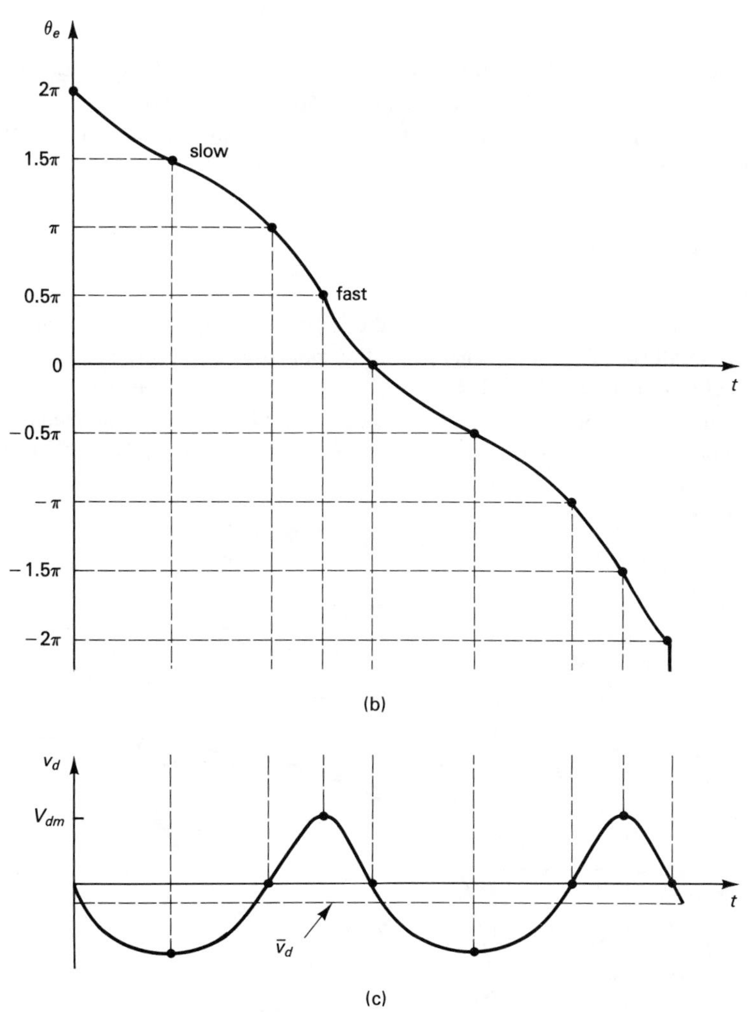

(c)

FIGURE 8–2 Beat note asymmetry

The expression for v_d as a function of time can be found by solving Eq. (8-3) together with Eq. (8-7). The average voltage $\bar{v}_d$ is called the *pull-in voltage* v_p. Richman [1] showed that

$$v_p = -K_d[\omega_c/K - \sqrt{(\omega_c/K)^2 - 1}] \tag{8-8}$$

for the case we are considering here: sinusoidal PD characteristic, active loop filter, $\omega_2 \leq K/4$, and $\theta_i = 0$. Equation (8-8) can be approximated by

$$v_p \approx -K_d K/2\omega_c \tag{8-9}$$

A plot of v_p and its approximation are given in Fig. 8–3. For $\omega_c > 2K$, the approximation is within 7%. For $\omega_c > K$, the approximation is conservative; it estimates less pull-in voltage than there actually is. For $\omega_c < K$, the PLL has locked, and pull-in voltage is undefined.

8–1–2 Pull-In Time T_p

For large frequency error ω_c, the asymmetry in v_d is slight, and the pull-in voltage v_p is small (see Fig. 8–3). Therefore, the capacitor is discharged slowly, and ω_c decreases slowly at first (see Fig. 8–4). As ω_c approaches K, v_p gets quite large, and the final descent of ω_c is more rapid. The pull-in is complete when ω_e can equal zero for some θ_e. According to Eqs. (8-4) and (8-6), this is when $\omega_c = K_o K_h V_{dm}$. This is called the *lock-in range* ω_L:

$$\omega_L = K_o K_h V_{dm} \tag{8-10}$$

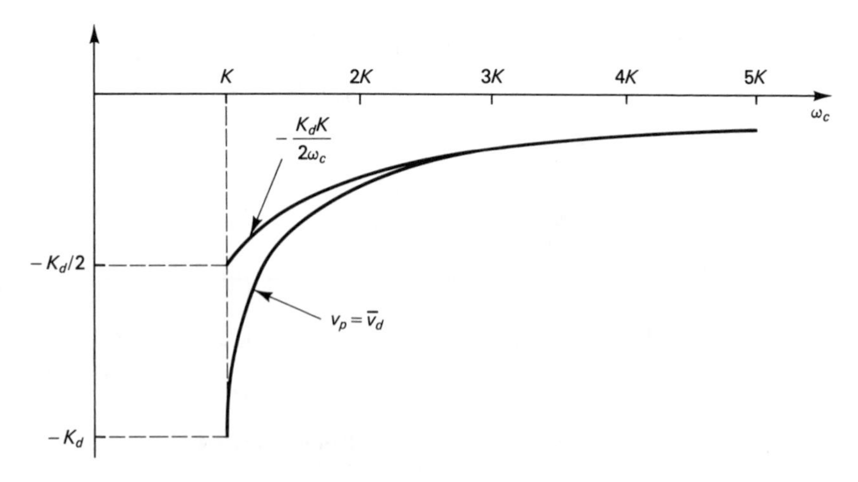

FIGURE 8–3 Pull-in voltage

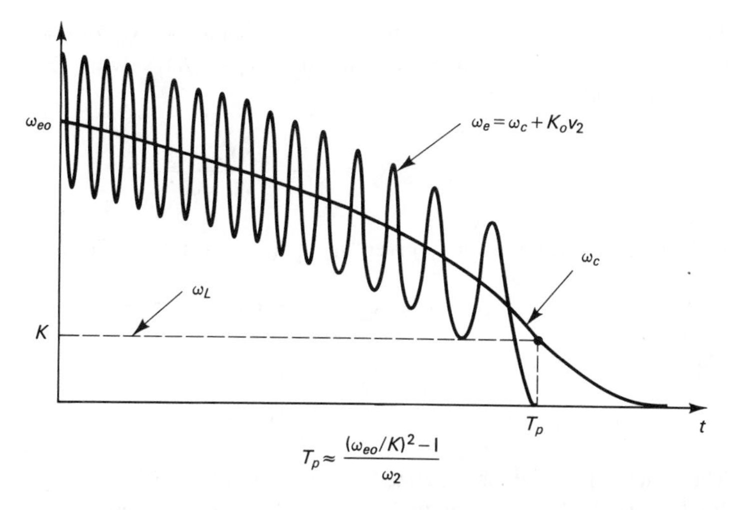

FIGURE 8–4 Pull-in time

An equivalent definition of lock-in range is the maximum frequency error ω_e for which acquisition is almost instantaneous; the PLL slips less than one cycle. (In section 7–3, the lock-in range was shown to be the maximum frequency step the PLL could handle and still maintain lock.) For a sinusoidal PD, $V_{dm} = K_d$, and from Eq. (8-10)

$$\omega_L = K \tag{8-10$'$}$$

From the approximation for v_p in Eq. (8-9), we can find an expression for the curve of ω_c versus t shown in Fig. 8–4. From this we can find the *pull-in time* T_p as a function of the initial frequency error ω_{eo}. Analyzing the circuit in Fig. 8–1, ω_c changes due to the capacitor being charged by the current $i = \bar{v}_d/R_1$:

$$\omega_c \equiv K_o v_3$$

$$\dot{\omega}_c = \mathrm{K_o}\ \dot{v}_3 = K_o i/C = K_o \bar{v}_d/R_1 C = K_o v_p K_h \omega_2 \tag{8-11}$$

where $K_h = R_2/R_1$ and $\omega_2 = 1/R_2C$. Then with Eq. (8-9),

$$\dot{\omega}_c \approx -K_o K_d K_h K \omega_2/2\omega_c = -K^2\omega_2/2\omega_c$$

$$2\ \omega_c\ d\omega_c \approx -K^2\omega_2\ dt$$

Integrating both sides gives

$$\omega_c^{\ 2} \approx \omega_{eo}^{\ 2} - K^2\omega_2 t \tag{8-12}$$

where $\omega_{eo}{}^2$ is the constant of integration such that $\omega_c = \omega_{eo}$ for $t = 0$. The pull-in time T_p is the time for ω_c to reach the lock-in range $\omega_L = K$. Solving Eq. (8-12) for t such that $\omega_c = K$ yields

$$T_p \approx \frac{(\omega_{eo}/K)^2 - 1}{\omega_2} \tag{8-13}$$

If the exact expression for v_p given in Eq. (8-8) is used in Eq. (8-11), the exact result is

$$T_p = [x^2 + x\sqrt{x^2 - 1} - \ell n(x + \sqrt{x^2 - 1}) - 1]/2\omega_2 \tag{8-13$'$}$$

where

$$x \equiv \omega_{eo}/K$$

The exact value of the normalized pull-in time $T_p\omega_2$ [Eq. (8-13′)] and the approximation for $T_p\omega_2$ [Eq. (8-13)] are plotted in Fig. 8–5. The approximation is conservative in that it gives too great a value. For $\omega_{eo} > 3\ K$, the approximation is in error by less than 8%.

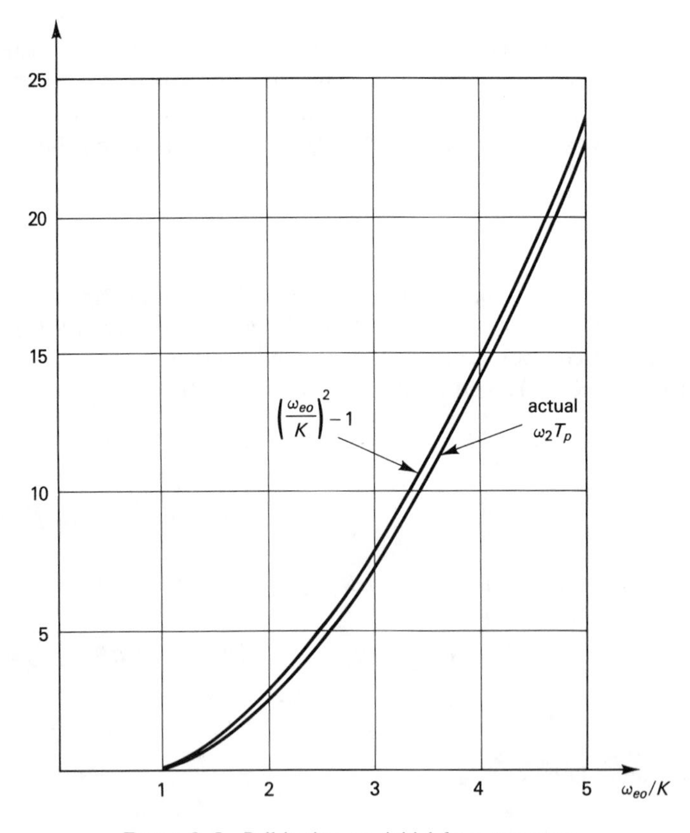

FIGURE 8–5 Pull-in time vs. initial frequency error

EXAMPLE 8–1

Given a PLL with $K = 8$ krad/s, $\omega_2 = 2$ krad/s, and a sinusoidal PD with $K_d = 2$ V/rad, find the pull-in time for initial frequency errors of $\omega_{eo} = 5$ krad/s, $\omega_{eo} = 100$ krad/s, and $\omega_{eo} = 2$ Mrad/s.

From Eq. (8-13), $T_p = 0$ for $\omega_{eo} = 5$ krad/s (ω_{eo} is within the lock-in range). $T_p = 78$ ms for $\omega_{eo} = 100$ krad/s, and $T_p = 31$ sec for $\omega_{eo} = 2$ Mrad/s. In the last case, it is probable that the PLL would actually never lock because the initial v_p is so small [Eq. (8-9) gives an initial $v_p = 4$ mV]. Such a small voltage could be overwhelmed by dc offset voltages, causing ω_o to move away from rather than toward ω_i.

8–1–3 Pull-In Range ω_p

The *pull-in range* of a PLL is defined as the largest frequency error ω_c for which the PLL will acquire lock. Suppose the PD has an offset voltage V_{do}. Then during pull-in, when the PLL is out of lock, the average voltage from the PD is

$$\bar{v}_d = V_{do} + v_p = V_{do} - K_d K/2\omega_c \tag{8-14}$$

(See Fig. 8–6.) For ω_c small enough, $\bar{v}_d$ is still negative, and the PLL pulls in. But for ω_c larger than some value, $\bar{v}_d$ is positive, and the frequency error ω_c actually increases. Then the pull-in range ω_p is that frequency error ω_c for which $\bar{v}_d = 0$. From Eq. (8-14)

$$\omega_p = K_d K/2V_{do} = MK/2 \tag{8-15}$$

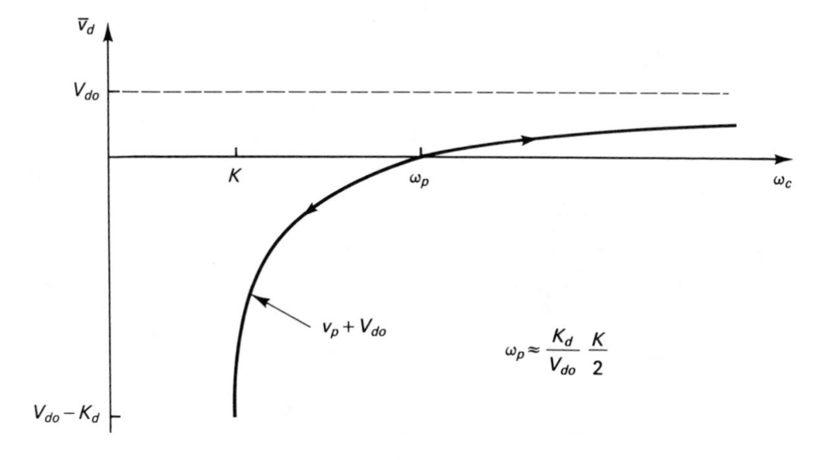

FIGURE 8–6 Pull-in range

where M is the PD figure of merit defined as $M \equiv K_d/V_{do}$. A typical value for M is 20, so ω_p can typically be $10 \times K$. With care, a PD can be made to have an M as great as 400. Then ω_p can be as great as $200 \times K$. (This is subject to limitation by ω_3, as shown in section 8.3.)

The analysis of pull-in time T_p and pull-in range ω_p has depended on Eq. (8-9) for v_p that holds for a sinusoidal PD characteristic. Meer [2] has carried out a similar analysis for triangular and sawtooth PD characteristics. The table in Fig. 8–7 compares the pull-in parameters for a sinusoidal PD, a triangular PD (such as an exclusive OR gate), and a sawtooth PD (such as a two-state PD). The pull-in times are within a factor of 7 of each other, and the pull-in ranges are also within a factor of 7 of each other. The expressions for v_p, T_p, and ω_p are approximations that hold for ω_c, ω_{eo}, and ω_p greater than $2\omega_L$. See Meer for the exact expressions.

8–2 SELF ACQUISITION: PASSIVE LOOP FILTER

For an active loop filter, the requirement for pull-in is simply that $\bar{v}_d$ be negative for positive frequency error, ω_c, and $\bar{v}_d$ be positive for negative ω_c. This is because the polarity of the charging current i depends only on $\bar{v}_d$. For a passive loop filter, however, i depends on both $\bar{v}_d$ and v_3, and v_3 depends on ω_c.

We will analyze the behavior of the passive loop filter in Fig. 8–8a during acquisition. The charging current is given by

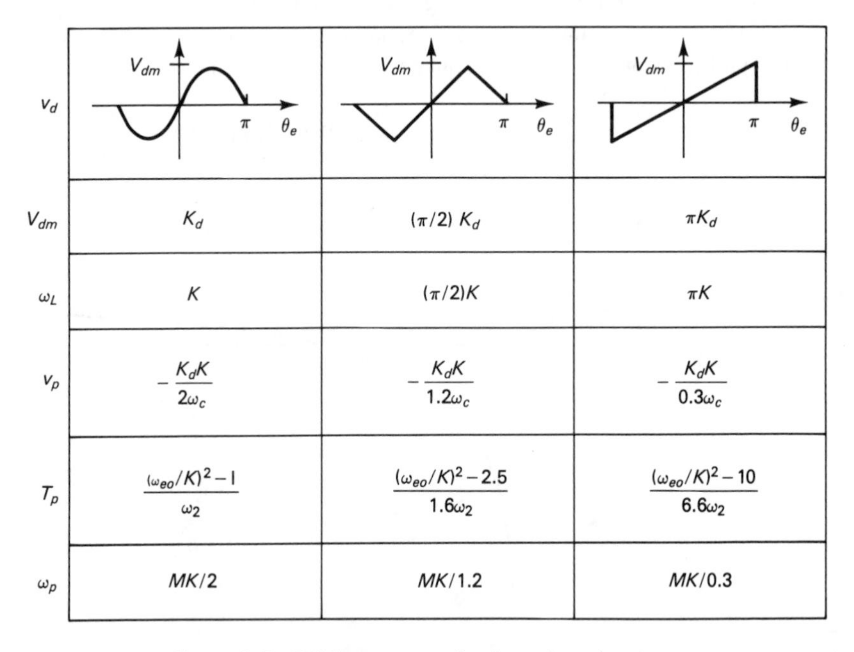

v_d	(sinusoidal: V_{dm}, π, θ_e)	(triangular: V_{dm}, π, θ_e)	(sawtooth: V_{dm}, π, θ_e)
V_{dm}	K_d	$(\pi/2)\,K_d$	πK_d
ω_L	K	$(\pi/2)K$	πK
v_p	$-\dfrac{K_d K}{2\omega_c}$	$-\dfrac{K_d K}{1.2\omega_c}$	$-\dfrac{K_d K}{0.3\omega_c}$
T_p	$\dfrac{(\omega_{eo}/K)^2 - 1}{\omega_2}$	$\dfrac{(\omega_{eo}/K)^2 - 2.5}{1.6\omega_2}$	$\dfrac{(\omega_{eo}/K)^2 - 10}{6.6\omega_2}$
ω_p	$MK/2$	$MK/1.2$	$MK/0.3$

FIGURE 8–7 Pull-in parameters for three phase detectors

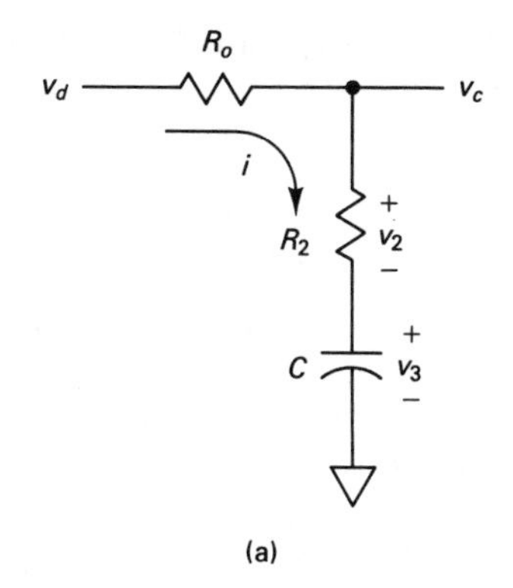

(a)

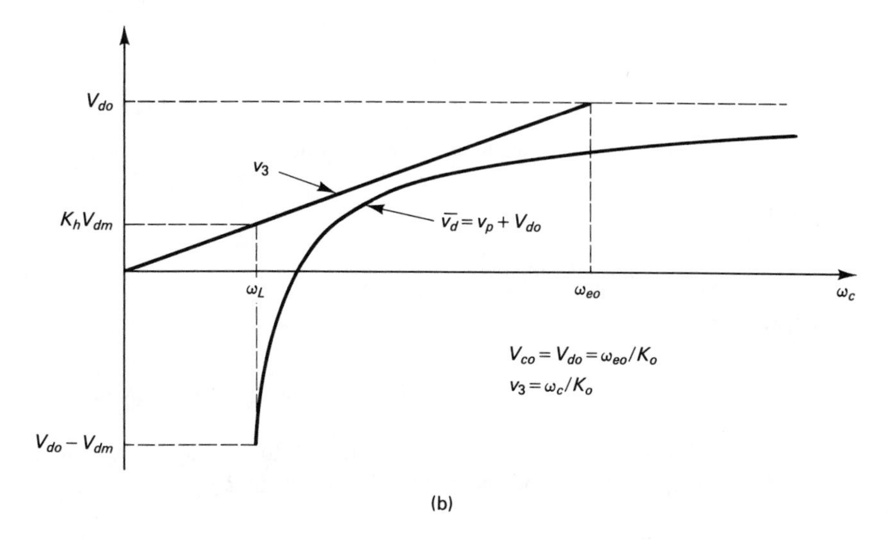

(b)

FIGURE 8–8 Pull-in with passive loop filter

$$i = (\overline{v}_d - v_3)/(R_0 + R_2)$$

Both $\overline{v}_d$ and v_3 are functions of ω_c during pull-in, as shown in Fig. 8–8b. We assume without loss of generality that $v_c = 0$ causes $\omega_e = 0$, and therefore $v_3 = 0$ causes $\omega_c = 0$. Initially the free-running voltage V_{do} of the PD charges the capacitor to $v_3 = V_{do}$, which is also the initial v_c since $v_2 = R_2 i = 0$. Then the initial frequency error is

$$\omega_{eo} = K_o V_{do}$$

For V_{do} positive, ω_{eo} is positive and i has the proper polarity (negative) for pull-in provided $\overline{v}_d$ is below v_3. This is true everywhere for the case in Fig. 8–8b, but at the point where the two curves get close, i becomes small, and the pull-in process becomes slow.

We find the pull-in time T_p for a PLL with a passive loop filter and a PD with a triangular characteristic.

$$\omega_c \equiv K_o v_3$$

$$\dot{\omega}_c = K_o \dot{v}_3 = K_o i/C = K_o(\bar{v}_d - v_3)/(R_0 + R_2)C$$
$$= K_o(\bar{v}_d - v_3)K_h\omega_2 = K_o(\bar{v}_d - \omega_c/K_o)K_h\omega_2$$

where $K_h = R_2/(R_o + R_2)$ and $\omega_2 = 1/R_2C$. From Fig. 8–7, $v_p = -K_dK/1.2\omega_c$ for a triangular PD characteristic. Then

$$\bar{v}_d = V_{do} - K_dK/1.2\omega_c = \omega_{eo}/K_o - K_dK/1.2\omega_c$$

and

$$\dot{\omega}_c = K_o(\omega_{eo}/K_o - K_dK/1.2\omega_c - \omega_c/K_o)K_h\omega_2$$
$$= \omega_{eo}K_h\omega_2 - K^2\omega_2/1.2\omega_c - K_h\omega_2\omega_c$$

Then

$$\omega_c\, d\omega_c/(-a + b\omega_c - c\omega_c^2) = dt \tag{8-16}$$

where

$$a = K^2\omega_2/1.2, \qquad b = K_h\omega_2\omega_{eo}, \qquad c = K_h\omega_2$$

As ω_c goes from ω_{eo} to $\omega_L = \pi K/2$, t goes from 0 to T_p. Integrating the left side of Eq. (8-16) from ω_{eo} to $\pi K/2$ and the right side from 0 to T_p yields

$$T_p = [\tan^{-1}(x/y - \pi/y) + \tan^{-1}(x/y)]\, x/(yK_h\omega_2)$$
$$- (1/2K_h\omega_2)\,\ell n(1 + 3K_h - 6K_hx/\pi) \tag{8-17}$$

where

$$x \equiv \omega_{eo}/K, \quad y \equiv \sqrt{\pi^2/3K_h - x^2} \tag{8-18}$$

This holds for a triangular PD characteristic. The normalized pull-in time $T_p\omega_2$ is plotted as a function of the normalized initial frequency error ω_{eo}/K in Fig. 8–9. Note that the pull-in time is also a function of the loop filter's high-frequency gain K_h. As ω_{eo} increases, the two curves in Fig. 8–8b eventually touch, and T_p goes to infinity. This can be seen in Fig. 8–9 for $\omega_{eo} = 4K$ when $K_h = 0.2$. The condition of $T_p = \infty$ arises in Eq. (8-17) when $y = 0$. Thus the pull-in range ω_p is found by solving Eq. (8-18) for the ω_{eo} such that $y = 0$:

$$\omega_p = \pi K/\sqrt{3K_h} \tag{8-19}$$

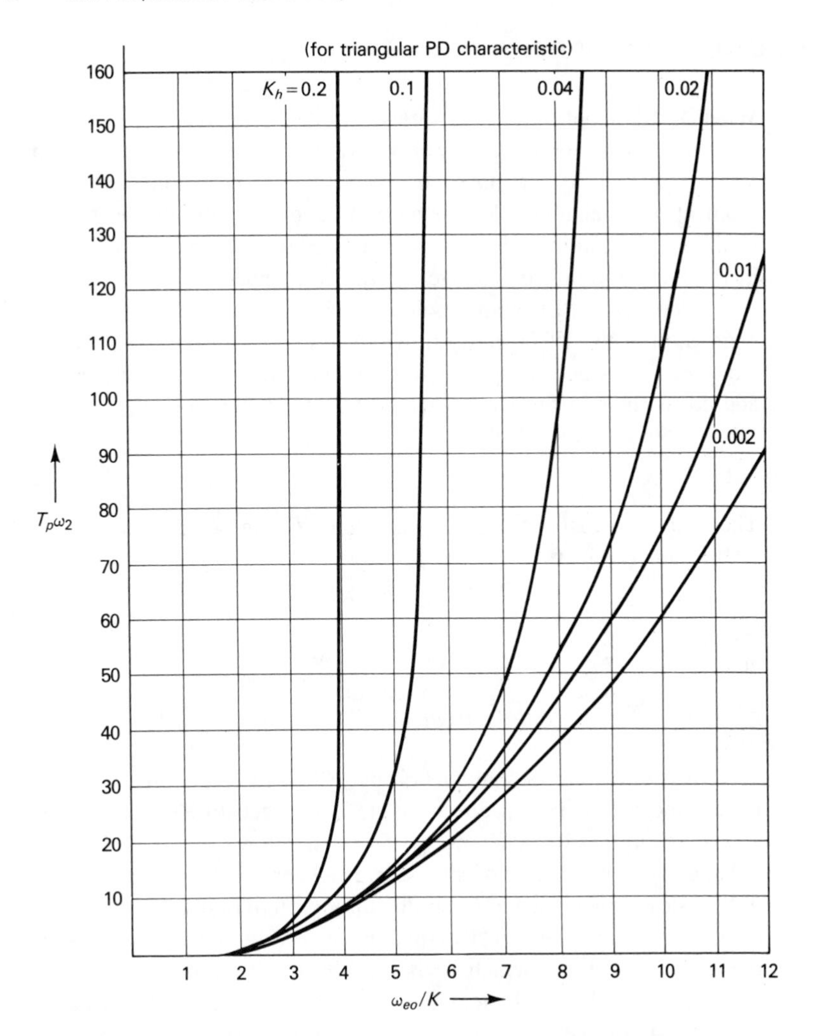

FIGURE 8–9 Pull-in time for passive loop filter

EXAMPLE 8–2

A PLL has a triangular PD with $K_d = 2$ V/rad, a VCO with $K_o = 0.2$ Mrad/s/V, and a passive loop filter with $K_h = 0.02$ and $\omega_2 = 2$ krad/s. Find the lock-in range, the pull-in range, and the pull-in time for $\omega_{eo} = 8$ krad/s, for $\omega_{eo} = 80$ krad/s, and for $\omega_{eo} = 800$ krad/s.

The bandwidth is $K = K_d K_h K_o = 8$ krad/s. From Fig. 8–7, the lock-in range is $\omega_L = \pi K/2 = \underline{12.5 \text{ krad/s}}$. From Eq. (8-19), the pull-in range is $\omega_p = \underline{103}$ krad/s. For $\omega_{eo} = 8$ krad/s, $T_p = 0$ because ω_{eo} is within ω_L. For $\omega_{eo} = 80$ krad/s, Fig. 8–9 gives $T_p\omega_2 = 109$, or $T_p = \underline{54 \text{ ms}}$. For $\omega_{eo} = 800$ krad/s, the PLL fails to lock because ω_{eo} exceeds ω_p.

8–3 ACQUISITION WITH A POLE AT ω_3

When a PLL is out of lock, v_d varies periodically, and a portion of this appears across R_2. It is this varying component v_2 of the control voltage that causes the pull-in. It causes the VCO frequency to vary in such a way that the beat note at v_d slows down at just the moment it is negative. (For negative frequency error, the beat note slows just as v_d is positive.) If the amplitude of v_2 is attenuated or if its phase is shifted by an additional pole in the loop, the pull-in voltage v_p will be reduced.

Consider a PLL with a pole at ω_3 in the loop. This may be due to the modulation bandwidth of the VCO, the bandwidth of the op amp in the active loop filter, or it may have been introduced purposely to smooth v_c when in lock. Let this pole be modeled by an additional filter in the loop with the transfer function

$$F'(s) = \omega_3/(s + \omega_3) \tag{8-20}$$

The frequency response of the magnitude $|F'|$ and of the phase ang(F'), shown in Fig. 8–10a, are given by

$$|F'| = \omega_3/(\omega_c^2 + \omega_3^2)^{1/2} \tag{8-21}$$

and

$$\text{ang}(F') = -\tan^{-1}(\omega_c/\omega_3) \tag{8-22}$$

It is clear that the attenuation by $|F'|$ will cause a proportional reduction in v_p. To understand the reduction due to ang(F') we need to look at some examples. The case shown in Fig. 8–10 is for $\omega_c = 5\omega_3$. Therefore, $|F'| \approx 1/5$, and ang$(F') = -1.37$ radians. This phase shift of almost $\pi/2$ radians between v_d and the VCO causes the beat note to slow down when v_d is passing through zero rather than when it is negative (see Fig. 8–10b). This has the effect of making v_d nearly symmetrical about the t axis, greatly reducing its average, which is the pull-in voltage.

Let the pull-in voltage with ω_3 present by v_p'. If ang(F') were $-\pi/2$, then v_p' would be reduced completely to zero. If ang(F') were $-\pi$ (it cannot be here), the beat note would slow down when v_d is positive, and $v_p' = -v_p$. This would be the opposite of the polarity that is needed to pull in, and "push-out" would occur, preventing lock of the PLL. These examples are a heuristic explanation of the fact that v_p' is further attenuated by a factor of cos[ang(F')], where $\cos(-\pi/2) = 0$, and $\cos(-\pi) = -1$.

The pull-in voltage reduced by both the magnitude and phase of F' is given by

$$v_p' = |F'| \cos[\text{ang}(F')]\, v_p \tag{8-23}$$

where v_p is the pull-in voltage with no F' in the loop. For our case of a single pole, Eqs. (8-21), (8-22), and (8-23) give

$$v_p' = \frac{\omega_3}{(\omega_c^2 + \omega_3^2)^{1/2}} \cos[\tan^{-1}(\omega_c/\omega_3)]\, v_p$$

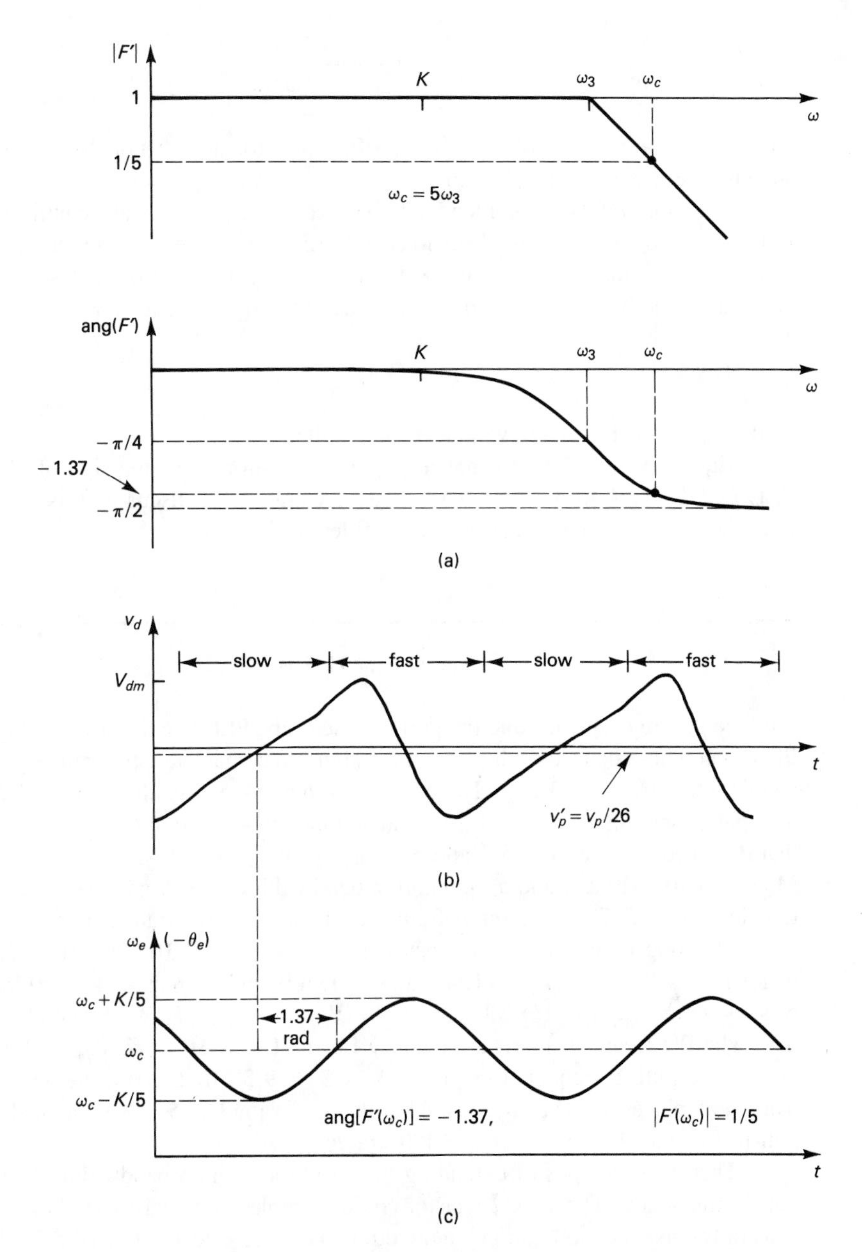

FIGURE 8–10 Effect of pole at ω_3 on pull-in

But $\cos[\tan^{-1}x] \equiv 1/(x^2 + 1)^{1/2}$. Therefore,

$$v_p' = \frac{\omega_3^{\,2}}{\omega_c^{\,2} + \omega_3^{\,2}}\, v_p \tag{8-24}$$

where v_p is the pull-in voltage given in Fig. 8–7 for no ω_3. For the case of $\omega_c = 5\omega_3$ shown in Fig. 8–10, $v_p' = v_p/26$.

Equation (8-24) is not intended to be used to design and analyze pull-in for $\omega_c > \omega_3$; v_p' becomes so small that neglected second-order effects become important. Rather, Eq. (8-24) suggests that the effect of ω_3 on pull-in is slight provided the designer keeps $\omega_3 > \omega_c$. Stated another way, in a practical sense the pull-in frequency is

$$\omega_p \approx \omega_3$$

or that given in Fig. 8–7, whichever is smaller.

Equation (8-23) shows that multiple poles can reverse the sign of v_p', causing the PLL to *false lock* to the wrong frequency. Gardner [3] discusses this effect when the multiple poles are introduced by an IF filter in the PLL.

EXAMPLE 8–3

Find the pull-in frequency and the pull-in time of the PLL circuit shown in Fig. 8–11. The PD is a double-balanced multiplier with a sinusoidal characteristic, gain $K_d = 0.4$ V/rad, and dc offset $V_{ao} = -1$ mV. The op amp has a gain-bandwidth product GBP $= 1$ MHz, an input offset voltage $V_{IO} = 1$ mV, and an input offset current $I_{IO} = 10$ nA. The VCO is that designed in Examples 5-3 and 5-5 for a gain $K_o = 3.7$ Mrad/s/V, a range from 95 Mrad/s to 105 Mrad/s, and a modulation bandwidth of 400 krad/s (see the VCO characteristic in Fig. 5–7b). The input frequency is $\omega_i = 100$ Mrad/s.

The loop filter's high-frequency gain is $K_h = 670\ \Omega/10\ \text{k}\Omega = 0.067$, so the PLL bandwidth is $K = K_dK_hK_o = 100$ krad/s. The reference voltage is $V_r = 0$ V, so the total offset given by Eq. (3-11a) is $V_{do} = (V_r - V_{ao}) + V_{IO} + I_{IO}R_1 = 1 + 1 + 0.1 = 2.1$ mV. The PD figure of merit is given by Eq. (4-14) as $M = K_d/V_{do} = 190$. From Eq. (8-15), the pull-in frequency is $\omega_p = M\,K/2 = 9.5$ Mrad/s. But the VCO modulation bandwidth places a pole at $\omega_3 = 400$ krad/s (see Example 5-5), so a practical value for the pull-in frequency is $\omega_p = \omega_3 = \underline{400\ \text{krad/s}}$.

There are methods of extending the VCO modulation bandwidth and therefore the pull-in frequency. The 15 kΩ resistor could be replaced by a 10-μH choke and a 600-Ω resistor to raise the modulation bandwidth to a double pole at 30 Mrad/s. In that case, the pole introduced by the op amp bandwidth becomes important. Equation (3-22) gives $\omega_3 = 2\pi\text{GBP}/(1 + K_h) = 5.9$ Mrad/s. This can also be improved by choosing an op amp with a GBP $= 10$ Mhz, raising the pole frequency to 59 Mrad/s. Since all poles are now greater than $M\,K/2 = 9.5$ Mrad/s, the improved pull-in frequency is $\omega_p = \underline{9.5\ \text{Mrad/s}}$.

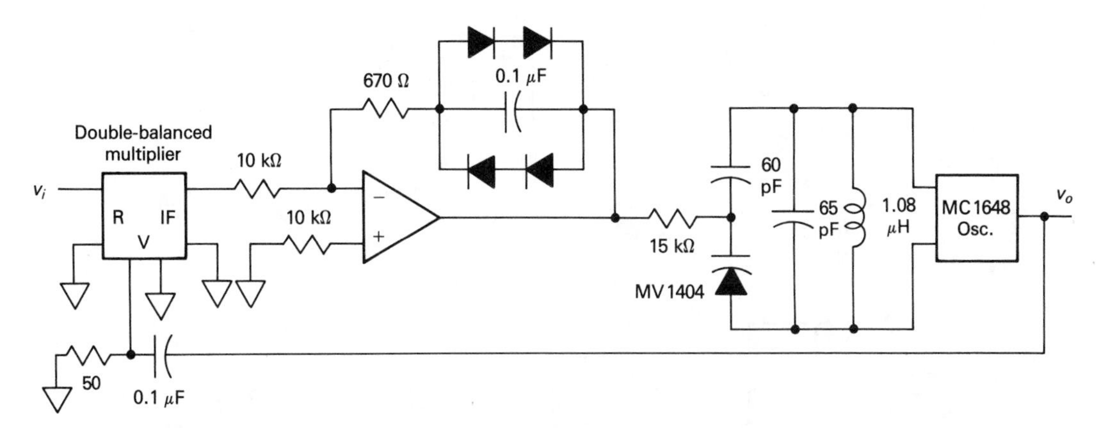

FIGURE 8–11 PLL for study of pull-in range

Before acquisition, the VCO sits at either 95 Mrad/s or 105 Mrad/s, while $\omega_i = 100$ Mrad/s. Therefore, the initial frequency error is $\omega_{eo} = 5$ Mrad/s, which is less than the improved ω_p. The loop filter's zero frequency is $\omega_2 = 1/R_2C = 15$ krad/s. Then from Eq. (8-13), the pull-in time is $T_p = (\omega_{eo}{}^2/K^2 - 1)/\omega_2 = \underline{167 \text{ ms}}$. If this is too long, an acquisition aid such as that described in section 8–6 is required.

8–4 ACQUISITION WITH A THREE-STATE PD

The advantage of a three-state PD in acquisition is that it acts as both a phase detector and a frequency detector. As we saw in section 4–8, v_d is always positive when $\omega_e < 0$ (when θ_e is increasing), and v_d is always negative when $\omega_e > 0$. From the characteristic in Fig. 4–8d, it would appear that $\bar{v}_d = V_{dm}/2$ if θ_e is increasing smoothly to the right. This is approximately correct. In fact, $\bar{v}_d$ becomes even greater as the rate of increase of θ_e is greater, as shown by the following discussion.

Consider the case when ω_o, the frequency of V, is slightly greater than ω_i, the frequency of R. Then the PD will eventually be cycling between State 1 and State 2 in Fig. 8–12a. A rising edge on V sets v_D high, and a rising edge on R resets v_D low. The waveforms are shown in Fig. 8–12b. Here R and V are represented as they would appear on an oscilloscope synchronized to R. Because ω_o is higher in frequency, V drifts in phase to the left. This causes the rising edge of v_D to drift to the left, increasing the duty cycle of V_D (many rising edges are shown in Fig. 8–12b). When the rising edge meets the falling edge (shown heavy here), it snaps back to a duty cycle of zero and beings drifting to the left again. If the phase drift is uniform, the average duty cycle is 50%, and $\bar{v}_d = -0.5V_{dm}$.

Now consider the case when ω_o is slightly greater that $2\omega_i$ as in Fig. 8–12c. Then V again drifts in phase to the left, causing the rising edge of v_D to drift to the left and

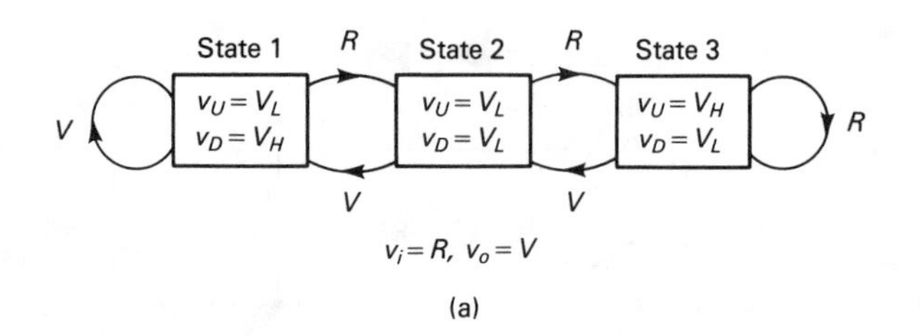

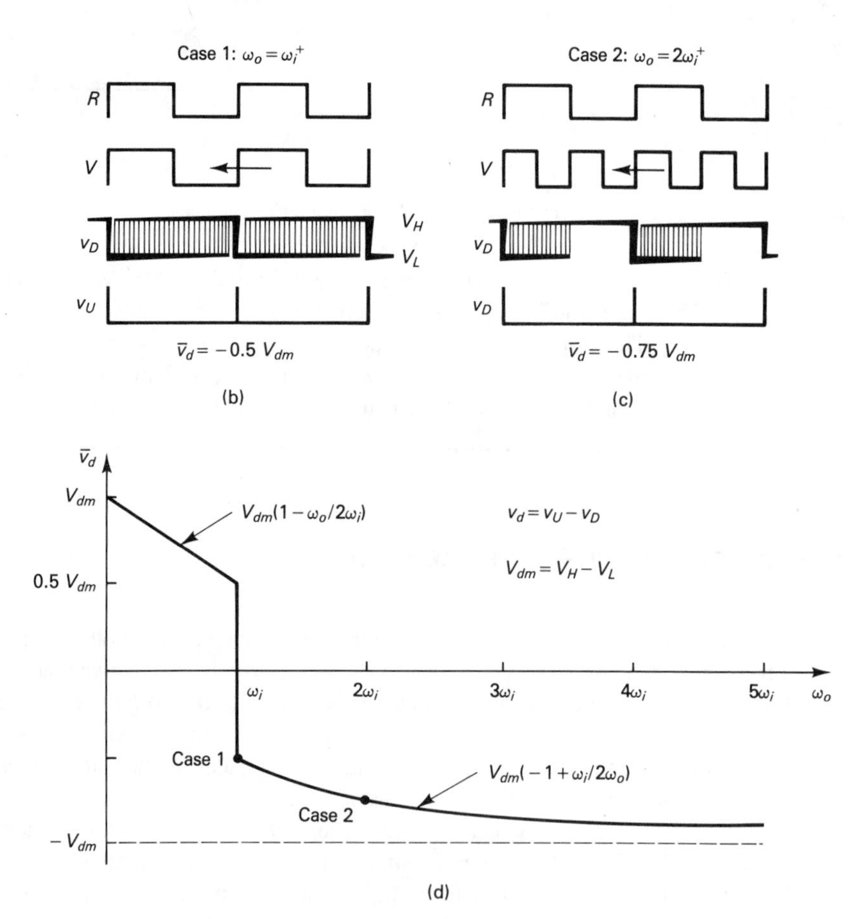

FIGURE 8–12 Pull-in with three-state PD

increasing the duty cycle of v_D. When the rising edge meets the falling edge, it snaps back to a duty cycle of 50% (not zero) and begins drifting to the left again. This time the average duty cycle is 75%, and $\bar{v}_d = -0.75V_{dm}$.

In general, the expression for $\bar{v}_d$ is given by

$$\bar{v}_d = V_{dm}(-1 + \omega_i/2\omega_o); \qquad \omega_o > \omega_i$$

$$\bar{v}_d = V_{dm}(1 - \omega_o/2\omega_i); \qquad \omega_o < \omega_i \tag{8-25}$$

which is plotted in Fig. 8–12d. For ω_o very close to ω_i (within 3 K), there is an additional $\overline{v_d}$ component (not shown) similar to that in Fig. 8–3 due to the asymmetry of the beat note. In the analysis of three-state PD pull-in behavior, we will neglect this small effect. For this PD the only limit to the pull-in range is the hold-in range (see section 7–1).

To get a simple expression for the pull-in time T_p, we will use a conservative approximation for Eq. (8-25):

$$\bar{v}_d \approx -0.5V_{dm} = -\pi K_d \tag{8-26}$$

for $\omega_o > \omega_i$, or $\omega_c > 0$. For a three-state PD we have

$$\omega_L = 2\pi K \tag{8-27}$$

The calculation of the pull-in time T_p is similar to that for the sinusoidal PD. Substituting Eq. (8-26) into Eq. (8-11), integrating, and solving for $t = T_p$ such that $\omega_c = \omega_L$, we obtain

$$T_p = \frac{\omega_{eo}/K - 2\pi}{\pi\omega_2} \tag{8-28}$$

In many applications, the range of the three-state PD is extended with $\div N$ frequency dividers, as was done for the two-state PD in Fig. 4–12. In that case, Eq. (8-26) becomes

$$\bar{v}_d \approx -0.5V_{dm} = -\pi N K_d \tag{8-26$'$}$$

Eq. (8-27) becomes

$$\omega_L = 2\pi N K \tag{8-27$'$}$$

and Eq. (8-28) becomes

$$T_p = \frac{\omega_{eo}/K - 2\pi N}{\pi N\omega_2} \tag{8-28$'$}$$

Comparing Eqs. (8-28) and (8-28′) with Eq. (8-13), we see that T_p increases only proportionally with ω_{eo}, rather than as its square. Note, however, that a three-state PD can be used only with strictly periodic signals. If there are any missing pulses, as in clock recovery applications or high-noise applications, then the three-state PD will not operate properly.

EXAMPLE 8–4

A PLL has $K = 8$ krad/s and $\omega_2 = 2$ krad/s. The initial frequency error is $\omega_{eo} = 2$ Mrad/s. The PD is a three-state phase detector. Find the pull-in time.

From Eq. (8-28), $T_p = 39$ ms. Compare this with the 31 seconds in Example 8-1 with similar conditions but a sinusoidal PD.

8–5 AIDED ACQUISITION WITH A THREE-STATE PD

If the acquisition time that a three-state PD can provide [see Eq. (8-28)] is not fast enough, a frequency detector (FD) must be added to the PLL. The purpose of the FD is to provide a strong pull-in signal in response to a frequency error ω_e (or ω_c if we ignore the small frequency fluctuations due to v_2 *across* R_2). All FDs used in conjunction with a PLL operate on the basis of sensing cycle slips between ω_i and ω_o. We have already seen these cycle slips as the beat note in v_d (see Fig. 8–2c). We also saw cycle slips in section 4–8 in the operation of three-state PDs. Referring to the state diagram of a three-state PD in Figure 8–12a, the looped arrow at the left corresponds to a cycle slip when $\omega_o > \omega_i$, where ω_o is the frequency of V and ω_i is the frequency of R. The looped arrow at the right corresponds to a cycle slip when $\omega_i > \omega_o$. A circuit to detect these cycle slips was presented in section 4–12 (see Fig. 4–16).

A PLL using cycle slip detectors to implement a FD is shown in Fig. 8–13a. A cycle slip is shown in the waveforms in Fig. 8–13b for the case $\omega_o > \omega_i$. Whenever a rising edge on V is not accompanied by a rising edge on v_D, a slip has occurred, and a pulse appears on v'_D. This pulse, which occurs at a rate equal to the frequency error, charges the capacitor in the direction to reduce the frequency error. The amount of charge each slip pulse delivers determines the acquisition speed.

The op amp in the loop filter is referenced to a V_r halfway between the logic high V_H and the logic low V_L:

$$V_r = (V_H + V_L)/2 \tag{8-29}$$

Assuming v_2 across R_2 is negligible, $v_x \approx V_r$. During a slip pulse on v'_D, $v'_D = V_H$, and the charging current is

$$\begin{aligned} i_3 &= (v_x - v'_D)/R_3 = (V_r - V_H)/R_3 \\ &= -(V_H - V_L)/2R_3 \end{aligned}$$

where the diode voltage drop has been neglected. Similarly, during a slip pulse on v'_U, $v'_U = V_L$, and the charging current is

$$i_3 = (v_x - v'_U)/R_3 = (V_H - V_L)/2R_3$$

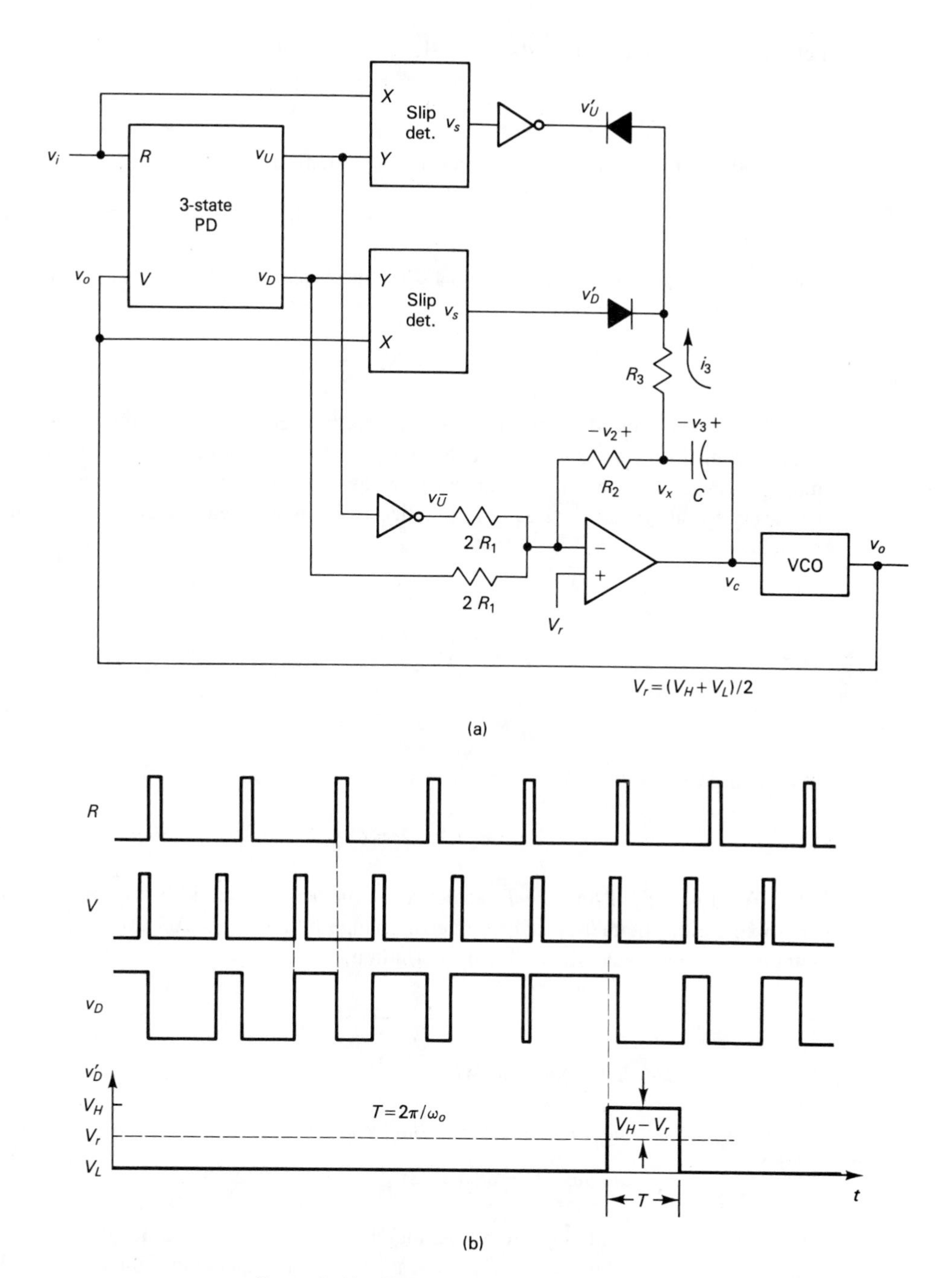

FIGURE 8–13 Three-state PD with aided acquisition

Let I be the magnitude of i_3 during a slip pulse. Then

$$I = (V_H - V_L)/2R_3 \tag{8-30}$$

The width of a slip pulse is the period of the signal at V:

$$T = 2\pi/\omega_o \approx 2\pi/\omega_i \tag{8-31}$$

The change in capacitor voltage due to a slip pulse is $|\Delta v_3| = IT/C$, and the corresponding change in frequency error is

$$|\Delta\omega_c| = K_o|\Delta v_3| = K_o IT/C \tag{8-32}$$

Then with every cycle slip, the frequency error is reduced by $\Delta\omega_c$, and the acquisition rate is proportional to $\Delta\omega_c$. But if $\Delta\omega_c$ is too large, the PLL may never reach steady-state; it may always jump over the lock-in range ω_L as it approaches $\omega_c = 0$. Therefore, it is necessary to satisfy $|\Delta\omega_c| < 2\omega_L$. In practice, it is good to leave a factor-of-two margin and choose I in Eq. (8-32) so that

$$|\Delta\omega_c| = \omega_L \tag{8-33}$$

For a three-state PD, $\omega_L = 2\pi K$, so

$$|\Delta\omega_c| = 2\pi K \tag{8-33$'$}$$

is a good design value for the circuit in Fig. 8–13a. Then Eqs. (8-32) and (8-33′) give

$$I = 2\pi KC/K_oT \tag{8-34}$$

What is the pull-in time T_p achieved by the acquisition aid in Fig. 8–13? We will approximate the frequency error's rate of change by $d\omega_c/dt \approx \Delta\omega_c/\Delta t$, where Δt is the interval between cycle slips. But by definition,

$$\Delta t \equiv 2\pi/|\omega_c|$$

Since ω_c and $\Delta\omega_c$ have opposite signs,

$$d\omega_c/dt \approx \Delta\omega_c/\Delta t = |\omega_c|\Delta\omega_c/2\pi = -\omega_c|\Delta\omega_c|/2\pi$$

$$d\omega_c/\omega_c = -dt|\Delta\omega_c|/2\pi \tag{8-35}$$

As ω_c goes from an initial ω_{eo} to a final lock frequency of ω_L, the time goes from 0 to T_p. Then integrating the left side of Eq. (8-35) from ω_{eo} to ω_L and the right side from 0 to T_p yields

$$T_p = (2\pi/|\Delta\omega_c|)\,\ell n(\omega_{eo}/\omega_L) \tag{8-36}$$

For a practical maximum of $|\Delta\omega_c| = 2\pi K$ as in Eq. (8-33′), the minimum T_p is

$$T_p = (1/K)\,\ell n(\omega_{eo}/2\pi K) \tag{8-37}$$

The waveforms during acquisition are illustrated in Fig. 8–14a. The cycle-slip pulses v_D' produces steps of $\Delta\omega_c$ in ω_c, which get farther apart in time as ω_c decreases. At the same time, the sawtooth waveform of v_2 also affects the total frequency error $\omega_e = \omega_c + K_o v_2$ (which happens to be a smooth curve for $|\Delta\omega_c| = \omega_L$). When ω_c finally jumps within the lock range ω_L, there are no more cycle slips, and the frequency acquisition is said to be complete.

The pull-in time T_p may be reduced somewhat by increasing $|\Delta\omega_c|$. But if $|\Delta\omega_c| > 2\omega_L$, then ω_c may never jump to within the lock-in range, as illustrated in Fig. 8–14b.

EXAMPLE 8–5

A PLL has $K = 8$ krad/s, $K_o = 0.2$ Mrad/s/V, a three-state PD, and a loop filter with $R_1 = 50$ kΩ, $R_2 = 5$ kΩ, and $C = 0.1$ μF. The input frequency is $\omega_i = 10$ Mrad/s. The digital logic levels of the PD are $V_H = 5$ V and $V_L = 0$ V. Choose R_3 in the FD for minimum pull-in time. Find the pull-in time for $\omega_{eo} = 2$ Mrad/s.

From Eq. (8-29), the reference voltage is $V_r = (V_H + V_L)/2 = 2.5$ V. From Eq. (8-31), the input signal period is $T = 2\pi/\omega_i = 628$ ns. From Eq. (8-34), $I = 2\pi KC/K_oT = 40$ mA. But the most load current that the PD can handle is 10 mA. One way to scale down I is to reduce C by a factor of four. In order to preserve $F(s)$ of the loop filter, all of its impedances have to be increased by a factor of four: $R_1' = 200$ kΩ, $R_2' = 20$ KΩ, $C' = 0.025$ μF. Then $I' = 10$ mA, and from Eq. (8-30) $R_3 = (2.5\text{ V})/I' = 205$ Ω. If a diode drop of 0.7 V is taken into account, then $R_3 = (1.8\text{ V})/I' = \underline{180\ \Omega}$.

I could be reduced further by scaling up the impedances of the loop filter further. Eventually, R_1 becomes so large that the input offset current of the op amp creates too much offset voltage V_{do} (see section 3–4).

For $\omega_{eo} = 2$ Mrad/s, Eq. (8-37) gives $T_p = (8\text{ krad/s})^{-1} \times \ell n(2000/50) = \underline{0.46\text{ ms}}$. Compare this with the 39 ms for similar conditions and an unaided three-state PD in Example 8–4.

8–6 ROTATIONAL FREQUENCY DETECTOR

We have seen that an unaided three-state PD does a good job of acquisition, and with the aid of an FD, it does an excellent job. But there are applications where a three-state PD can't be used. Since it doesn't forgive a missing pulse in R or V, it can't be used with data or in high-noise applications. The FD described in this section does work with data and with high-noise applications. Richman [4] described the original version, which he called

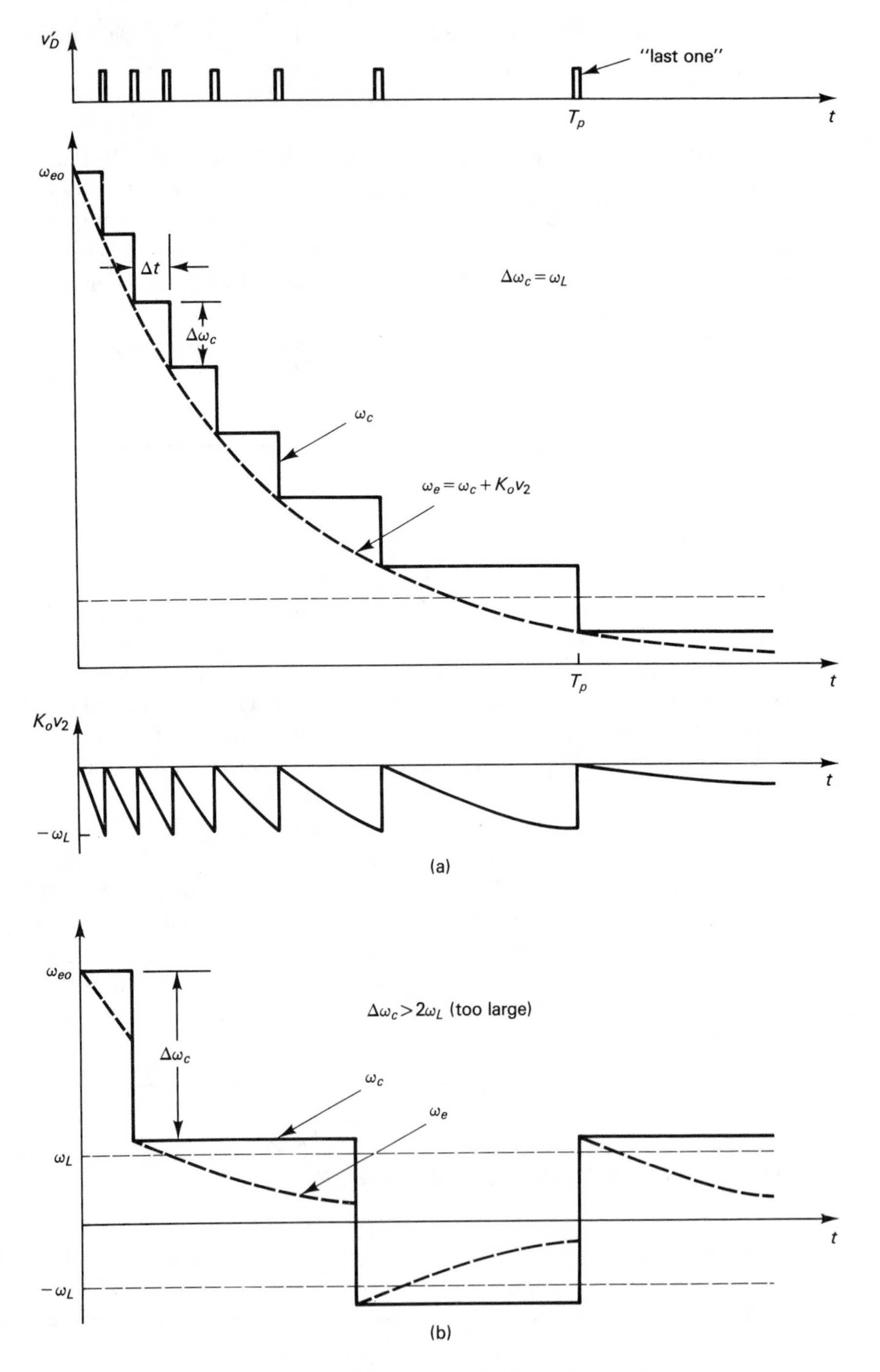

FIGURE 8–14 Pull-in waveforms with aided acquisition

a "quadricorrelator." Messerschmitt [5] described a digital version which he called a "rotational FD." At about the same time, Afonso et al. [6] published a circuit realizing such a frequency detector.

The rotational FD circuit in Fig. 8–15a generates pulses corresponding to cycle slips—negative pulses at v'_U for $\omega_o < \omega_i$ and positive pulses at v'_D for $\omega_o > \omega_i$. These charge or discharge the capacitor in the loop filter to bring ω_o equal to ω_i. As in Eq. (8-30), the magnitude of the charging current i_3 is $I = (V_H - V_L/2r_3$. To avoid jumping over the lock-in range, we again choose $|\Delta\omega_c| = \omega_L$ as in Eq. (8-33). Then in a similar development to that of Eq. (8-34), we have here

$$I = \omega_L C/K_o\tau \tag{8-38}$$

where ω_L depends on the particular PD that is used. The designer is free to choose τ here, but it must be bounded by

$$\tau \leq 2\pi/\omega_{eo} \tag{8-39}$$

so the pulses at v'_U or v'_D don't overlap.

The basis of detecting cycle slips is to divide the phase of the VCO signal v_o into four quadrants and sample the quadrants with the pulses of the input signal v_i. The sequence of quadrants tells the direction of cycle slips. The clock v_o is delayed by 90 deg. to form the signal v'_o. These two signals define four quadrants of phase in which v_o and v'_o are 1,1, then 0,1, then 0,0, then 1,0, where high and low logic levels are represented by 1 and 0. The signal v_i in Fig. 8–15b is RZ data with a "mark" represented by a pulse and "space" represented by a missing pulse. If ω_o is greater than ω_i ($\omega_c > 0$), then the rising edges of v_i advance through the quadrants, causing A and B (the sampled quadrant) to be 1,1, then 0,1, then 0,0, then 1,0, etc. (see the sequence in Fig. 8–15b). This is clockwise rotation through the quadrants in Fig. 8–15c. We wish to generate a pulse at v'_D every time the heavy line in Fig. 8–15c is crossed in a clockwise direction (a "D cycle slip"). This corresponds to v_i sampling the quadrant 0,0 immediately after the quadrant 0,1. The circuit in Fig. 8–15a remembers the previous quadrant as C and D (the previous values of A and B). Then a D cycle slip corresponds to the state, A,B,C,D = 0,0,0,1, or the truth of the Boolean expression $\overline{A}\cdot\overline{B}\cdot\overline{C}\cdot D$. This is realized by an AND gate in Fig. 8–15a which produces a pulse at F for every D cycle slip. A monostable multivibrator stretches each pulse to a width of τ at v'_D. Similarly, a counterclockwise (or U) cycle slip produces a pulse at v'_U.

If the difference ω_c between ω_i and ω_o is too great or if there are too many sequential pulses missing from the data signal v_i, it is possible for the circuit to miss some v'_D pulses when v_i jumps over either the 0,1 quadrant or the 0,0 quadrant. Even worse, the sequence of the sampled quadrants may appear to go backwards, generating some spurious v'_U pulses. Let f_U be the average rate of v'_U pulses, and let f_D be the average rate of v'_D pulses. For the case of all-ones data (no v_i pulses are missing), $f_U - f_D$ is given by the function

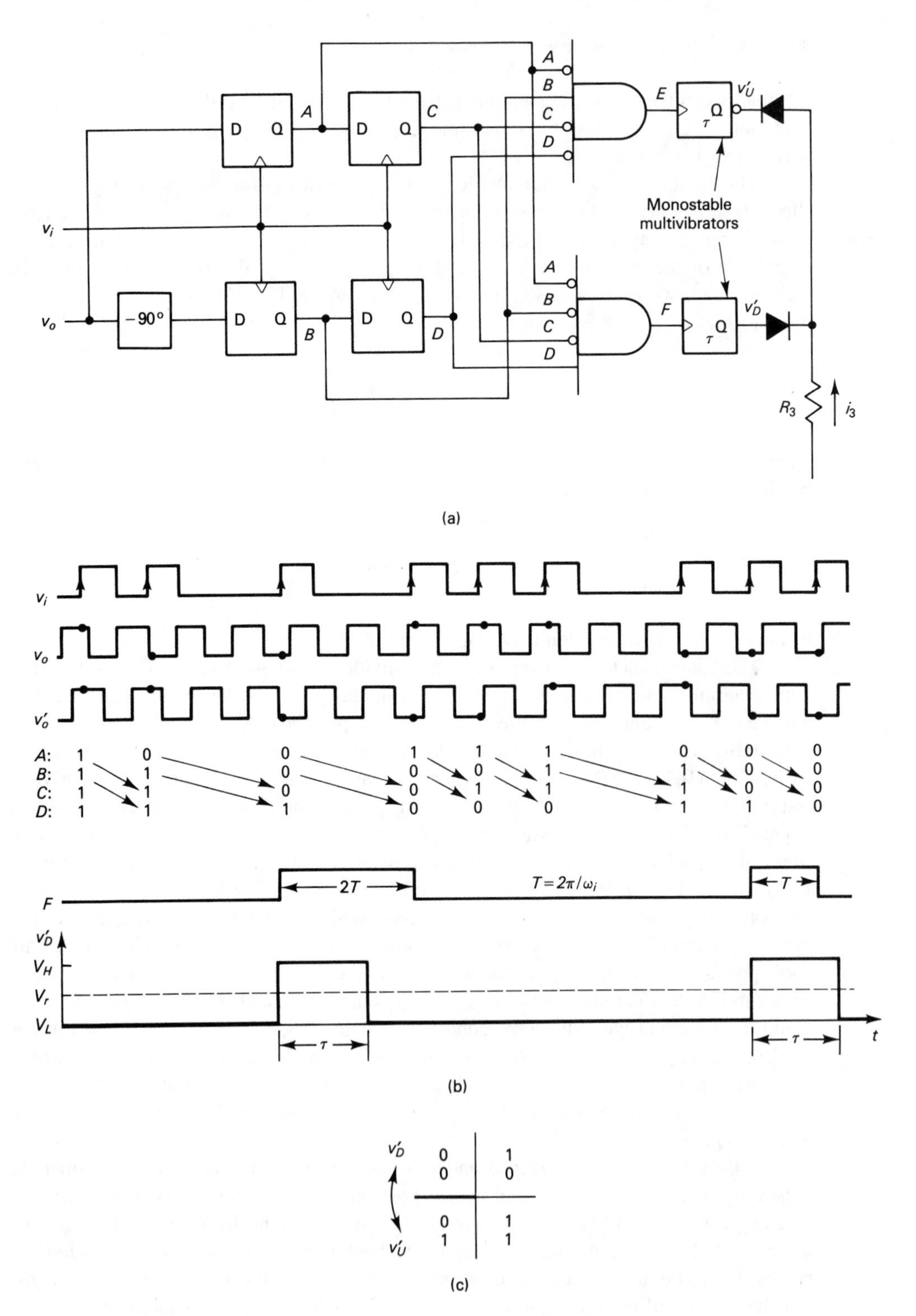

FIGURE 8–15 Acquisition aid for clock recovery

$g(\omega_c)$ in Fig. 8–16. For $0 < \omega_c < \omega_i/4$, all D cycle slips are detected, and there are no U cycle slips.

$$g(\omega_c) \equiv f_U - f_D = -f_D$$
$$= -\omega_c/2\pi; \qquad 0 < \omega_c < \omega_i/4 \tag{8-40a}$$

As ω_c goes beyond $\omega_i/4$, the probability of jumping over the 0,1 quadrant or the 0,0 quadrant increases (f_D decreases) until no cycle slips are detected for $\omega_c = \omega_i/2$:

$$g(\omega_c) = -\omega_i/4\pi + \omega_c/2\pi; \qquad \omega_i/4 \leq \omega_c \leq \omega_i/2 \tag{8-40b}$$

For $\omega_c > \omega_i/2$, aliasing makes the sequence of sampled quadrants appear to go backwards, and pulses start occurring on v'_U($f_D = 0$, and f_U increases). $g(\omega_c)$ becomes positive, and the FD pushes out rather than pulling in. Therefore, the pull-in range for all-ones data is

$$\omega_p = \omega_i/2 \tag{8-41}$$

To find the pull-in time for all-ones data, we can proceed as in the development of Eq. (8-35) by approximating the time derivative with

$$d\omega_c/dt \approx |\Delta\omega_c|(f_U - f_D) = \omega_L g(\omega_c)$$

Using the expressions for $g(\omega_c)$ given in Eqs. (8-40a) and (8-40b), choosing $|\Delta\omega_c| = \omega_L$, and integrating between the appropriate limits as in the previous section, we find

$$T_p = (2\pi/\omega_L)\,\ell n(\omega_{eo}/\omega_L); \qquad 0 \leq \omega_{eo} \leq \omega_i/4 \tag{8-42a}$$

$$T_p = (2\pi/\omega_L)\,\ell n[\omega_i^2/8\omega_L(\omega_i - 2\omega_{eo})]; \qquad \omega_i/4 \leq \omega_{eo} \leq \omega_i/2 \tag{8-42b}$$

All-ones data is not interesting in itself, but it models a sinusoidal carrier. Even for a noisy carrier, the FD misses few of the cycles (pulses), and the $g(\omega_c)$ characteristic in Fig. 8-16 applies. Messerschmitt [7] used a rotational FD to acquire lock with good results for signal-to-noise ratios as low as 15 dB.

When v_i is random data, the effect of not detecting cycle slips and detecting apparently reversed cycle slips is aggravated due to the many missing pulses. For all-ones data [to which $g(\omega_c)$ applies] the spacing of the v_i pulses is $T = 2\pi/\omega_i$. For random data, the probability is 1/2 that the spacing is T. The probability is 1/4 that the spacing is $2T$, which has the effect of a frequency error of $2\omega_c$. The probability is 1/8 that the spacing is 3T, which has the effect of a frequency error of $3\omega_c$, etc. Therefore, $f_U - f_D$ for random data is given by

$$g'(\omega_c) = \sum_{n=1}^{\infty} 2^{-n-1}\, g(n\omega_c) \tag{8-43}$$

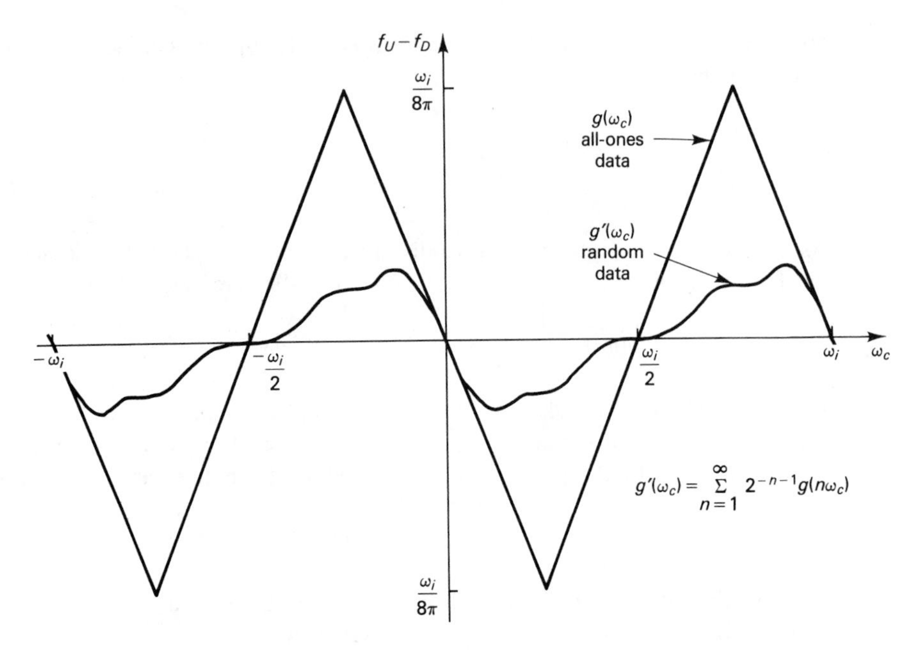

FIGURE 8–16 Net pulse rate for acquisition aid in Fig. 8–15

This function, plotted in Fig. 8–16, agrees with the experimental data of Afonso et al. [8]. For $\omega_c < \omega_i/20$, practically no cycle slips are missed, and $g'(\omega_c) \approx -\omega_c/2\pi$. But for $\omega_c = \omega_i/4$, for example, a net of only 22% of the cycle slips are effective, and $g'(\omega_c) = 0.22\omega_c/2\pi$. Beyond $\omega_c = \omega_i/3$, the pull-in effect is so small as to be not reliable or useful. Therefore, the pull-in range for random should be practically taken as

$$\omega_p \approx \omega_i/3 \tag{8-44}$$

In order to get an expression for the pull-in time for random data, we make a (conservative) piecewise linear approximation:

$$g'(\omega_c) \approx -\omega_c/2\pi; \qquad 0 \le \omega_c \le \omega_i/20 \tag{8-45a}$$

$$g'(\omega_c) \approx -\omega_i/40\pi; \qquad \omega_i/20 \le \omega_c \le \omega_i/3 \tag{8-45b}$$

To find the pull-in time for random data, we again begin with the approximation

$$d\omega_c/dt \approx |\Delta\omega_c|(f_U - f_D) = \omega_L g'(\omega_c)$$

and substitute into it the expressions for $g'(\omega_c)$ in Eq. (8-45). Integrating between the appropriate limits yields

$$T_p \approx (2\pi/\omega_L)\,\ell n(\omega_{eo}/\omega_L); \qquad 0 \le \omega_{eo} \le \omega_i/20 \qquad (8\text{-}46a)$$

$$T_p \approx (2\pi/\omega_L)\,\ell n(\omega_i/20\omega_L) + (40\pi\omega_{eo}/\omega_i - 2\pi)/\omega_L; \qquad \omega_i/20 \le \omega_{eo} \le \omega_i/3 \quad (8\text{-}46b)$$

EXAMPLE 8–6

A PLL is used to recover a clock from random data. The phase detector is a two-state PD with a sawtooth characteristic, and $K = 8$ krad/s. The VCO has a gain of $K_o = 0.2$ Mrad/s/V, and the capacitor in the active loop filter is 0.1 μF. For $\omega_i = 10$ Mrad/s and $\omega_{eo} = 2$ Mrad/s, choose τ and I, and find the pull-in time.

For a two-state PD, $\omega_L = \pi K = 25$ krad/s. In accord with Eq. (8-39), we maximize τ by choosing $\tau = 2\pi/\omega_{eo} = 3.14$ μs. Then from Eq. (8-38) we get the minimum T_p [given by Eq. (8-46)] by designing for $I = \omega_L C/K_o\tau = 4$mA. For $\omega_i = 10$ Mrad/s, $\omega_{eo} =$ 2 Mrad/s $> \omega_i/20$, and Eq. (8-46b) gives $T_p = 0.75 + 0.75 = \underline{1.5 \text{ ms}}$.

REFERENCES

[1] D. Richman, "Color-carrier Reference Phase Synchronization Accuracy in NTSC Color Television," *Proc. IRE,* vol. 43, pp. 108–33, January 1954.

[2] S. A. Meer, "Analysis of Phase-Locked Loop Acquisition: A Quasi Stationary Approach," *IEEE International Convention Record,* vol. 14, pt. 7, pp. 85–106, 1966.

[3] F. M. Gardner, *Phaselock Techniques,* Wiley: New York, 1979, pp. 151–56.

[4] Richman, "Color-carrier Reference Phase Synchronization."

[5] D. G. Messerschmitt, "Frequency Detectors for PLL Acquisition in Timing and Carrier Recovery," *IEEE Trans. on Communications,* vol. COM-27, pp. 1288–295, September 1979.

[6] J. A. Afonso, A. J. Quiterio, and D. S. Arantes, "A Phase-Locked Loop with Digital Frequency Comparator for Timing Signal Recovery," *IEEE National Telecommunications Conference,* pp. 14.4.1–14.4.5, November 28–29, 1979.

[7] Messerschmitt, "Frequency Detectors."

[8] Afonso, et al., "Phase-Locked Loop."

CHAPTER

9

Modulation and Demodulation

Some PLL applications that involve modulating and demodulating a carrier were described briefly in Chapter 1. In this chapter, we will look more in depth at the following: phase modulation, phase demodulation, frequency modulation, and frequency demodulation. In each case the proper PLL bandwidth K will be determined, and the necessary ranges of the PD and the VCO will be considered. Until now we have ignored the high-frequency component of the PD output (the difference between v_d and $\bar{v}_d$). We will see that it can cause spurious phase and frequency modulation.

9–1 PHASE MODULATION

Let the modulating signal be $m(t)$ with a bandwidth B_m, where B_m is in Hz. The objective is to modulate the phase θ_o of a carrier so that $\theta_o(t) = \alpha\, m(t)$, where α is some constant. This can be done by adding the signal $m(t)$ into a PLL after the PD, as in Fig. 9–1a. The input to the PLL is a carrier $v_i = \sin(\omega_i t)$ with no phase modulation ($\theta_i = 0$). If the PLL bandwidth is great enough, v_d can follow $m(t)$ to effectively cancel it [$v_d \approx -m(t)$]. But this v_d must be produced by a proportional θ_o from the VCO, producing the desired phase modulation.

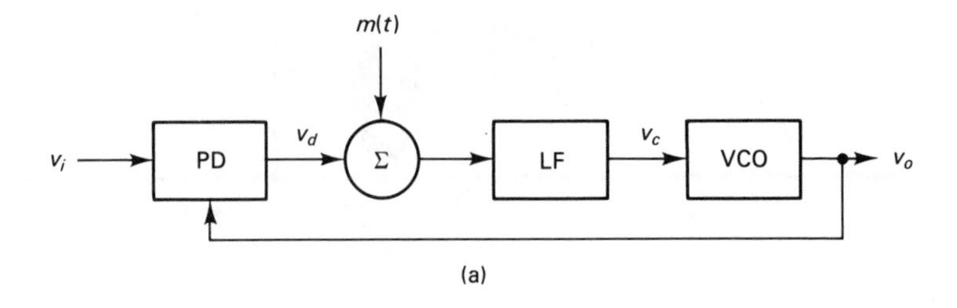

(a)

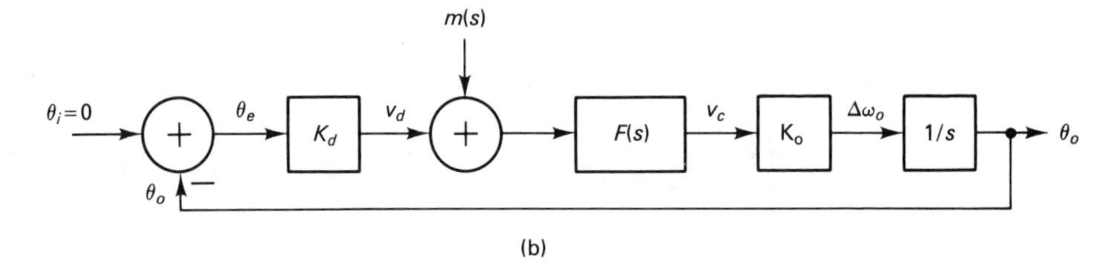

(b)

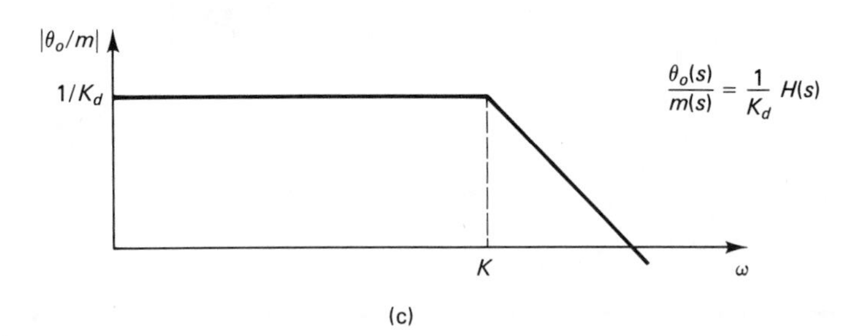

(c)

FIGURE 9–1 Phase modulation

9–1–1 Bandwidth, Phase and Frequency Ranges

A signal flow graph of the PLL is shown in Fig. 9–1b. We need to ensure that the transfer function from $m(s)$ to $\theta_o(s)$ has a flat frequency response of sufficient bandwidth. Let the forward gain from m to θ_o be $A \equiv F(s)K_o/s$, and let the feedback from θ_o to m be $B \equiv K_d$. Then from control theory (see Phillips and Harbor [1] for example),

$$\frac{\theta_o(s)}{m(s)} = \frac{A}{1 + AB} = \frac{F(s)K_o/s}{1 + K_dF(s)K_o/s} =$$

$$= \frac{G(s)/K_d}{1 + G(s)} = \frac{1}{K_d} H(s) \approx \frac{K/K_d}{s + K} \qquad (9\text{-}1)$$

where $H \equiv G/(1 + G)$. $H(s)$ has unity gain with a high-frequency cutoff at $\omega = K$. Therefore, $\theta_o(s)/m(s)$ has a gain of $1/K_d$ with a high-frequency cutoff at $\omega = K$, as shown in Fig. 9–1c. Then if we design the PLL to have

$$K \geq 2\pi B_m \tag{9-2}$$

all the spectral components of $m(t)$ will be passed with a gain of $1/K_d$, and

$$\theta_o(t) = (1/K_d)m(t) \tag{9-3}$$

as desired.

Since the input phase θ_i is zero, all of the output phase θ_o appears as phase difference θ_e across the PD. Therefore, the output θ_o must not exceed the linear range of the PD:

$$|\theta_o| \leq \theta_{em} \tag{9-4}$$

For example, if a three-state PD is used, the peak θ_o can't be greater than 2π. If a larger θ_o is desired, then the range of the PD must be extended by one of the methods discussed in section 4–11 or 4–12. The most common means is to precede the PD with two $\div N$ frequency dividers, as shown in Fig. 9–2a. The dividers are realized by digital counters—binary counters when N is a power of 2, or programmable counters for other values. The dividers extend the phase range by a factor of N; the extended-range PD has a range of $\theta_{em} = 2\pi N$ (see Fig. 9–2b). Then for this extended-range PD, the maximum phase modulation is

$$|\theta_o| \leq 2\pi N \tag{9-4$'$}$$

As $m(t)$ modulates θ_o, it is also modulating $\Delta\omega_o$ through the relationship $\Delta\omega_o \equiv \dot{\theta}_o$. Since $\omega_o = \omega_i + \Delta\omega_o$, the VCO must have a range of at least

$$\omega_o = \omega_i \pm \dot{\theta}_o \tag{9-5}$$

9–1–2 Spurious Modulation

The disadvantage of using $\div N$ dividers is that the high-frequency component of $\tilde{v}_d$ is lowered, and it may result in excessive spurious phase modulation of θ_o. [This is an undesired modulation not in response to $m(t)$.] The connection between $\tilde{v}_d$ and θ_o is illustrated in Fig. 9–2c. After the $\div N$, the frequency into the three-state PD is ω_i/N. The low-frequency component of the PD output $\tilde{v}_d$ is v_d, which is proportional to θ_e. But there is also a high-frequency component $\tilde{v}_d - v_d$ with a frequency.

$$\omega_d = \omega_i/N \tag{9-6}$$

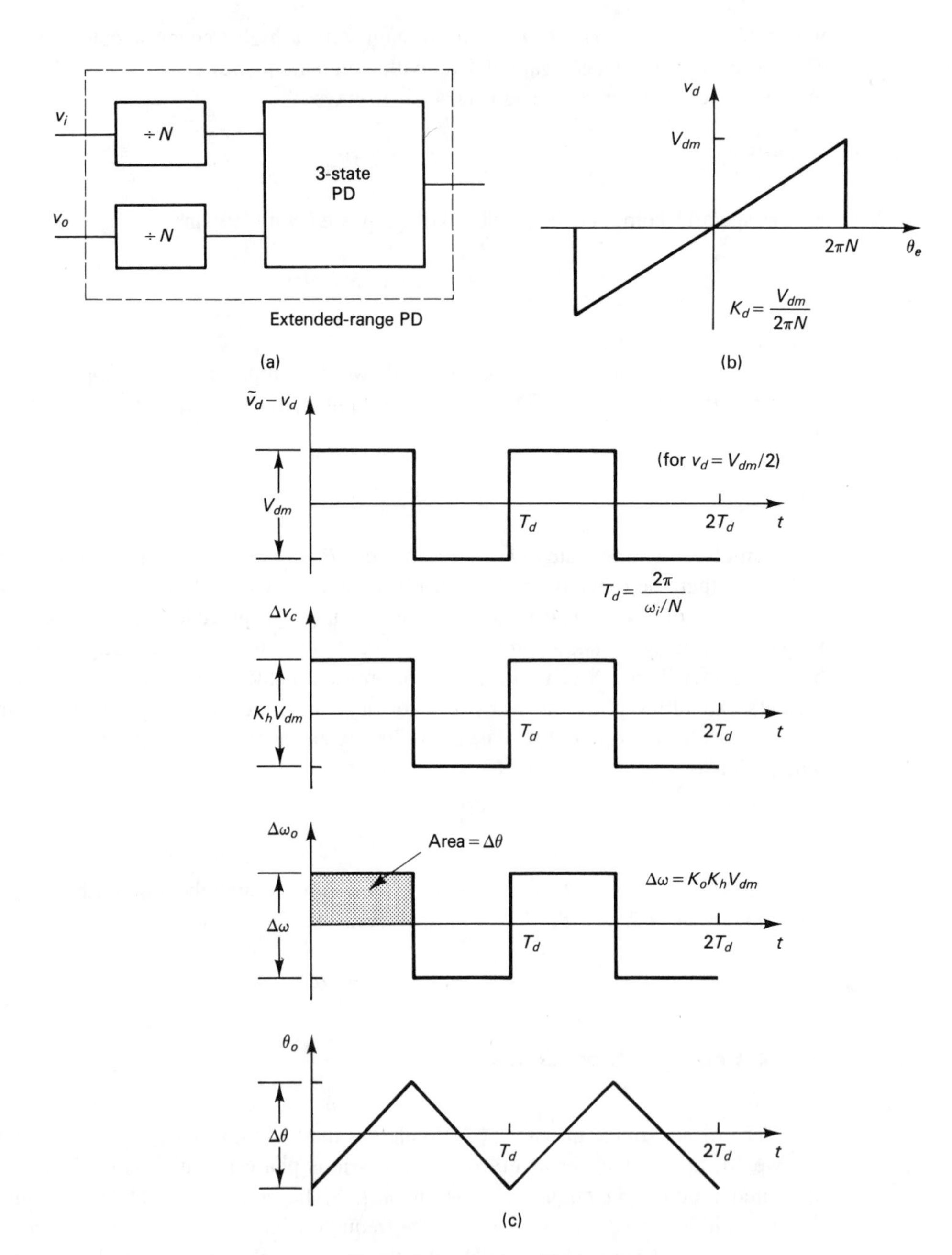

FIGURE 9–2 Spurious FM and PM

The period is $T_d = 2\pi/\omega_d$, or

$$T_d = 2\pi N/\omega_i \tag{9-7}$$

The most severe spurious modulation is when $\tilde{v}_d - v_d$ has a 50% duty cycle, as in Fig. 9–2c. This corresponds to $v_d = V_{dm}/2$ (see section 4–8 on the three-state PD). In practice, ω_d is always high enough ($>> \omega_2$) that the gain of the loop filter is essentially K_h. Then the deviation of the VCO control voltage is $\Delta v_c = K_h(\tilde{v}_d - v_d)$ with a peak-to-peak amplitude of $K_h V_{dm}$. The spurious modulation of the VCO frequency is $\Delta\omega_o = K_o \Delta v_c$ with a peak-to-peak amplitude $\Delta\omega = K_o K_h V_{dm}$. But for a three-state PD with $\div N$ frequency dividers,

$$V_{dm} = 2\pi N K_d \tag{9-8}$$

(see the characteristic in Fig. 9–2b). Then the amplitude of the spurious FM is

$$\Delta\omega = 2\pi N K_o K_h K_d = 2\pi N K \tag{9-9}$$

The spurious phase modulation θ_o is the integral of $\Delta\omega_o$. Therefore, its peak-to-peak rise $\Delta\theta$ is the area under a positive half-cycle of $\Delta\omega_o$:

$$\Delta\theta = (\Delta\omega/2)\,(T_d/2) = (\pi N)^2 K/\omega_i \tag{9-10}$$

The resulting triangular modulation of θ_o shown in Fig. 9–2c is present while $m(t)$ is also modulating θ_o. The spurious $\Delta\theta$ will vary as $m(t)$ increases and decreases the duty cycle of $\tilde{v}_d - v_d$, but it will never be larger than that given by Eq. (9-10).

9–1–3 Spurious Modulation with a Pole at ω_3

The spurious modulation can be attenuated by adding a pole to the response $F(s)$ of the loop filter as shown in Fig. 9–3a. This can be realized by an additional capacitor as in Fig. 3–14. For stability, we require

$$\omega_3 \geq 4\,K \tag{9-11}$$

The roll-off of $|F|$ beyond $\omega = \omega_3$ reduces the amplitude of the unwanted waveforms with frequency ω_d when $\omega_d > \omega_3$. For $\omega > \omega_3$, the loop filter acts essentially like an integrator:

$$F(s) \approx K_h\omega_3/s; \qquad |s| > \omega_3 \tag{9-12}$$

(see Fig. 9–3a).

The worst-case $\tilde{v}_d - v_d$ is shown in Fig. 9–3b with 50% duty cycle, peak-to-peak amplitude V_{dm}, and period $T_d = 2\pi N/\omega_i$. This unwanted signal is integrated by the transfer function in Eq. (9-12) to deviate the VCO control voltage:

$$\Delta v_c = K_h\omega_3 \int (\tilde{v}_d - v_d)dt$$

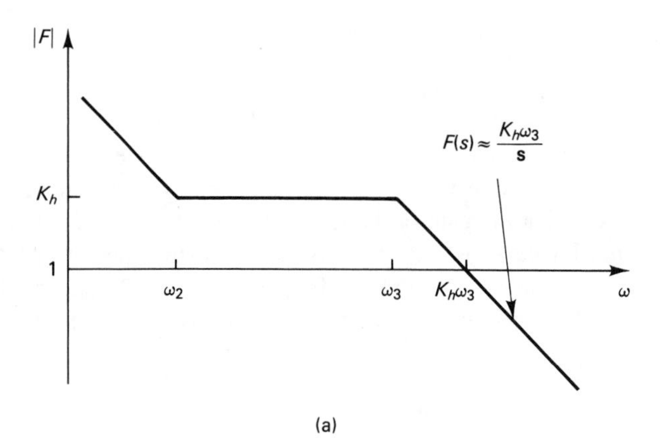

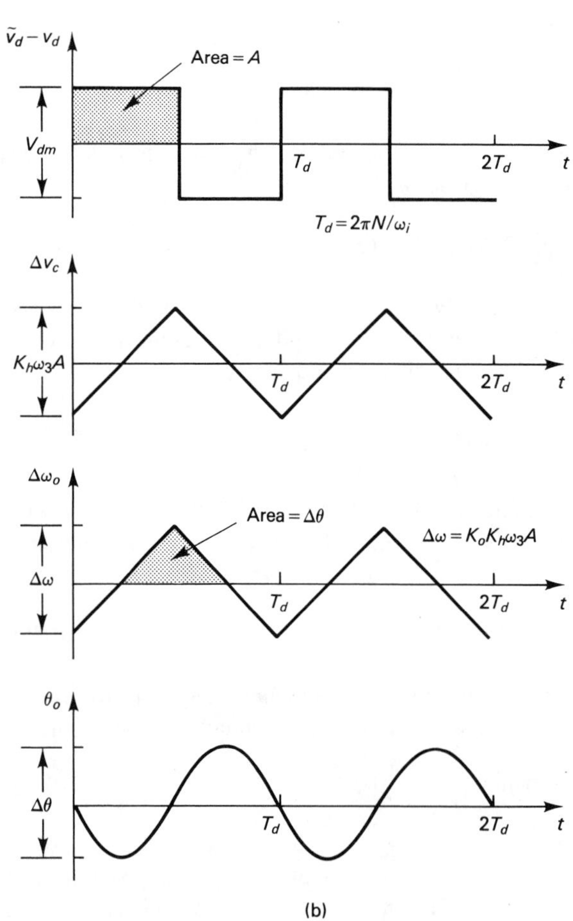

FIGURE 9–3 Spurious FM and PM with pole at ω_3

Then the peak-to-peak amplitude of Δv_c is $K_h\omega_3 A$, where A is the area under a positive half-cycle of $(\tilde{v}_d - v_d)$: $A = V_{dm}T_d/4 = (\pi N)^2 K_d/\omega_i$ [see Eqs. (9-7) and (9-8)]. The spurious FM is $\Delta\omega_o = K_o\Delta v_c$ with a peak-to-peak amplitude

$$\Delta\omega = K_oK_h\omega_3 A = K_oK_h\omega_3(\pi N)^2 K_d/\omega_i$$

$$= (\pi N)^2 K\omega_3/\omega_i \tag{9-13}$$

This amplitude is minimized by choosing the smallest ω_3 within the bounds of Eq. (9-11): $\omega_3 = 4K$. Then the amplitude of the spurious FM is

$$\Delta\omega = (2\pi NK)^2/\omega_i; \qquad \omega_3 = 4K \tag{9-13'}$$

The spurious phase modulation θ_o is the integral of $\Delta\omega_o$. Therefore, its peak-to-peak rise $\Delta\theta$ is the area under a positive half-cycle of $\Delta\omega_o$:

$$\Delta\theta = (\Delta\omega/2)\ (T_d/2)/2 = (\pi N)^3 K\omega_3/4\omega_i^2 \tag{9-14}$$

$$= (\pi N)^3 (K/\omega_i)^2; \qquad \omega_3 = 4K \tag{9-14'}$$

Further reduction of $\Delta\omega$ and $\Delta\theta$ is possible by making a multiple pole at ω_3. For each additional pole, the spurious modulation is reduced by an additional factor of $\omega_3/\omega_d = N\omega_3/\omega_i$. For a total of n poles at ω_3, Eqs. (9-13) and (9-14) become

$$\Delta\omega = \pi^2 NK(N\omega_3/\omega_i)^n \tag{9-15}$$

$$\Delta\theta = (\pi^3 N^2 K/4\omega_i)\ (N\omega_3/\omega_i)^n \tag{9-16}$$

With multiple poles, ω_3 must be kept *more* than a factor of four away from K. Otherwise, the in-band phase of the multiple poles will cause instability. The phase at $\omega = K$ due to the multiple poles at ω_3 should be kept less than about 0.5 radian (or 29 deg.).

EXAMPLE 9–1

A carrier with frequency $\omega_i = 10$ Mrad/s is to be phase-modulated by $m(t) = V_m \sin(\omega_m t)$ to produce $\theta_o(t) = 7\pi \sin(\omega_m t)$, where $\omega_m = 10$ krad/s. The phase detector is a three-state PD with $V_{dm} = 2.5$ V together with two $\div N$ frequency dividers. Choose N, V_m, and the PLL bandwidth K. Find the range required of the VCO, and find the spurious PM amplitude $\Delta\theta$. Add a pole at ω_3 in the loop filter if necessary to keep $\Delta\theta$ less than 1% of the peak θ_o.

Equation (9-4′) requires a minimum $N = \underline{4}$. Then $K_d = V_{dm}/2\pi N = 0.1$ V/rad. For a peak $\theta_o(t)$ of 7π radians, Eq. (9-3) requires a peak $m(t)$ of $V_m = 7\pi K_d = \underline{2.2\text{ V}}$. Equation (9-2) requires $K > 2\pi B_m = \omega_m = 10$ krad/s. Choose $K = \underline{20\text{ krad/s}}$. The frequency modulation is $\Delta\omega_o = \dot{\theta}_o = 7\pi\omega_m \cos(\omega_m t) = (220 \text{ krad/s}) \cos(\omega_m t)$. Then Eq. (9-5) requires a VCO range of $\omega_o = \underline{10 \pm 0.22 \text{ Mrad/s}}$.

From Eq. (9-10), the spurious PM is $\Delta\theta = (\pi N)^2 K/\omega_i = (\pi 4)^2(20 \text{ krad/s})/10 \text{ Mrad/s} = \underline{0.316 \text{ radians}}$. This is slightly greater than 1% of the peak θ_o of 22 radians. With a pole at $\omega_3 = 4K = 80$ krad/s, Eq. (9-14′) gives $\Delta\theta = 2(\pi N)^3(K/\omega_i)^2 = 2(\pi 4)^3(0.02/10)^2 = \underline{0.016 \text{ radians}}$.

9–2 PHASE DEMODULATION

In phase demodulation, the modulated phase θ_i of a carrier is converted back into a voltage $m(t) = \alpha\theta_i(t)$, where α is some constant. The demodulation is preformed by locking a narrow-bandwidth PLL to the modulated carrier $v_i = \sin(\omega_i t + \theta_i)$. If the bandwidth is small enough, the VCO phase θ_o won't follow the phase modulation, and θ_o serves as a reference against which to compare θ_i. The PD compares θ_i against θ_o, and the PD output v_d is the demodulated output $m(t)$ (see Fig. 9–4a). A low-pass filter (LPF) is usually necessary to remove high-frequency components from $\tilde{v}_d$, leaving $v_d = m(t)$.

A signal flow graph of the PLL is shown in Fig. 9–4b. The low-pass filter has a transfer function $F'(s)$ with an in-band gain of unity and a cutoff at ω_{LP}. For an active loop filter, the transfer function from θ_i to θ_e is $H_e(s) = s^2/(s^2 + Ks + K\omega_2)$, and the transfer function from θ_e to $m(s)$ is $K_d F'(s)$. Then

$$\frac{m(s)}{\theta_i(s)} = K_d H_e(s) F'(s) = \frac{K_d s^2}{(s^2 + Ks + K\omega_2)} F'(s) \tag{9-17}$$

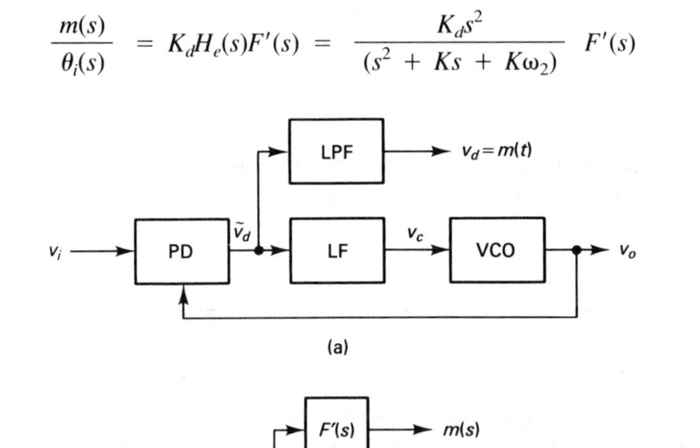

FIGURE 9–4 Phase demodulation

The frequency response of $|m/\theta_i|$ is illustrated in Fig. 9–5c. It has a mid-band gain of K_d, a low-frequency cutoff at K, and a high-frequency cutoff at ω_{LP} (compare the response $|H_e|$ in Fig. 3–16b).

A typical time waveform for $\theta_i(t)$ and its spectrum are shown in Fig. 9–5a. The bandwidth of $\theta_i(t)$ is from B_{m1} to B_{m2}. The whole PD output $\tilde{v}_d$ is actually a pulse-width-modulated signal with a "carrier" frequency ω_d and low-frequency component v_d, as shown in Fig. 9–5b. The spectrum $\tilde{v}_d(\omega)$ is also shown; it has upper and lower sidebands around ω_d and a baseband component that is the spectrum of v_d. To recover the baseband component, it is clear from the response $|m/\theta_i|$ in Fig. 9–5c that both of the following bounds must be satisfied:

$$K \leq 2\pi B_{m1} \tag{9-18}$$

$$2\pi B_{m2} \leq \omega_{LP} < \omega_d - 2\pi B_{m2} \tag{9-19}$$

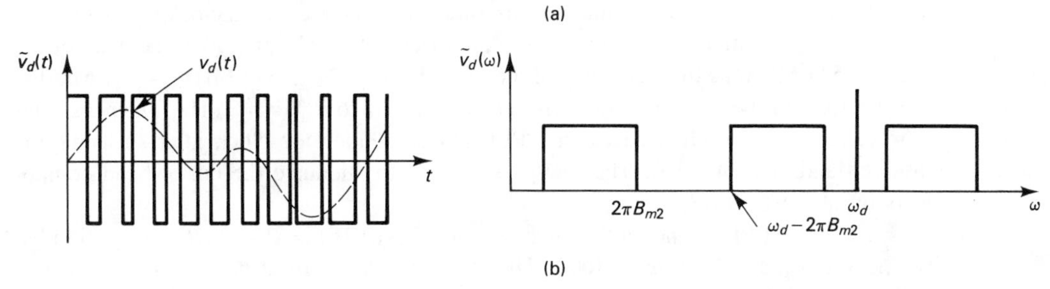

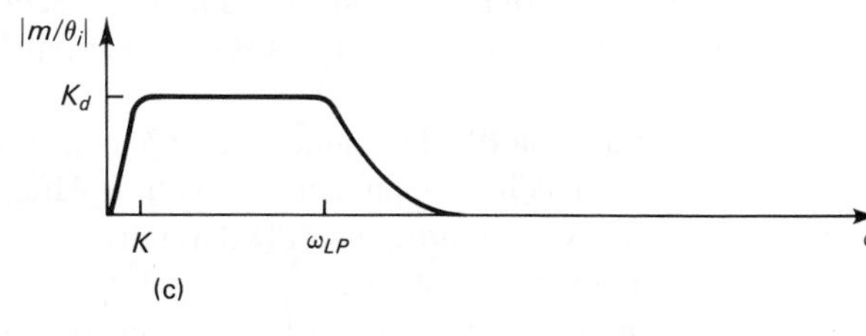

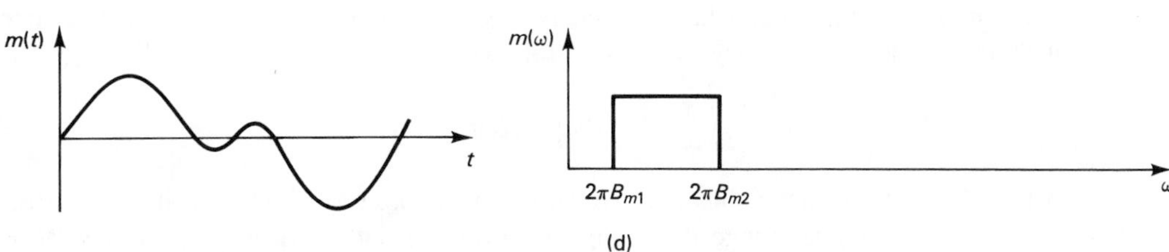

FIGURE 9–5 Phase demodulation waveforms and spectra

In some applications, the low-pass filter must be of high order (such as a fifth-order Butterworth) if the bounds in Eq. (9-19) are tight. If Eqs. (9-18) and (9-19) are satisfied, then

$$m(t) = v_d(t) = K_d\theta_i(t) \tag{9-20}$$

as desired.

Since $\theta_o \approx 0$, $\theta_e(t) \approx \theta_i(t)$. Then for θ_e to stay in the linear range of the PD, we must satisfy

$$|\theta_i(t)| < \theta_{em} \tag{9-21}$$

As with phase modulation in the previous section, it may be necessary to extend the PD range if $\theta_i(t)$ is large. Leave enough margin to avoid false lock during acquisition if a three-state PD is used (see section 4–13).

EXAMPLE 9–2

A digital data signal has accumulated phase jitter, and the recovered clock signal has this same jitter $\theta_i(t)$. (Jitter and clock recovery are discussed in Chapter 10.) The baud of the data is 1.544 Mb/s, so the frequency of the clock is $\omega_i = 2\pi(1.544 \text{ MHz}) = 9.7$ Mrad/s. The spectrum of the jitter θ_i extends from $B_{m1} = 10$ Hz to $B_{m2} = 40$ kHz. The peak jitter is 5 cycles, or 10π radians. Design a PLL to demodulate the clock jitter so it may be observed and characterized. The low-pass filter is to attenuate the out-of-band components by at least 50 dB.

The largest PLL bandwidth which satisfies Eq. (9-18) is $K = 2\pi B_{m1} = \underline{62.8 \text{ rad/s}}$. This is very small: $K = \omega_i/154000$. Therefore, the VCO must be a crystal-controlled VCXO to avoid injection problems (see section 5–9). This greatly restricts the VCO range, but for phase demodulation the required VCO frequency range is vanishingly small.

Equation (9-21) requires that $\theta_{em} > 10\pi$. For a three-state PD, θ_{em} is only 2π. We will try extending the phase range with $\div N$ frequency dividers, as in Fig. 9–2a. To avoid false lock with a three-state PD during acquisition, it is necessary to make the PD range even greater than that given by Eq. (9-21). For our application, Eq. (4-47) gives the bound $|\theta_e| < \pi$ to avoid false lock when there are no frequency dividers. This may be extended to include the use of frequency dividers as follows:

$$|\theta_e| < N\pi$$

where $\theta_e \approx \theta_i$ for our application. Therefore, we choose $\underline{N = 11}$ so that $N\pi > 10\pi$.

For the extended-range PD in Fig. 9–2a, the detector frequency is $\omega_d = \omega_i/N = (9.7 \text{ Mrad/s})/11 = 882$ krad/s. Now, $2\pi B_{m2} = 251$ krad/s. Then Eq. (9-19) requires the low-pass filter bandwidth to satisfy 251 krad/s $\leq \omega_{LP} <$ 631 krad/s. If $\omega_{LP} =$ 251 krad/s, a seventh-order filter will provide an attenuation of $(631/251)^7 =$ 56 dB at 631 krad/s.

We will now try a design that doesn't require such a high-order low-pass filter. Let the phase detector range be extended by using an n-state PD. According to Eq. (4-34), θ_{em} for an n-state PD is $(n - 1)\pi$, and Eq. (9-21) requires $(n - 1)\pi > 10\pi$. But it is necessary to make n even greater than this to avoid false lock. The condition for avoiding false lock given by Eq. (4-47) can be extended to an n-state PD as follows:

$$|\theta_e| < (n - 2)\pi$$

where $\theta_e \approx \theta_i$ for our application. Then to satisfy $(n - 2)\pi > 10\pi$, we choose $n = 13$. A 13-state PD requires a 10-stage shift register (compare the six-state PD in Fig. 4–15). The advantage is that the detector frequency is higher now: $\omega_d = \omega_i =$ 9.7 Mrad/s. Then Eq. (9-12) requires the low-pass filter bandwidth to satisfy 251 krad/s $\leq \omega_{LP} <$ 9.45 Mrad/s. If $\omega_{LP} =$ 251 krad/s, a second-order filter will provide an attenuation of $(9450/251)^2 =$ 63 dB at 9.45 Mrad/s. This design trades off a more complex PD for a simpler low-pass filter.

9–3 PHASE DEMODULATION WITH NO CARRIER

For some phase modulation, the carrier (the spectral component at ω_i) actually disappears. Then there is nothing for the PLL to lock onto, and the phase demodulation process described in the previous section won't work.

A simple example of phase modulation for which there is no carrier is $\theta_i(t) = x \sin \omega_m t$, where $x =$ 2.4 radians and ω_m is some modulation frequency (see Fig. 9–6a). Then the input signal is

$$\begin{aligned} v_i(t) &= \sin(\omega_i t + \theta_i) = \sin(\omega_i t + x \sin \omega_m t) \\ &= J_0(x) \sin \omega_i t + \sum_{n \neq 0} J_n(x) \sin(\omega_i + n\omega_m t) \end{aligned} \tag{9-22}$$

(See Lathi [2] for a discussion of the Bessel functions J_n.) For $x =$ 2.4 radians, $J_0(x) = 0$, and the carrier at ω_i disappears. Figure 9–7 shows the spectrum of v_i for $x =$ 2.4 radians; there is no component at ω_i—only the sideband components from the second term in Eq. (9-22). The phasor representation of v_i in Fig. 9–6b gives some understanding of why the carrier disappears. The many phasors show $\boldsymbol{v}_i$ at many instants of time, but the carrier (with fixed frequency and phase) is the average of these positions over time. By symmetry, the average points neither up nor down, and for $\theta_{i\ max} =$ 2.4 radians, the average points neither left nor right. Therefore, the average is zero (no carrier).

A more common example of phase modulation for which there is no carrier is phase-shift keying (PSK). For binary PSK, data is transmitted with $\theta_i = \pi/2$ representing

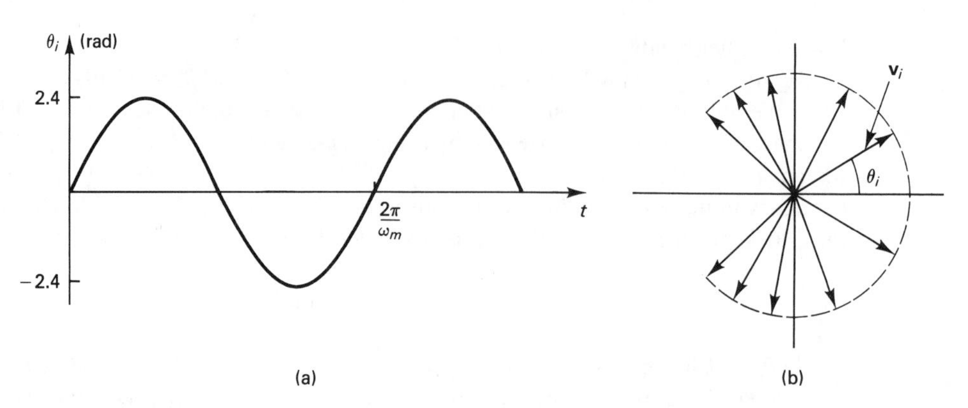

FIGURE 9–6 Sinusoidal PM with no carrier

a "mark" and $\theta_i = -\pi/2$ representing a "space." Figure 9–8a shows an example of $\theta_i(t)$ for random data. The corresponding phasors for v_i are shown in Fig. 9–8b. If the probabilities of a mark and a space are equal, the average of the phasor over time is zero (no carrier).

If a phase-modulated $\mathbf{v}_i$ like that in Fig. 9–6b or Fig. 9–8b is applied to the input of a PLL, the output phasor $\mathbf{v}_o$ will flow $\mathbf{v}_i$ if the bandwidth K is large enough. But if the bandwidth is very narrow (as for phase demodulation), $\mathbf{v}_o$ tries to follow the average of $\mathbf{v}_i$, which doesn't exist. The PLL fails to lock, and phase demodulation is not possible using the simple techniques in the previous section. However, it is possible to generate a carrier by using a nolinearity—by squaring v_i.

9–3–1 SQUARING LOOP

The modulated signal for the PSK in Fig. 9–8 can be represented by

$$v_i = \cos[\omega_i t - (\pi/2)\ m(t)] \tag{9-23}$$

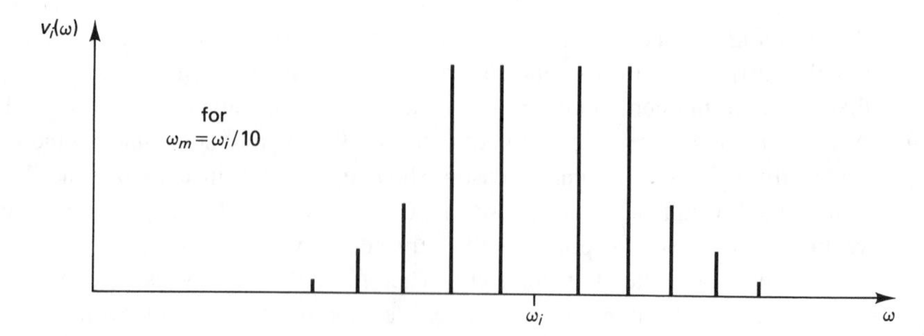

FIGURE 9–7 Spectrum of signal with PM in Fig. 9–6

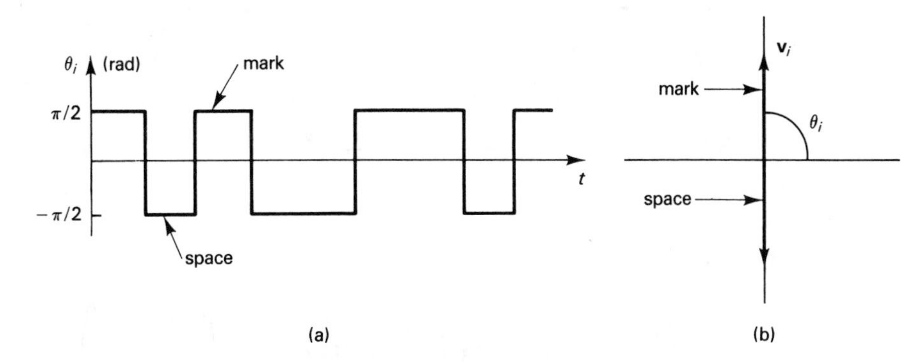

FIGURE 9–8 PSK modulation with no carrier

where $m(t)$ is the modulating data signal with values of ± 1 (see Fig. 9–9a). As m goes from $+1$ to -1, it changes the phase of the carrier by π radians. But this is just an inversion of the signal. Therefore, an equivalent representation of the phase modulation is

$$v_i = m(t) \sin \omega_i t \tag{9-24}$$

An example illustrating the phase reversals in v_i is shown in Fig. 9–9b. The product of any sinusoid with v_i given in Eq. (9-24) has zero average because $m(t)$ has a zero time average. Therefore, a PLL cannot lock to v_i.

If we square the modulated signal, we get

$$v_i^2 = m^2 \sin^2 \omega_i t = 0.5(m^2 - m^2 \cos 2\omega_i t) \tag{9-25}$$

(see Fig. 9–9c). It is the carrier portion of v_i^2 we are interested in, so we throw away the first term in Eq. (9-25) and use $v_i^2 = -m^2 \cos 2\omega_i t$. Since m^2 is always positive, it has a nonzero average, and there is a strong spectral component at $2\omega_i$—double the carrier frequency.

A scheme for recovering the carrier from v_i by using a squarer is shown in Fig. 9–10. After bandpass filtering to limit the noise, v_i is squared, doubling the frequency to $2\omega_i$ [see Eq. (9-25)]. A PLL locks onto this component, and provides $v_o = \sin(2\omega_i t - \theta_e)$, where $\theta_e = 0$ in steady-state (see the waveform in Fig. 9–9d). A $\div 2$ frequency divider reduces the $2\omega_i$ to the carrier frequency ω_i (see the waveform v_o' in Fig. 9–9e). This is the recovered carrier, which is used to demodulate v_i. The multiplier in Fig. 9–10a does the job of demodulation performed by the PD in Fig. 9–4a. The waveform of the product $v_i \times v_o' \equiv \tilde{m}$ is shown in Fig. 9–9f. Low-passing recovers the baseband component m. The cutoff frequency of the low-pass filter must be great enough to pass the signal m:

$$\omega_{LP} \geq 2\pi B_m \tag{9-26}$$

where B_m is the message bandwidth in Hz.

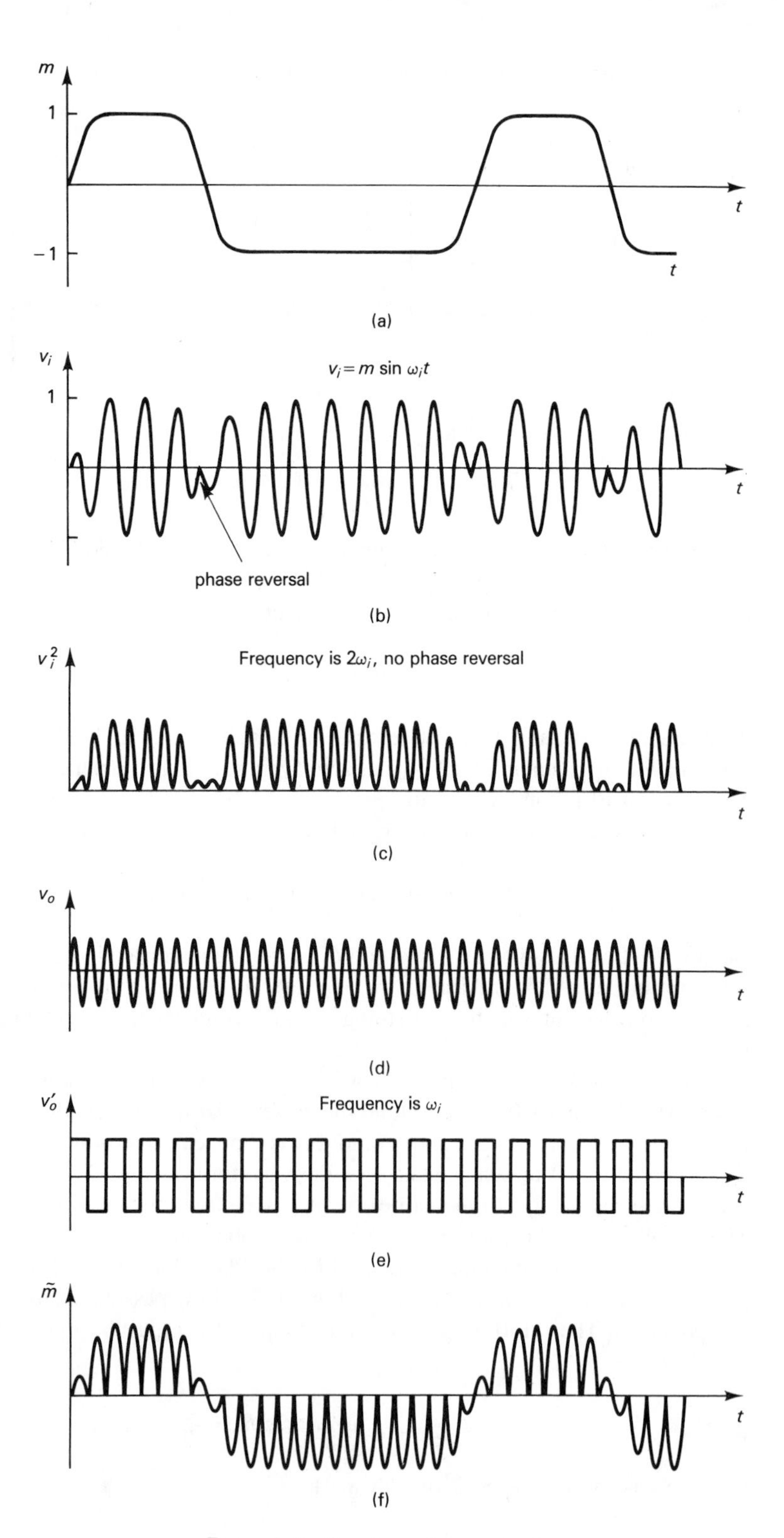

FIGURE 9–9 Waveforms for squaring loop

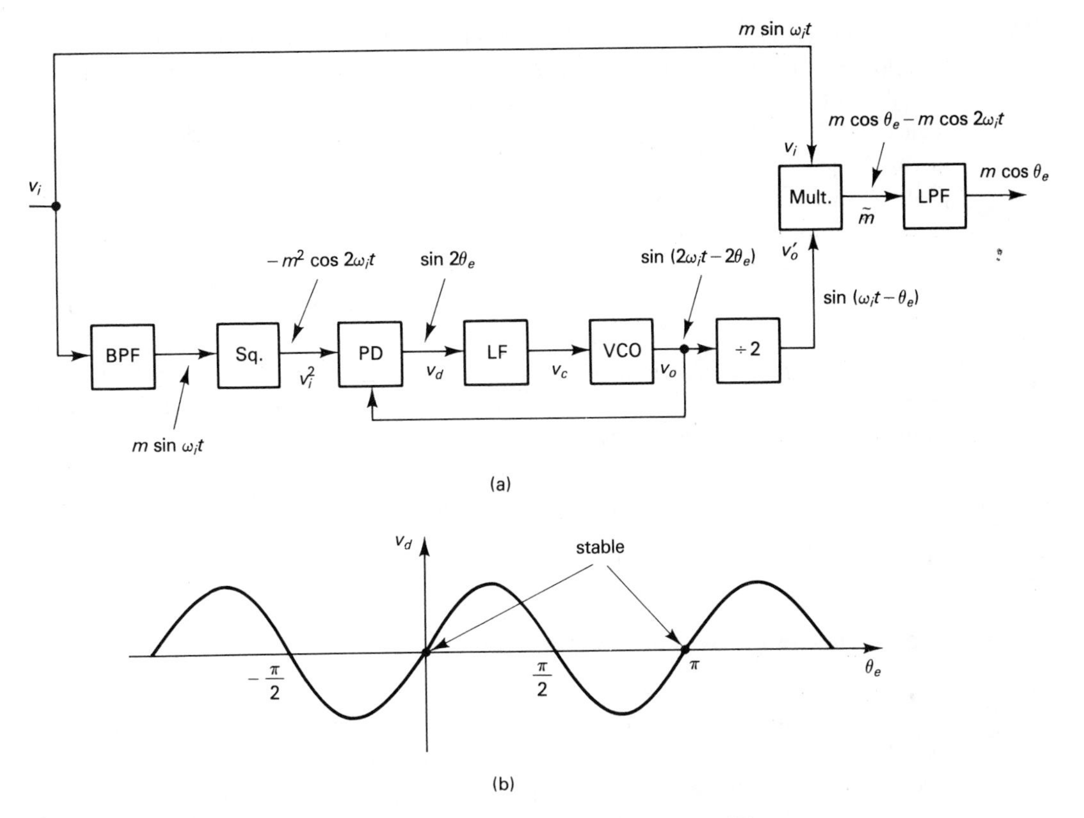

FIGURE 9–10 Squaring loop to demodulate PSK

Note that the PD output in Fig. 9–10a is $v_d = \sin 2\theta_e$. This PD characteristic is shown in Fig. 9–10b. We have been assuming that the PLL settles to the stable point at $\theta_e = 0$. But there is another stable point at $\theta_e = \pi$. If the PLL happens to settle at $\theta_e = \pi$, then the demodulated output $m \cos \theta_e$ is $-m$ rather than m. This ambiguity cannot be resolved by the demodulation circuit; some pattern in the data must tell which phasor in Fig. 9–8b is "up."

It is the nonlinearity of a PD characteristic that gives rise to the cycle slips discussed in section 6–8. The nonlinearity of the characteristic in Fig. 9–9b comes at half the θ_e compared with a normal sinusoidal characteristic (see Fig. 4–1c). Therefore, a squaring loop encounters cycle slips at lower noise levels. The guideline for negligible cycle slips was given in Eq. (6-66) as $\theta_{o\ \mathrm{rms}} \leq 0.3$ radians. For a squaring loop, the same performance requires

$$\theta_{o\ \mathrm{rms}} \leq 0.15 \text{ radians} \tag{9-27}$$

where θ_o is the phase of v'_o with noise present at the input.

In section 9–2, the bandwidth K of the PLL had to be small enough that θ_o didn't try to follow θ_i. In a squaring loop, there is no such requirement since the signal v_i^2 into the PLL doesn't have any phase modulation. However, it is still desirable to keep K small to reduce the θ_o due to noise. If squaring were not involved, then Eq. (6-29) would apply, giving the phase noise $\overline{\theta_o^2} = 2B_L N_o/V_i^2$. It can be shown [3] that the squaring causes the phase noise to be greater by a factor $(1 + B_i N_o/V_i^2)$:

$$\overline{\theta_o^2} = (2B_L N_o/V_i^2)\,(1 + B_i N_o/V_i^2) \tag{9-28}$$

$$= (2B_L N_o/V_i^2)\,(1 + 1/2\text{SNR}_i) \tag{9-29}$$

where B_i is the noise bandwidth of the bandpass filter at the input. If B_i could be reduced to zero, there would be no phase noise penalty for the squaring. But to pass the modulation, it is necessary to make

$$B_i \geq 2B_m \tag{9-30}$$

where B_m is the bandwidth of the baseband modulation.

9–3–2 Remodulator and Costas Loop

Two techniques similar to the squaring loop are the *remodulator* and the *Costas loop* [4] shown in Fig. 9–11. They have the advantage of not doubling the carrier frequency, which relaxes the speed requirements of the circuit components.

Consider first the remodulator in Fig. 9–11a, which is more similar to the squaring loop. Rather than multiply $m \sin \omega_i t$ by itself (squaring), it multiplies $m \sin \omega_i t$ by m. This gets the desired m^2 factor without doubling the frequency. The input to the PLL is $m^2 \sin \omega_i t$, which does have a spectral component at ω_i that the PLL can lock onto. The recovered carrier (after a phase shift of $-\pi/2$) is used to demodulate v_i exactly as with the squaring loop.

Before steady-state with $\theta_e = 0$ is reached, the recovered signal m at the output is actually $m \cos \theta_e$. This function of θ_e causes the PD output to be $v_d = \sin 2\theta_e$, and the PD characteristic is again that shown in Fig. 9–10b. Therefore, $\theta_e = \pi$ is also a stable point, producing a demodulated signal $-m$ at the output. As with the squaring loop, this ambiguity is unavoidable.

Although the remodulator doesn't involve squaring, it has been shown [5] that the phase noise at the PLL output is still that given in Eq. (9-28).

The bandpass filter at the input limits the noise. From Eqs. (9-28) and (9-30), the optimum bandwidth is $B_i = 2B_m$. But this small a bandwidth is often not attainable due to physical limitations on the Q of a resonant circuit, where $Q = \omega_i/2B_i$. This limitation is avoided by the Costas loop.

The Costas loop shown in Fig. 9–11b is mathematically identical to the remodulator. The order of multiplication has merely been reversed by putting the PD ahead of the multiplier (the PD is actually a multiplier too). The advantage is that the bandpass filter

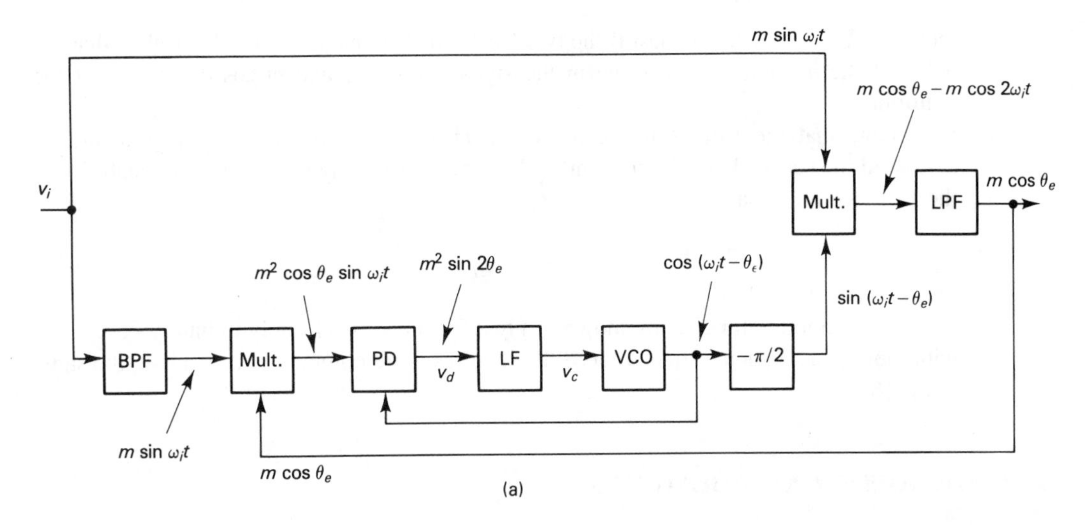

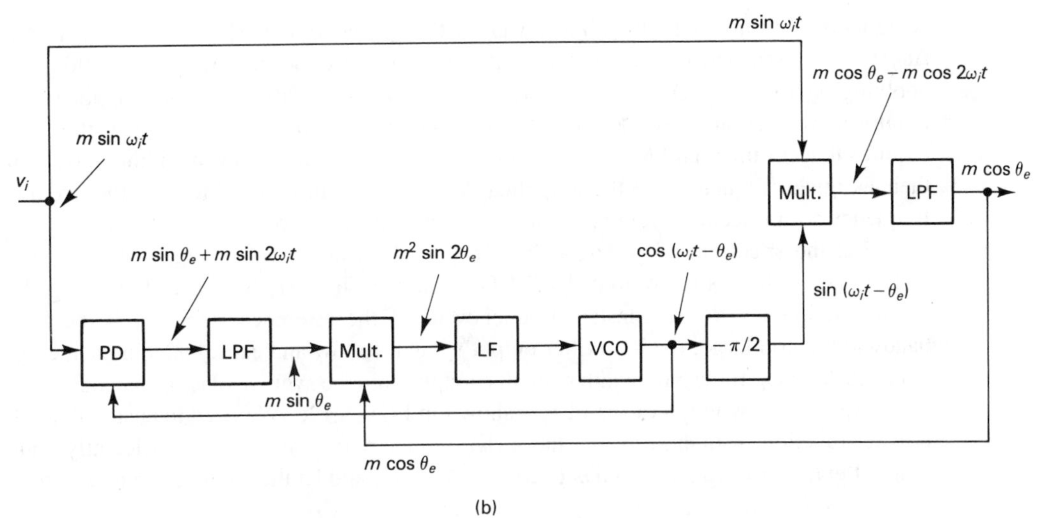

FIGURE 9–11 Remodulator (a), and Costas loop (b)

can be replaced by a lowpass filter (LPF) after the PD brings the signal down to the baseband. The LPF must pass the modulation, so its cutoff frequency must satisfy the same constraint given by Eq. (9-26) for the LPF at the output:

$$\omega_{LP} \geq 2\pi B_m$$

Since the LPF is taking the role of the BPF in the remodulator, $2\omega_{LP}$ replaces B_i in the expression for the phase noise:

$$\overline{\theta_o^{\,2}} = (2B_L N_o/V_i^2)\ (1 + 2\ \omega_{LP} N_o/V_i^2) \tag{9-31}$$

[compare Eq. (9-28)]. It is best if the two LPF cutoff frequencies are identical so that the delay of the m factor is the same in the signals $m \sin \theta_e$ and $m \cos \theta_e$ arriving at the multiplier.

Note that the LPF is inside the PLL. Therefore, it constitutes a pole at ω_3, as discussed in section 3–7. Consequently, the same rules apply to ω_{LP} as to ω_3: stability of the PLL requires that

$$K \leq \omega_{LP}/4 \tag{9-32}$$

The demodulator circuits shown in Figs. 9–9a and 9–11 apply to binary PSK. PSK with more phases such as quaternary PSK requires more complex circuits along the same lines. [6]

9–4 FREQUENCY MODULATION

A signal m is to modulate the *frequency* ω_o of the carrier so that $\omega_o = \omega_i + \Delta\omega_o$, where $\Delta\omega_o(t) = \alpha m(t)$, and α is some constant. This can be done in simple applications by applying $m(t)$ to a VCO (not in a PLL) so that $\Delta\omega_o = K_o m(t)$. But some applications require more accurate control of the average frequency (carrier). For example, the FCC requires that commercial FM stations maintain their carrier frequency to within 0.001% of their assigned frequency. In these applications, a PLL can be used to lock the average frequency to an accurate reference such as a crystal oscillator.

Let the spectrum of m extend from B_{m1} to B_{m2}, as shown in Fig. 9–12a. Figure 9–12b shows the configuration of a VCO modulated by $m(t)$ with a PLL locking the average frequency to the constant frequency ω_i of the reference signal v_i. If the PLL bandwidth is too large, the loop filter output v_{cd} will follow and cancel $m(t)$ in its attempt to match ω_o to ω_i. If the bandwidth is small enough, only the average of ω_o is matched to ω_i.

A signal flow graph of the PLL is shown in Fig. 9–12c. We need to ensure that the transfer function from $m(s)$ to $\Delta\omega_o$ has a flat frequency response over a sufficiently wide range. Let the forward gain from m to $\Delta\omega_o$ be $A = K_o$, and let the feedback from $\Delta\omega_o$ to m be $B = K_d F(s)/s$. Then

$$\frac{\Delta\omega_o(s)}{m(s)} = \frac{A}{1 + AB} = \frac{K_o}{1 + K_d F(s) K_o / s}$$

$$= \frac{K_o}{1 + G(s)} = K_o H_e(s) \approx \frac{K_o s^2}{(s + \omega_2)(s + K)} \tag{9-33}$$

where $H_e \equiv 1/(1 + G)$. But $H_e(s)$ has unity gain with a low-frequency cutoff at $\omega = K$. Therefore, $\Delta\omega_o/m$ has a gain of K_o with a cutoff at $\omega = K$, as shown in Fig. 9–12d. Then if we design the PLL to have a bandwidth that satisfies

$$K \leq 2\pi B_{m1} \tag{9-34}$$

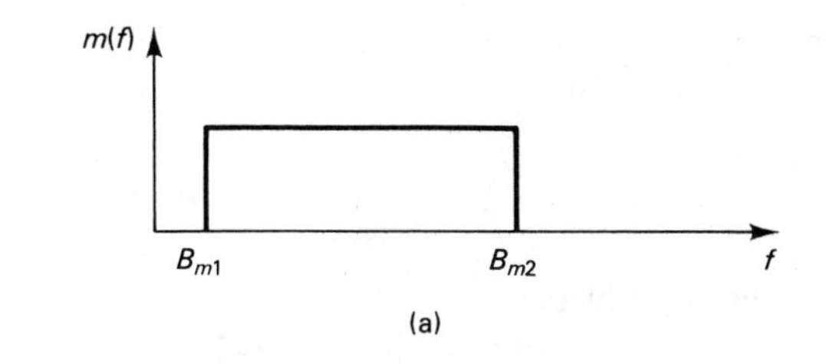

(a)

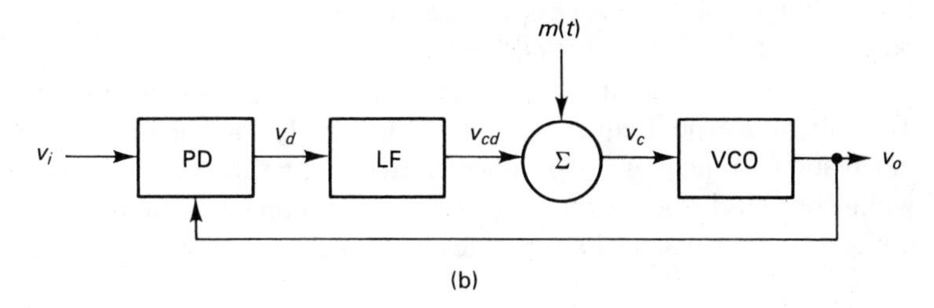

(b)

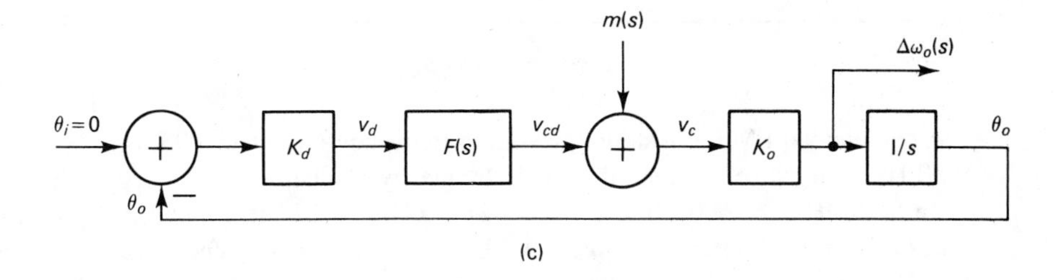

(c)

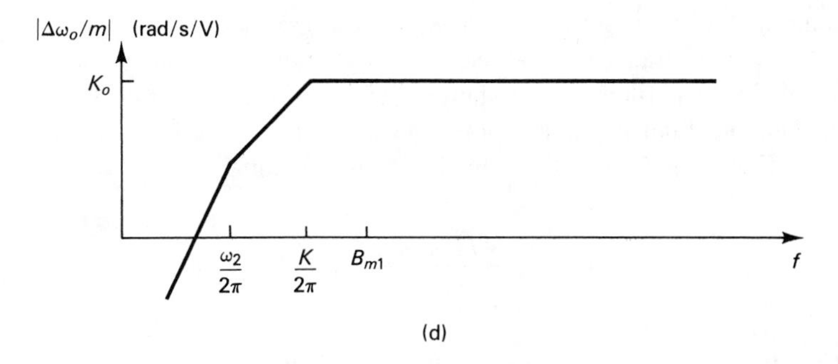

(d)

FIGURE 9–12 Frequency modulation

the spectrum of m will be passed, and

$$\Delta\omega_o(t) = K_o m(t) \tag{9-35}$$

as desired.

By definition,

$$\theta_o(t) \equiv \int \Delta\omega_o(t)\, dt$$

Since for our application here $\theta_i = 0$, then $\theta_e = -\theta_o$. Then to stay in the linear range of the PD, we need to satisfy

$$|\theta_e(t)| = |\int \Delta\omega_o(t)\, dt| \leq \theta_{em} \tag{9-36}$$

where θ_{em} is the limit of the linear PD range. If the range needs to be extended, $\div N$ frequency dividers can be used as in Fig. 9–2a. For large N, the spurious frequency modulation $\Delta\omega$ may be excessive [see Eq. (9-9)]. A multiple pole in the loop filter at ω_3 helps reduce $\Delta\omega$ [see Eq. (9-15)].

Another way to extend θ_{em} to meet Eq. (9-36) is to use an n-state PD, as in Fig. 4–13. Then there are no frequency dividers ($N = 1$), and there is no spurious frequency modulation. [Equation (9-9) indicates $\Delta\omega = 2\pi K$ for $N = 1$, but this variation is all within one cycle of ω_i since $\omega_d = \omega_i$.] Sometimes a combination of $\div N$ frequency dividers and an n-state PD provides adequate performance with the simplest circuitry.

EXAMPLE 9–2

A commercial FM radio station wishes to frequency-modulate a 95.5-MHz carrier with a 50-Hz sine wave so that the peak frequency deviation is 75 kHz. That is, $\omega_i = 2\pi(95.5\text{MHz}) = 600$ Mrad/s, $\omega_m = 2\pi(50\text{ Hz}) = 314$ rad/s, and $\Delta\omega_o = 2\pi(75\text{ kHz}) \cos \omega_m t = (471\text{ krad/s}) \cos \omega_m t$. A PLL performs the frequency modulation using two $\div N$ frequency dividers to extend the range of a three-state PD as in Fig. 9–2a. A crystal-controlled VCO is available with $K_o = 100$ krad/s/V, and a resonant VCO is available with $K_o = 5$ Mrad/s/V. Find the necessary modulation voltage $m(t)$. Design K_h, ω_2, and ω_3 of the loop filter so the spurious modulation is at least 60 dB below the desired modulation. Find the acquisition time T_p for $\omega_{eo} = 0.05\ \omega_i = 30$ Mrad/s.

From Eq. (9-36), the effective phase modulation is

$$\theta_e(t) = \frac{471\text{ krad/s}}{\omega_m} \sin \omega_m t = 1500 \sin \omega_m t$$

Then the range of the three-state PD must be extended to $2\pi N > 1500$, requiring $N = \underline{239}$. To satisfy Eq. (9-34) with the highest K possible, let $K = 314$ rad/s. This is a very low bandwidth: $K = \omega_i/1{,}900{,}000$. Only a crystal-controlled VCO (or VCXO) can avoid injection problems under these conditions [see Eq. (5-33)]. However, a peak-to-peak deviation of 2×471 krad/s $= 942$ krad/s for $\Delta\omega_o$ is 1570 ppm of $\omega_i = 600$ Mrad/s, which exceeds the linear range of most VCXOs (see Fig. 5–13 for example).

The rule of thumb for a resonant VCO is $K > \omega_i/10{,}000$ (see Eq. (5-33)]. We can improve on this rule by eliminating the $\div 239$ that divides the 600-Mrad/s v_i down to 2.5105 Mrad/s (see Fig. 9–2a) and providing a 2.5105-Mrad/s v_i directly to the three-state PD. This eliminates the 600 Mrad/s signal that was "leaking" into the VCO and causing the injection problems. However, the 239th harmonic of the 2.5105-Mrad/s square wave is 600 Mrad/s, and this component causes some injection in the VCO. For a square wave, the nth harmonic is $1/n$ of the fundamental amplitude, so the injected signal V_I is 1/239 as large. This reduces the injection constant $K_I = (\omega_{oo}/2Q)(V_I/V_1)$ by 1/239, and our rule of thumb becomes $K > \omega_i/2{,}390{,}000$. This allows the design value $K = 314$ rad/s.

The resonant VCO has $K_o = 5$ Mrad/s/V. From Eq. (9-35), the voltage necessary to produce the modulation $\Delta\omega_o = 2\pi(75 \text{ kHz}) \cos \omega_m t$ is $m(t) = \Delta\omega_o/K_o = (94 \text{ mV}) \cos \omega_m t$.

Let the three-state PD use ECL devices with $V_H = -0.8$ V and $V_L = -1.8$ V. Then $V_{dm} = V_H - V_L = 1$ V, and $K_d = V_{dm}/2\pi N = 1/480\pi = 663\ \mu V/\text{rad}$ (see Fig. 9–2b). To realize the bandwidth K, we need $K_h = K/K_d K_o = 314/(663\ \mu \times 5\text{ M}) = \underline{0.095}$.

To reduce the spurious modulation, we place a pole at $\omega_3 = 4K = \underline{1256 \text{ rad/s}}$. Then from Eq. (9-13′), the spurious frequency modulation is $\Delta\omega = (2\pi NK)^2/\omega_i = 373$ rad/s. This is only 0.04% (or -68 dB) of the peak-to-peak $\Delta\omega_o$ modulation 2×471 krad/s $= 942$ krad/s. Further reduction would require another pole at ω_3 or an n-state PD.

The pull-in time for a three-state PD with $\div N$ frequency division is given by Eq. (8-28′):

$$T_p = (\omega_{eo}/K - 2\pi N)/\pi N\omega_2 \tag{9-37}$$

To minimize T_p, we pick ω_2 as large as possible: $\omega_2 = K/4 = \underline{78 \text{ rad/s}}$. Then for $\omega_{eo} = 30$ Mrad/s, $T_p = \underline{1.6 \text{ sec}}$.

9–5 FREQUENCY DEMODULATION

The frequency of a carrier has been modulated with some variation $\Delta\omega_i(t)$ that is limited to a bandwidth B_m (see Fig. 9–13a). The objective of demodulation is to produce a proportional voltage $m(t) = \alpha\Delta\omega_i(t)$, where α is some constant. The configuration of a PLL used as a frequency demodulator is shown in Fig. 9–13b. If the PLL bandwidth is great enough, the VCO frequency will follow the input frequency, and $\Delta\omega_o \approx \Delta\omega_i$. But $\Delta\omega_o = K_o v_c$, so $v_c \approx \Delta\omega_i/K_o$, which is the desired demodulation $m(t)$.

We can get this result more formally from the signal flow graph in Fig. 9–13c. The transfer function from $\Delta\omega_i$ to v_c can be seen in three stages: $\theta_i/\Delta\omega_i = 1/s$, $\theta_o/\theta_i \equiv H(s)$, and $v_c/\theta_o = s/K_o$. Then the product of the three stages is

$$v_c/\Delta\omega_i = (1/K_o)\, H(s)$$

with a flat gain of $1/K_o$ out to $\omega = K$ (see Fig. 9–13d). To pass the whole spectrum of $\Delta\omega_i$ we require

$$K \geq 2\pi B_m \tag{9-38}$$

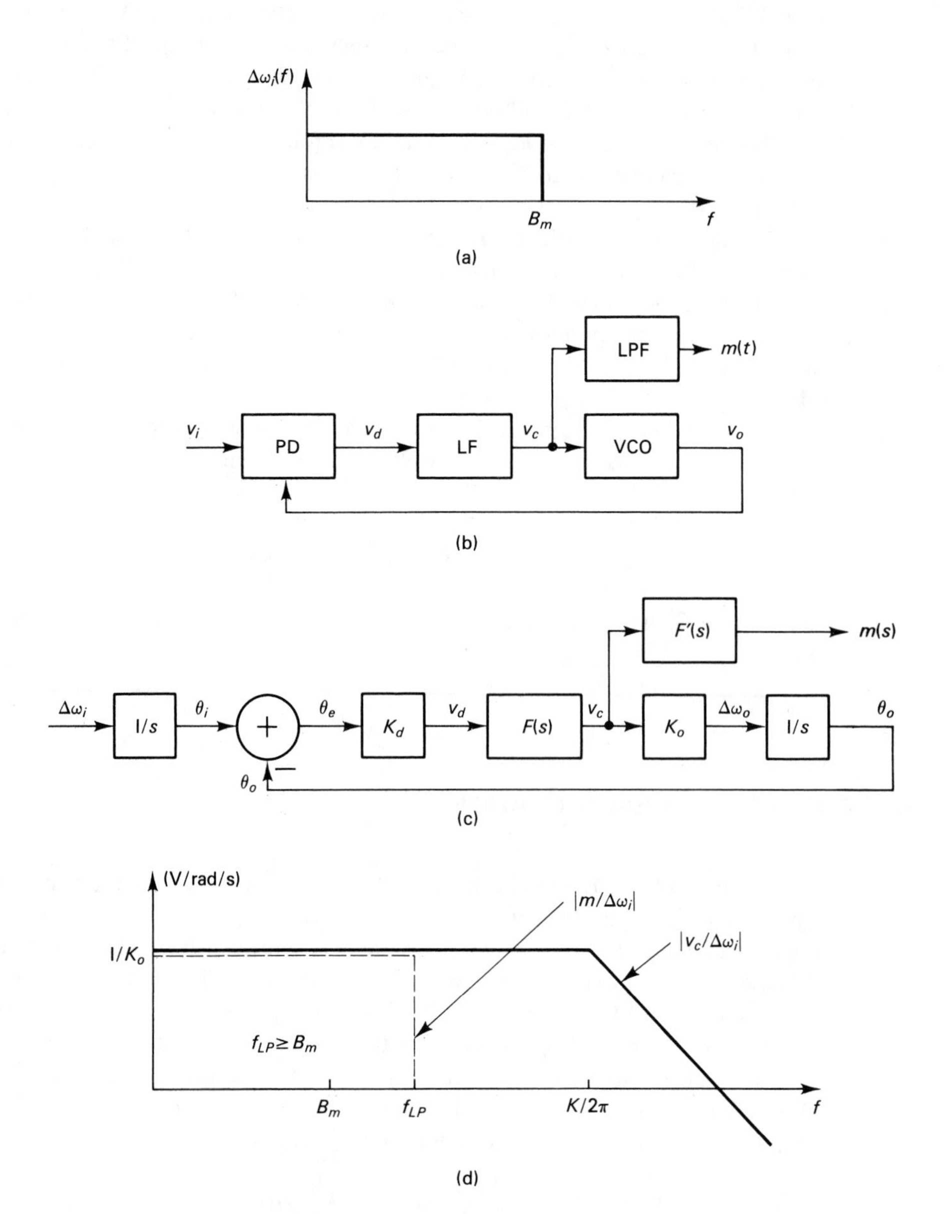

FIGURE 9–13 Frequency demodulation

But to maintain lock in the presence of random FM, an even greater K is required:

$$K \geq \frac{7\pi\, B_m\, \Delta\omega_{i\ \mathrm{rms}}}{\sqrt{3}\ \omega_2} \tag{9-39}$$

for a sinusoidal PD characteristic [see Eq. (7-34a)]. Therefore, a lower cutoff at f_{LP} is provided by a low-pass filter to better limit noise. Let the transfer function of this filter be $F'(s)$. Then

$$\frac{m(s)}{\Delta\omega_i(s)} = \frac{1}{K_o} H(s)F'(s) \approx \frac{K/K_o}{s + K} F'(s) \tag{9-40}$$

A frequency response for $|m/\Delta\omega_i|$ is shown in Fig. 9–13d for the case of a high-order low-pass $F'(s)$ with a sharp cutoff at f_{LP}. As in the bound for K in Eq. (4-38), we also require

$$f_{LP} \geq B_m \tag{9-41}$$

Then all the frequency components of $\Delta\omega_i$ are passed, and

$$m(t) \approx (1/K_o)\Delta\omega_i(t) \tag{9-42}$$

as desired.

A noise component $n(t)$ at the PLL input causes random frequency modulation that we will call the *frequency noise* $\omega_n(t)$. The standard practice is to limit this noise with a bandpass filter before the PLL. The spectral density of $n(t)$ has power density N_o and noise bandwidth B_i as shown in Fig. 9–14a. The noise power is given by $\overline{n^2} = N_oB_i$. As shown in section 6–8, the power density of the phase noise θ_n is $\Phi_{\theta n} = 2N_o/V_i^2$, where V_i is the amplitude of the carrier v_i. Since $\omega_n(s) = s\theta_n(s)$, the power density spectrum of the frequency noise is

$$\Phi_{\omega n} = |s|^2\Phi_{\theta n} = (2\pi f)^2 2N_o/V_i^2 \tag{9-43}$$

out to $B_i/2$ (see plot in Fig. 9–14c). The portion of $\overline{\omega_n^2}$ that gets through the low-pass filter is the area under the curve out to $f = f_{LP}$. This area is minimized by the lower bound of Eq. (9-41): $f_{LP} = B_m$. Integrating Eq. (9-43) from $f = 0$ to $f = B_m$,

$$\overline{\omega_n^2} = 26.3\ B_m^3 N_o/V_i^2 = 13.2\ B_m^3/B_i\mathrm{SNR}_i \tag{9-44}$$

where the carrier-to-noise ratio is

$$\mathrm{SNR}_i \equiv \overline{v_i^2}/\overline{n^2} = V_i^2/2N_oB_i \tag{9-45}$$

The signal-to-noise ratio after demodulation is

$$\mathrm{SNR}_o \equiv (\Delta\omega_{i\ \mathrm{rms}})^2/\overline{\omega_n^2} \tag{9-46}$$

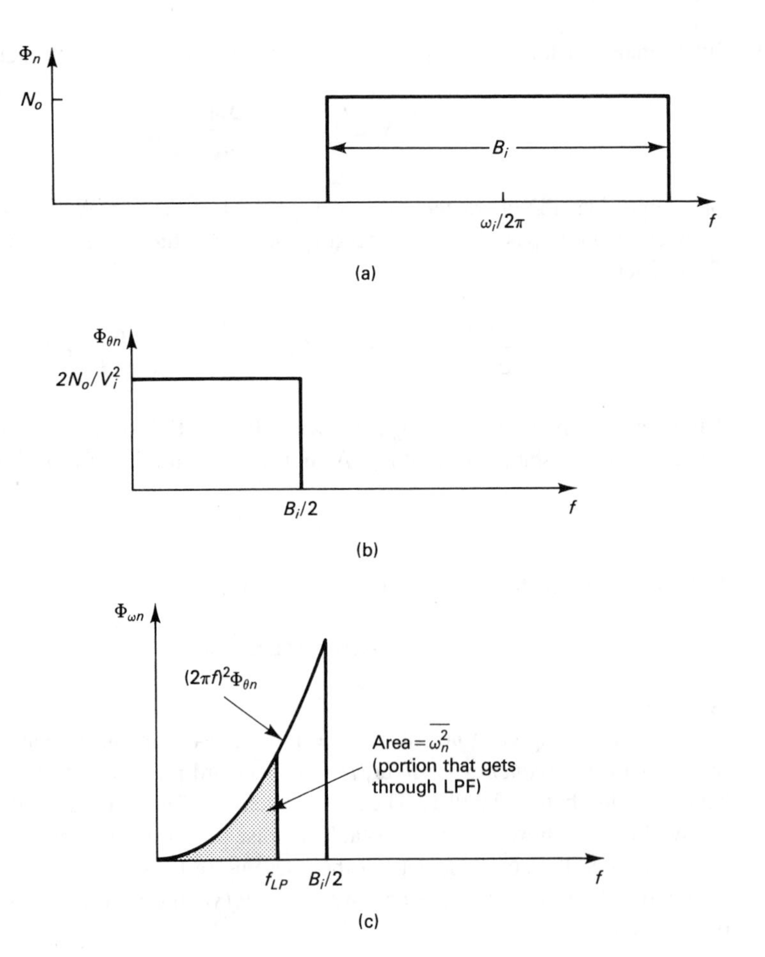

FIGURE 9–14 FM frequency noise

EXAMPLE 9–3

A carrier is frequency-modulated with a signal that has a bandwidth B_m = 15 kHz and causes an effective peak frequency deviation of $3.5\Delta\omega_{i\ rms} = 2\pi$(75 kHz), or $\Delta\omega_{i\ rms}$ = 135 krad/s. A PLL is to demodulate the signal. The noise bandwidth of the bandpass filter at the PLL input is B_i = 200 kHz, and the carrier-to-noise ratio is SNR_i = 20 dB = 100. Find K and the signal-to-noise ratio after demodulation.

Equation (9-38) requires $K > 94$ krad/s. But to maintain lock, Eq. (9-39) requires K = 320 krad/s and ω_2 = 80 krad/s (see Example 7-3). Satisfying the lower bound of Eq. (9-40), we set $f_{LP} = B_m$ = 15 kHz. Then Eq. (9-44) gives $\overline{\omega_n^2}$ = 2.23 (krad/s)^2, and Eq. (9-46) gives $\text{SNR}_o = (135)^2/2.23 = 8173$ = 78 dB.

REFERENCES

[1] C. L. Phillips and R. D. Harbor, *Feedback Control Systems,* Prentice-Hall: Englewood Cliffs, NJ, 1988.

[2] B. P. Lathi, *Modern Digital and Analog Communication Systems,* HRW: Philadelphia, 1989, section 4.13.

[3] F. M. Gardner, *Phaselock Techniques,* Wiley: New York, 1979, Appendix B.

[4] J. P. Costas, "Synchronous Communications," *Proc. IRE,* vol. 44 (December 1956), pp. 1713–1718.

[5] S. A. Butman and J. R. Lesh, "The Effects of Bandpass Limiters on *n*-Phase Tracking Systems," *IEEE Trans. Commun.*, vol. COM-25 (June 1977), pp. 569–576.

[6] Butman and Lesh, "Effects of Bandpass Limiters."

CHAPTER

10

Clock Recovery

In digital communication systems, information is conveyed by a series of bits—1's and 0's. A typical binary data signal v_i is shown in Fig. 10–1a. The bit sequence here is 1,1,0,1,0,0,1,1, where a pulse represents a 1, and the absence of a pulse represents a 0. The bit rate is called the *baud* f_B. To process the data correctly, the receiver usually synchronizes a clock to the data so the clock frequency f_o equals f_B. The process of synchronizing the frequency and phase of the clock is called *clock recovery,* and it is usually accomplished by a PLL. Applications requiring clock recovery include compact disk players, floppy disk readers, and satellite data links.

The application of PLLs to clock recovery has some special design considerations. Because of the random nature of data, the choice of phase detectors is restricted. In particular, three-state PDs won't work, and other means must be used to aid acquisition. One useful method is the rotational frequency detector described in section 8–6. The random data also cause the PLL to introduce undesired phase variation in the recovered clock. This is called *timing jitter,* and it is the principal topic of this chapter. Through proper design of the PLL, this jitter can be minimized. Trischitta and Varma [1] provide a comprehensive reference on jitter sources, effects, and standards.

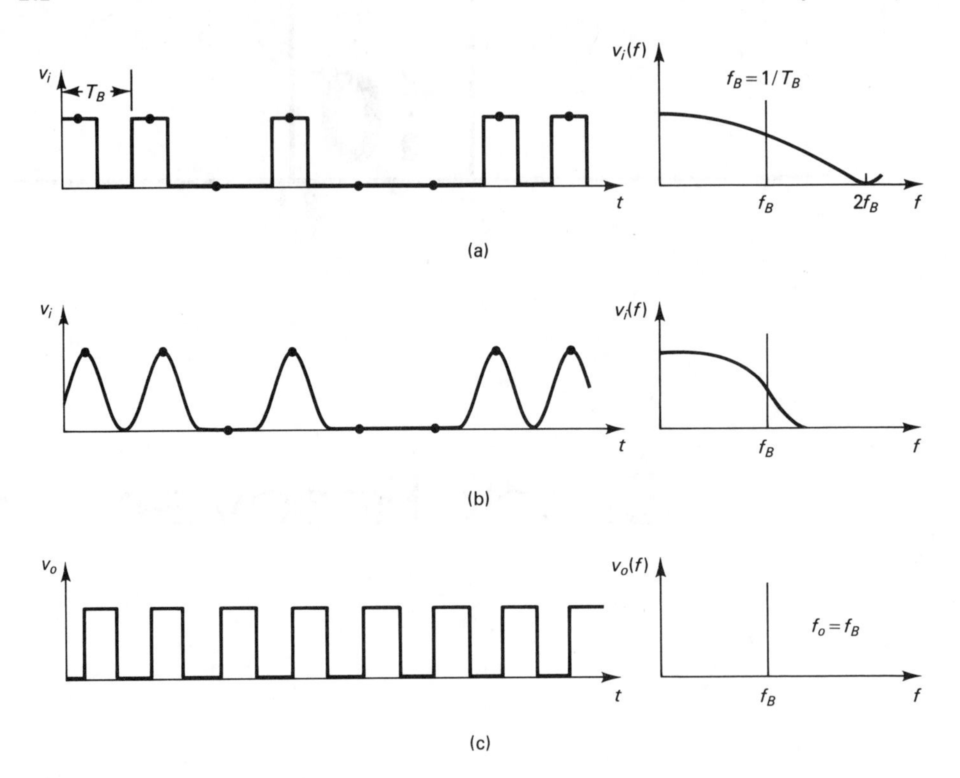

FIGURE 10–1 RZ data and clock

10–1 DATA FORMATS AND SPECTRA

Consider the data signal in Fig. 10–1a, where the first two bits are adjacent 1's. Because the signal goes to zero between adjacent pulses, this format is called *return-to-zero* (RZ) data. The result is a signal that is a periodic square wave with period T_B but with some of the pulses missing. Therefore, the spectrum $v_i(\omega)$ has a line component at $f_B = 1/T_B$. The spectrum also has a continuous component that extends beyond f = $2f_B$; this corresponds to the random pattern of missing pulses.

In communication applications, the data is filtered by a low-pass filter to eliminate as much noise as possible, as in Fig. 10–1b. The result is a rounded waveform v_i in the time domain and a narrower spectrum $v_i(\omega)$ in the frequency domain. Note that the filtering is broad enough that the signal still returns to zero between pulses, and there is still a line component at f_B in the spectrum. Clock recovery amounts to extracting this line component as either a sine wave or a square wave v_o (see Fig. 10–1c). The desired phase is to position the rising edges of v_o at the center of the v_i pulses. Then the clock can sample the data at the optimum time (see the dots on the v_i waveforms) to determine whether the bit is a 1 or a 0.

Another data format is the non-return-to-zero (NRZ) data v_i' shown in Fig. 10–2a. Again the bit sequence is 1,1,0,1,0,0,1,1, but the signal does not go to zero between

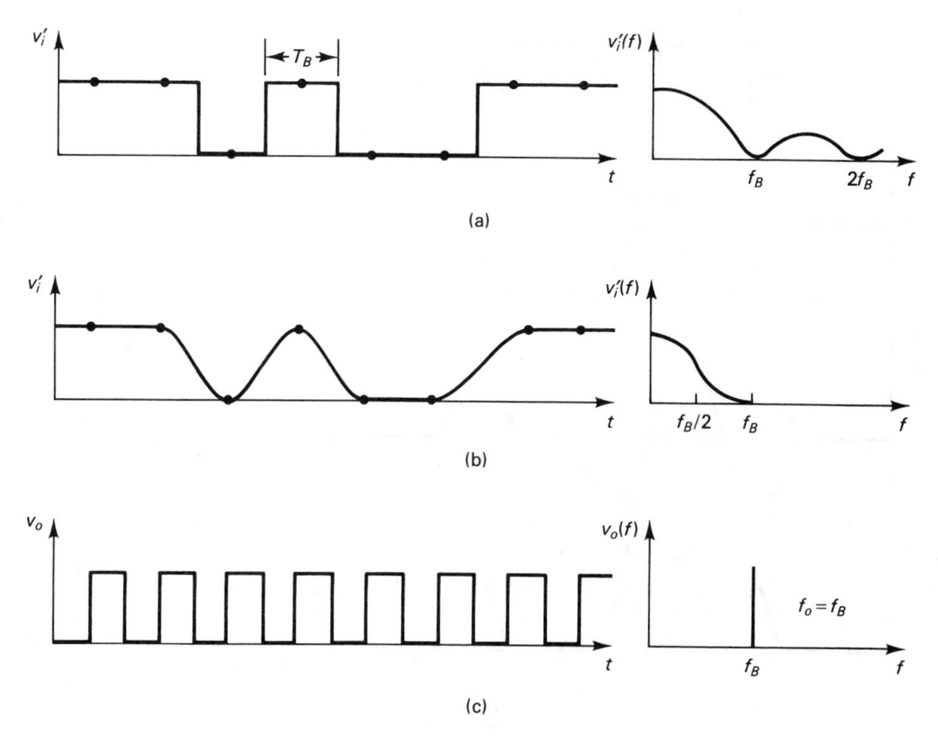

FIGURE 10–2 NRZ data and clock

adjacent pulses representing 1's. That is, the pulse width equals the pulse spacing T_B. It can be shown [2] that the corresponding spectrum v_i' (ω) has no line component at f_B (see Fig. 10–2a). In fact, the continuous part of the spectrum actually goes to zero at f_B. The band-limited form of the NRZ data is shown in Fig. 10–2b. Most of the spectrum of this signal lies below $f_B/2$, a result of filtering to reject noise.

Since NRZ data has no component at f_B, a PLL will not lock to the data to produce the clock signal. As we found in section 9–3–1, a nonlinear process can create a line component at f_B, and a PLL can recover the desired clock signal v_o, as in Fig. 10–2c. When in lock, the PLL usually phases the clock so that its rising edges are centered on the data pulses (see the dots on the v_i' waveforms). If the PLL aligns the *falling* edge of v_o in the center of the v_i pulses, the complement of the clock can be used for data sampling.

10–2 CONVERSION FROM NRZ TO RZ DATA

One way to recover clock from NRZ data is to convert it to an RZ-like data signal that has a line component at f_B, and then recover clock from that RZ data with a PLL. The conversion process for band-limited NRZ data is illustrated in Fig. 10–3. Since the phase

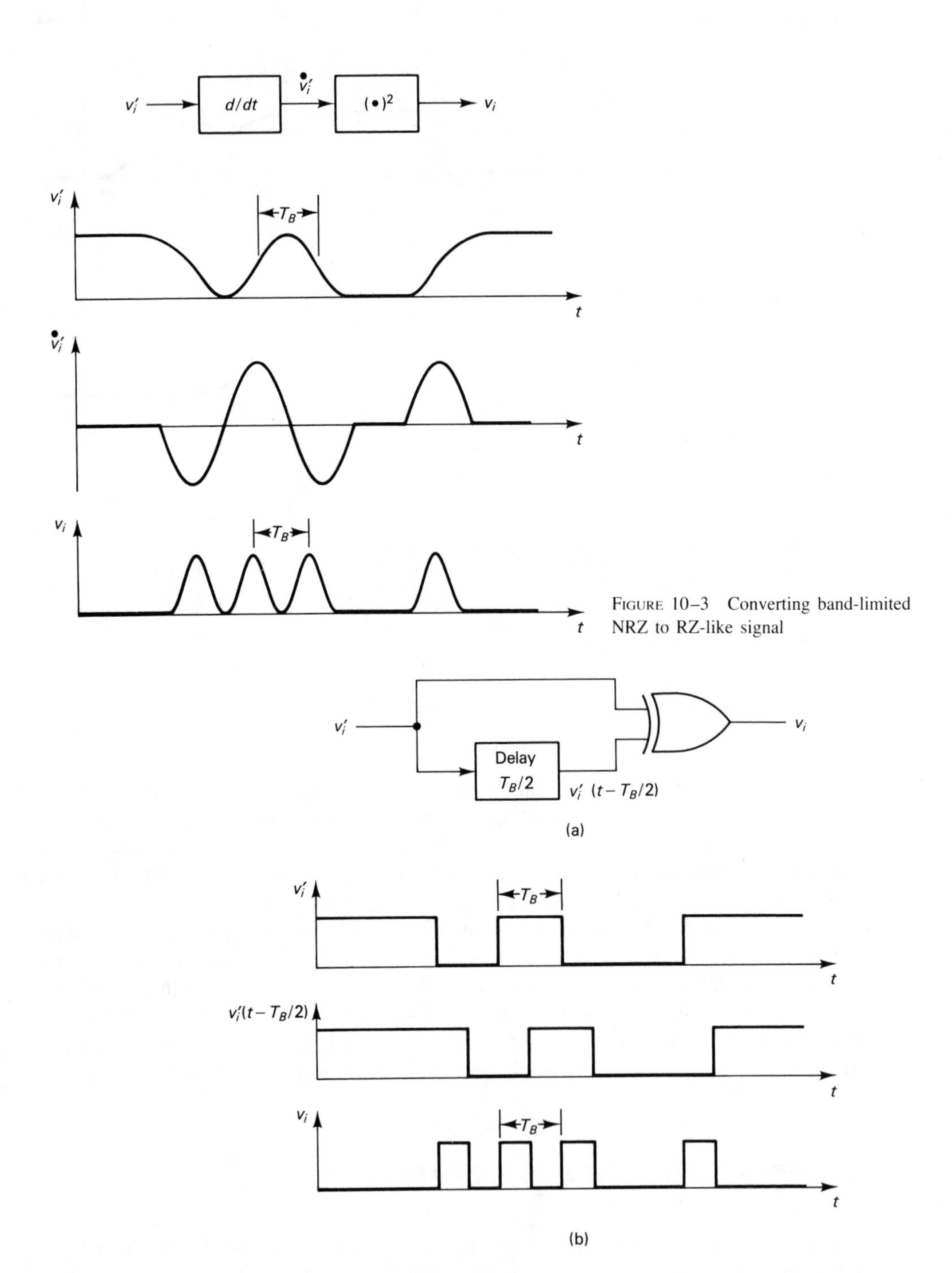

FIGURE 10–3 Converting band-limited NRZ to RZ-like signal

FIGURE 10–4 Converting NRZ logic signal to RZ-like signal (high-data rates)

information is in the transitions of the data v_i', the data is first differentiated to generate a pulse (either positive or negative) corresponding to each transition (from 0 to 1 or from 1 to 0). These pulses are made to be all positive by squaring the differentiated signal $\dot{v}_i'$. (Squaring produces fewer harmonics and reduces the in-band noise compared with full-wave rectification.) The result is signal v_i that looks just like RZ data; pulses are spaced at intervals of T_B, and some pulses are missing. But here a pulse stands for a transition rather than a 1.

In low-noise applications, other options are available for converting NRZ data to RZ data. A method useful at high data rates ($f_B > 20$ Mb/s) is shown in Fig. 10–4. The data v_i' is delayed by $T_B/2$ and compared with itself by an exclusive-OR gate. Each time that $v_i'(t)$ and $v_i'(t - T_B/2)$ are different (after a transition), the exclusive-OR generates a pulse. For example, at a data rate of $f_B = 50$ Mbit/sec the bit spacing is $T_B = 20$ ns, and a delay of $T_B/2 = 10$ ns is needed. Since signals propagate through delay lines at about 0.2 m/ns, this requires 2 m of cable. Since logic devices decrease the effective signal-to-noise ratio, this method is not used in recovering clock in low signal-to-noise applications. But for clock recovery from a logic signal, the circuitry here is simpler than that in Fig. 10–3.

For low data rates, the cable length to realize T_B delay may be excessive. In that case, the method for converting NRZ to RZ data shown in Fig. 10–5 is simpler. Each

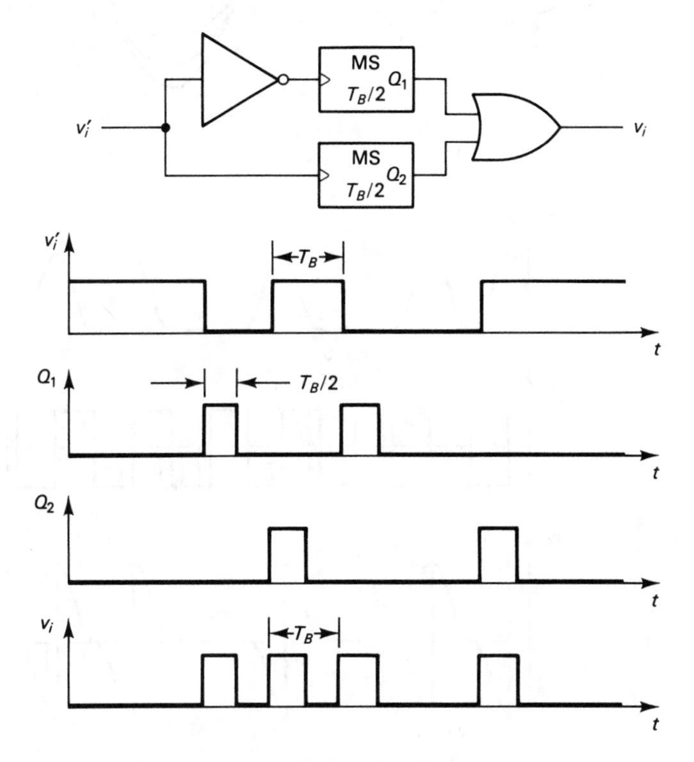

FIGURE 10–5 Converting NRZ logic signals to RZ-like signal (low-data rates)

rising and falling edge of the NRZ data triggers a monostable multivibrator with pulse width $T_B/2$, and an OR gate combines the pulses.

10–3 PHASE DETECTORS FOR RZ DATA

In high-noise applications, a multiplier is usually used as a phase detector, as in Fig. 10–6a. (Multipliers as phase detectors were discussed in section 4–1.) For the case shown here, the data pulses v_i are positive, and the clock signal v_o goes both positive and negative. In steady-state with $\theta_e = 0$, each falling edge of v_o splits a data pulse (the solid v_o waveform in Fig. 10–6c). Then the product $\tilde{v}_d$ consists of pulses with equal positive and negative areas, and v_d, the average, is zero. For positive phase error θ_e, the clock is delayed slightly

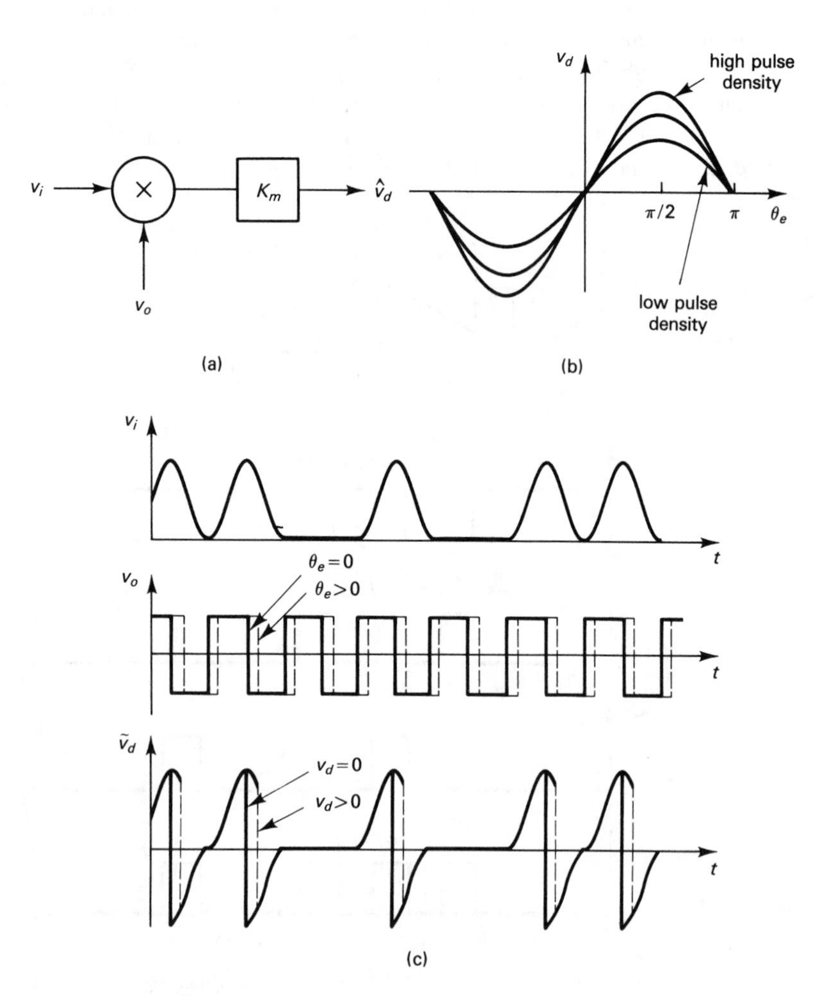

FIGURE 10–6 Phase detector for high-noise RZ data

compared to the data (the dashed v_o waveform), the $\tilde{v}_d$ pulses have more positive than negative area, and v_d is positive. The resulting PD characteristic of v_d versus θ_e is shown in Fig. 10–6b. The characteristic is sinusoidal with a maximum value that depends on the density of data pulses. If half the pulses are missing on average, the characteristic is half the height it would have with no missing pulses.

In low noise applications, a simple exclusive-OR gate can be used as a multiplier, as in Fig. 10–7a. Again, the falling edges of the clock v_o split the data pulses evenly in steady-state, and v_d, the average of $\tilde{v}_d$, is zero, (see Fig. 10–7c). For positive θ_e, the clock v_o is delayed (dashed waveform) relative to the data v_i, and the average of $\tilde{v}_d$ is positive. Note that during the time that there are no v_i pulses, the average of $\tilde{v}_d$ is zero even for nonzero θ_e. Then for $\theta_e = \pi/2$, $\tilde{v}_d$ will be $V_H - V_L$ half the time and an average of zero half the time if the pulse density is 50%. Therefore, $v_d = (V_H - V_L)/2$ corresponds to

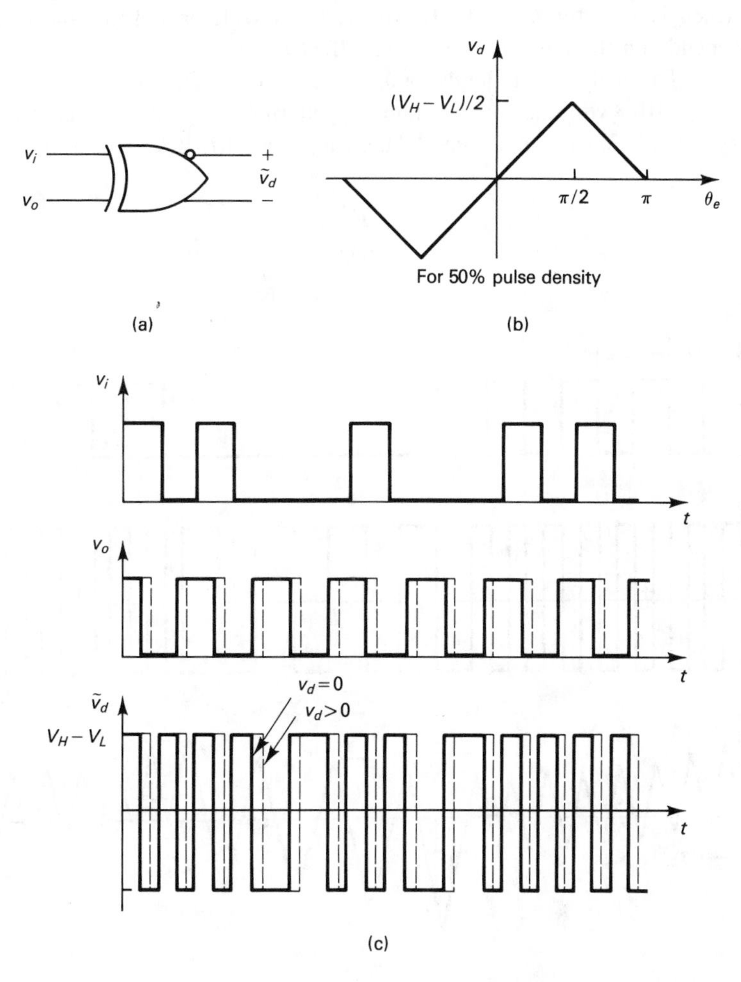

FIGURE 10–7 Phase detector for low-noise RZ data

$\theta_e = \pi/2$ for a 50% pulse density (see the PD characteristic in Fig. 10–7b). The corresponding PD gain is

$$K_d = (V_H - V_L)/\pi; \qquad 50\% \text{ pulse density} \tag{10-1}$$

10–4 PATTERN-DEPENDENT JITTER

We have been assuming that the PLL responds to v_d, the average of $\tilde{v}_d$, because of the low-pass nature of the PLL transfer function $H(s)$. When the bandwidth K is much less than the baud f_B, it is reasonable to consider the long-term average of $\tilde{v}_d$ in Fig. 10–7c, which is zero for $\theta_e = 0$. However, for a wide-band PLL, the pattern of $\tilde{v}_d$, which depends on the data, does have an effect on θ_o.

The analysis here holds for an exclusive-OR PD. Consider the data pattern v_i shown in Fig. 10–8 with the corresponding $\tilde{v}_d$ pattern for $\theta_e = 0$. We can analyze the effect of $\tilde{v}_d$ by considering it a phase-modulation input at $m(t)$ in Fig. 9–1. From Eq. (9-1),

$$\frac{\theta_o(s)}{\tilde{v}_d(s)} \approx \frac{K/K_d}{s + K} \tag{10-2}$$

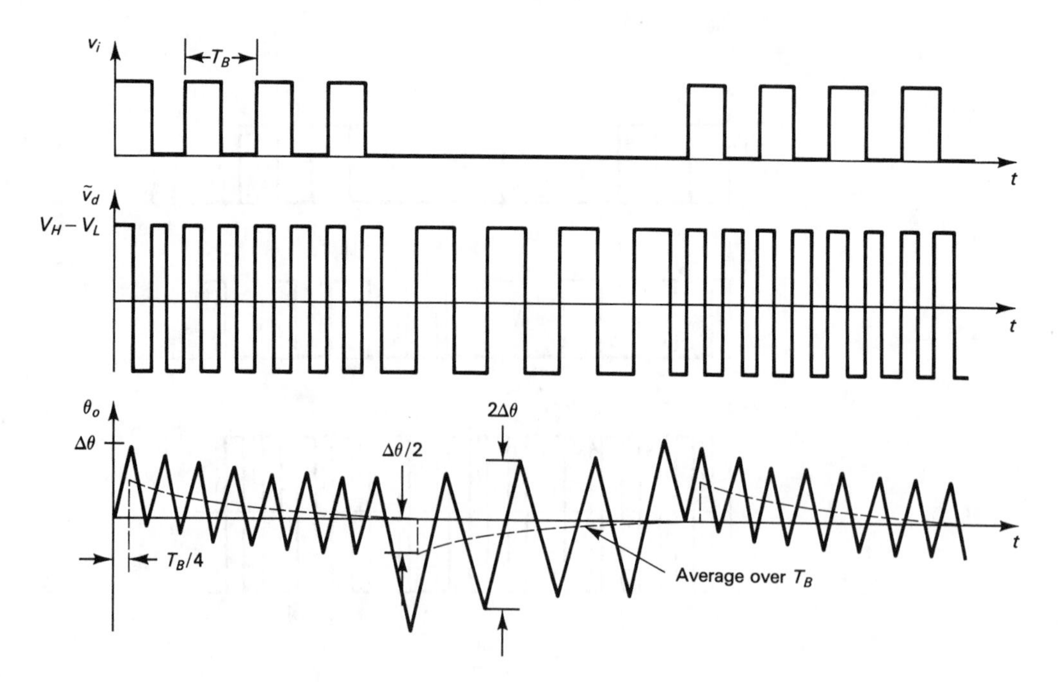

FIGURE 10–8 Pattern-dependent jitter with XOR PD

The result of this low-pass transfer function is shown by the $\theta_o(t)$ waveform in Fig. 10–8. During pulses on v_i, θ_o is a triangular wave with a peak-to-peak amplitude of $\Delta\theta$. When there are no pulses on v_i, θ_o is a triangular wave with a peak-to-peak amplitude of $2\Delta\theta$. There is also a decaying exponential transient when the data pattern changes; this transient is the short-term average of θ_o. The amplitude of the transient is $\Delta\theta/2$.

The value of $\Delta\theta$ can be obtained easily in terms of K and T_B as follows. For frequencies on the order of f_B (and therefore greater than K), Eq. (10-2) can be approximated by

$$\frac{\theta_o(s)}{\tilde{v}_d(s)} \approx \frac{K/K_d}{s}$$

Rearranging, and taking the inverse Laplace transform, we get

$$s\ \theta_o(s) \approx \tilde{v}_d(s)\ K/K_d$$

$$\dot{\theta}_o(t) \approx \tilde{v}_d(t)\ K/K_d$$

When a data pulse is present on v_i, then $\tilde{v}_d = V_H - V_L$ for an interval $T_B/4$ (see Fig. 10–8). Then the change in θ_o during this interval is

$$\Delta\theta = \dot{\theta}_o\ T_B/4 = (V_H - V_L)K/K_d)T_B/4$$

But from Eq. (10-1), the PD gain is $K_d = (V_H - V_L)/\pi$ for 50% 1's density in the data pattern. Therefore

$$\Delta\theta = (\pi/4)K\ T_B \tag{10-3}$$

When pulses are not present in the data pattern v_i, the interval during which θ_o ramps becomes $T_B/2$, and the change becomes $2\Delta\theta$ (see Fig. 10–8). Since the exponential transients have amplitudes of $+\Delta\theta/2$ and $-\Delta\theta/2$, the peak-to-peak phase jitter of θ_o is the $\Delta\theta$ given by Eq. (10-3).

The specific form of the transient behavior is of course dependent on the data pattern. If the pattern is alternate 1's and 0's, then the jitter is virtually nonexistent. But if there are long strings of 1's and 0's, then the jitter amplitude given in Eq. (10-3) holds.

EXAMPLE 10–1

A PLL with an exclusive-OR phase detector is used to recover clock from RZ data. The PLL bandwidth is one-tenth the baud; that is, $K = 0.1 \times 2\pi f_B$. Find the pattern-dependent jitter.

Since $f_B = 1/T_B$, we have $K = 0.2\pi/T_B$, and from Eq. (10-3), $\Delta\theta = 0.2\pi^2/4 = 0.5$ radian. This is a significant 8% of a bit interval!

The pattern-dependent jitter that accompanies an exclusive-OR PD can be avoided by using a two-state PD. The performance of the two-state PD was analyzed in section 4–7 for both v_i and v_o periodic. Here, v_i is RZ data applied to the two-state PD, as in Fig. 10–9a. The result is a waveform v_y of positive pulses with width proportional to the phase difference between the data v_i and the recovered clock v_o (see Fig. 10–9b). For $\theta_e = 0$, the pulse width equals that of the RZ data, and $v_y = -v_x$. Then $\tilde{v}_d = 1/2(v_x + v_y) = 0$, and there is no pattern at the output of the PD for $\theta_e = 0$. Therefore, there is no pattern-dependent jitter. Another advantage of the two-state PD is that the phase range is $\pm\pi$, or double that of an exclusive-OR PD (see Fig. 10–9c).

10–5 PHASE DETECTORS FOR NRZ DATA

To recover clock from NRZ data, the NRZ data is usually converted to RZ-like data, as in Fig. 10–4, and then applied to an RZ phase detector, as in Fig. 10–7 or Fig. 10–9.

It is also possible to compare the phase of NRZ data directly with a clock, as in Fig. 10–10a. This circuit behaves essentially like a two-state PD, with a phase range from $-\pi$ to π, as shown in Fig. 10–10b. The data v_i is sampled by the rising clock edge v_o, and the sampled data Q_1 is compared with the data by an exclusive-OR gate. The result is a signal $\tilde{v}_b$ with pulses whose width goes from zero to T_B as θ_e goes from $-\pi$ to π (see Fig. 10–10c). For $\theta_e = 0$, the pulses have width $T_B/2$, but the average of $\tilde{v}_b$ depends on the data transition density (the number of $\tilde{v}_b$ pulses). Therefore, the waveform $\tilde{v}_a$ is needed as a reference. It maintains (independent of θ_e) the waveform that $\tilde{v}_b$ would have for $\theta_e = 0$. Then $\tilde{v}_d = \tilde{v}_b - \tilde{v}_a$ always has a zero average for $\theta_e = 0$. This corresponds to $v_d = 0$ for $\theta_e = 0$ in the PD characteristic (see Fig. 10–10b). The maximum value of the characteristic depends on the transition density; for a 50% density, the maximum is $(V_H - V_L)/4$, where V_H is the logic high level, and V_L is the logic low level.

10–6 OFFSET JITTER

When the PD has dc offset V_{do}, another kind of pattern-dependent phase jitter arises. A brief explanation is that the offset voltage causes the clock phase to drift when there are no data pulses (no phase information). The drift will be longer or shorter according to the variations of the data pattern. To be quantitative in our analysis of this effect, we need a model for the PD in the presence of a data pattern.

Let an RZ data signal v_i be considered the product of two waveforms:

$$v_i(t) = \delta(t)\ v_{im}(t)$$

where δ is the corresponding NRZ waveform and v_{im} is a square wave to reduce the last half of each RZ pulse to zero (see Fig. 10–11b). The periodic signal v_{im} is one that we

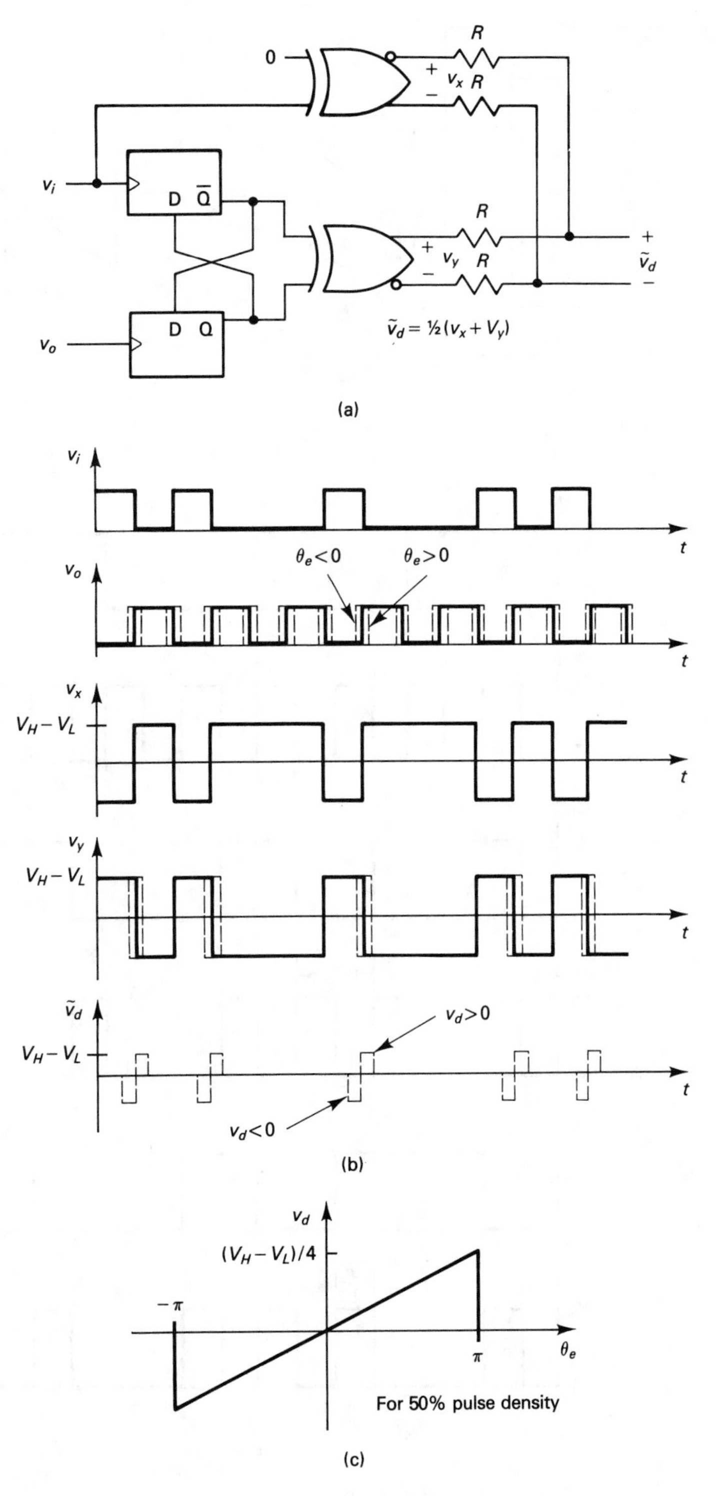

FIGURE 10–9 Two-state PD with data input

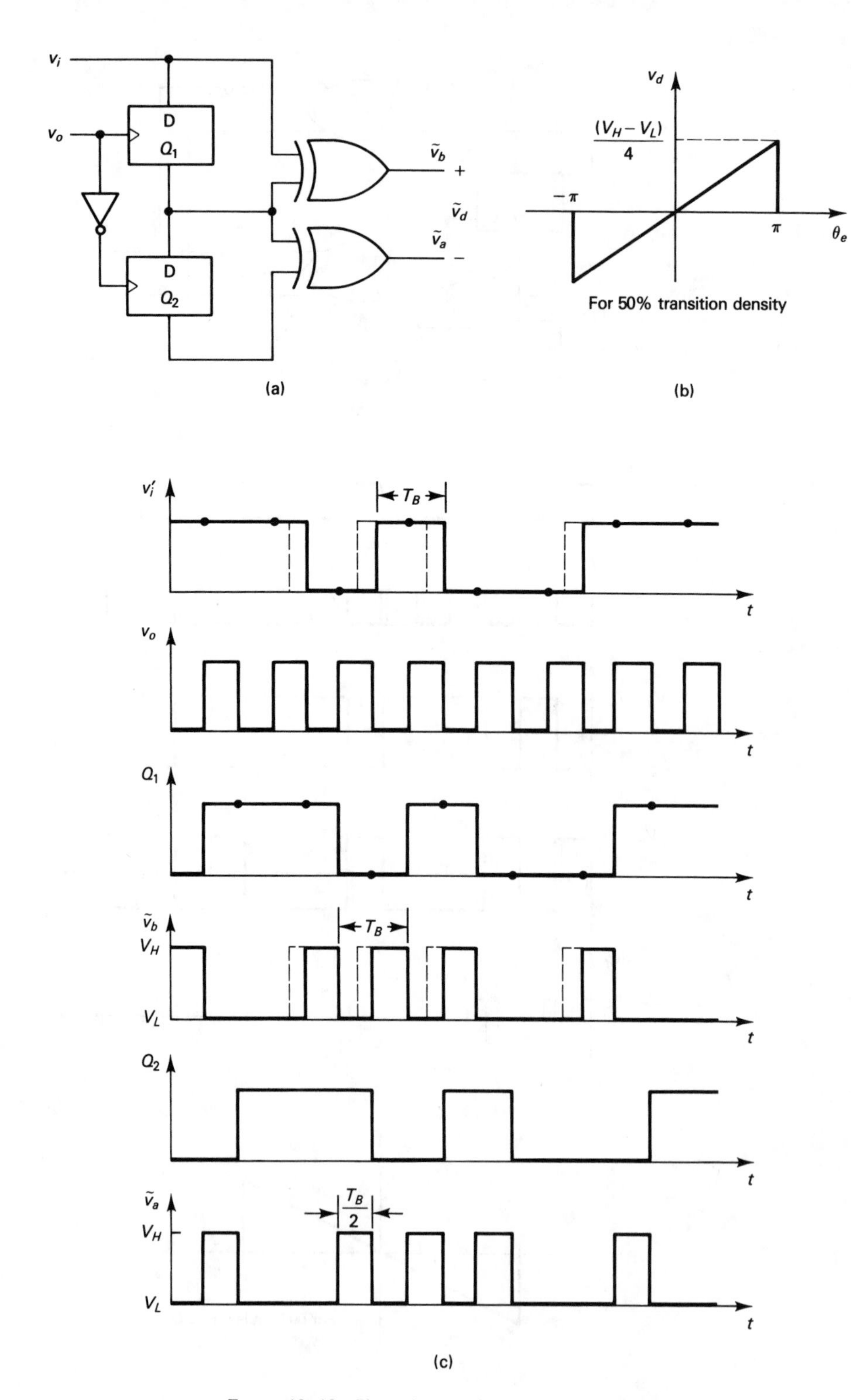

FIGURE 10–10 Phase detector for square-wave NRZ

have worked with before in section 4–6, and we will be able to apply some of that analysis.

The development here holds for all phase detectors, but to be clear, let the PD be an exclusive-OR gate comparing the phase of v_i with that of a clock v_o to produce $\tilde{v}_d$, as in Fig. 10–11a. We define v'_d as the average of $\tilde{v}_d$ while δ is constant (either 1 or 0). When $\delta = 1$ (no missing pulse), the analysis in section 4–6 applies, and $v'_d = K_{dm}\theta_e$, where

$$K_{dm} \equiv V_{dm}/(\pi/2) \tag{10-4}$$

When $\delta = 0$, $v'_d = 0$ (see the waveforms in Fig. 10–11b). This can be expressed as

$$v'_d = \begin{cases} K_{dm}\theta_e; & \delta = 1 \\ 0; & \delta = 0 \end{cases}$$

or

$$v'_d = \delta K_{dm}\theta_e \tag{10-5}$$

This relationship is incorporated into a signal flow graph for the PLL in Fig. 10–12a. The input phase θ_i is assumed to be zero so we can concentrate on the effects of the data pattern. The input variable to the loop here can be considered δ, which is the data pattern.

Unfortunately, the loop in Fig. 10–12a involves a multiplier, which makes it a nonlinear system and difficult to analyze. Therefore, our first task is to approximate the behavior of the multiplier by an adder as follows. Let the inputs δ and θ_e to the multiplier be resolved into dc and ac components:

$$\delta(t) \equiv \delta_o + \tilde{\delta}(t)$$
$$\theta_e(t) \equiv \theta_{eo} + \tilde{\theta}_e(t)$$

where the subscript o indicates the dc component (or average), and the "$\tilde{}$" indicates the ac component with zero mean. Typical waveforms for these variables are shown in Fig. 10–12b. The variables δ_o and $\tilde{\delta}(t)$ are of the same magnitude, but we can make the approximation

$$\tilde{\theta}_e(t) << \theta_{eo}$$

Then the product in Fig. 10–12a can be approximated by

$$\begin{aligned} \delta(t)\theta_e(t) &= [\delta_o + \tilde{\delta}(t)]\theta_e(t) = \delta_o\theta_e(t) + \tilde{\delta}(t)\,[\theta_{eo} + \tilde{\theta}_e(t)] \\ &\approx \delta_o\theta_e(t) + \tilde{\delta}(t)\theta_{eo} \end{aligned} \tag{10-6}$$

Neither of the products on the right involves two time functions; the constants δ_o and θ_{eo} can be considered "gain" factors. Therefore, the product on the left is approximated by a

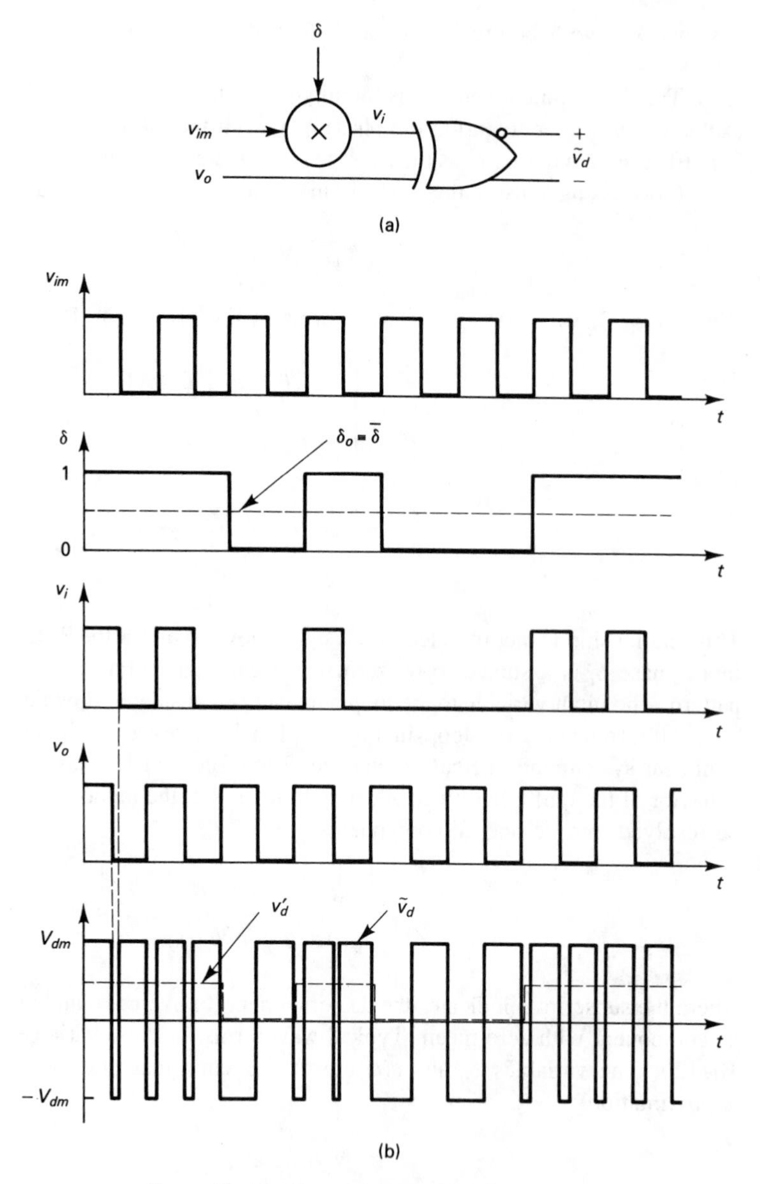

FIGURE 10–11 Decomposition of RZ waveform v_I

sum on the right. Equation (10-6) can also be put in the form $\delta\theta_e \approx \delta_o(\theta_e + \delta\theta_{eo}/\delta_o)$. The flow graph in Fig. 10–12c substitutes this sum for the product in Fig. 10–12a. The input variable is now $\tilde{\delta}$, a zero-mean signal determined by the data pattern (see Fig. 10–13a). The form of the flow graph in Fig. 10–12c is the same as that of a PLL with a PD gain of

$$K_d = \delta_o K_{dm} \tag{10-7}$$

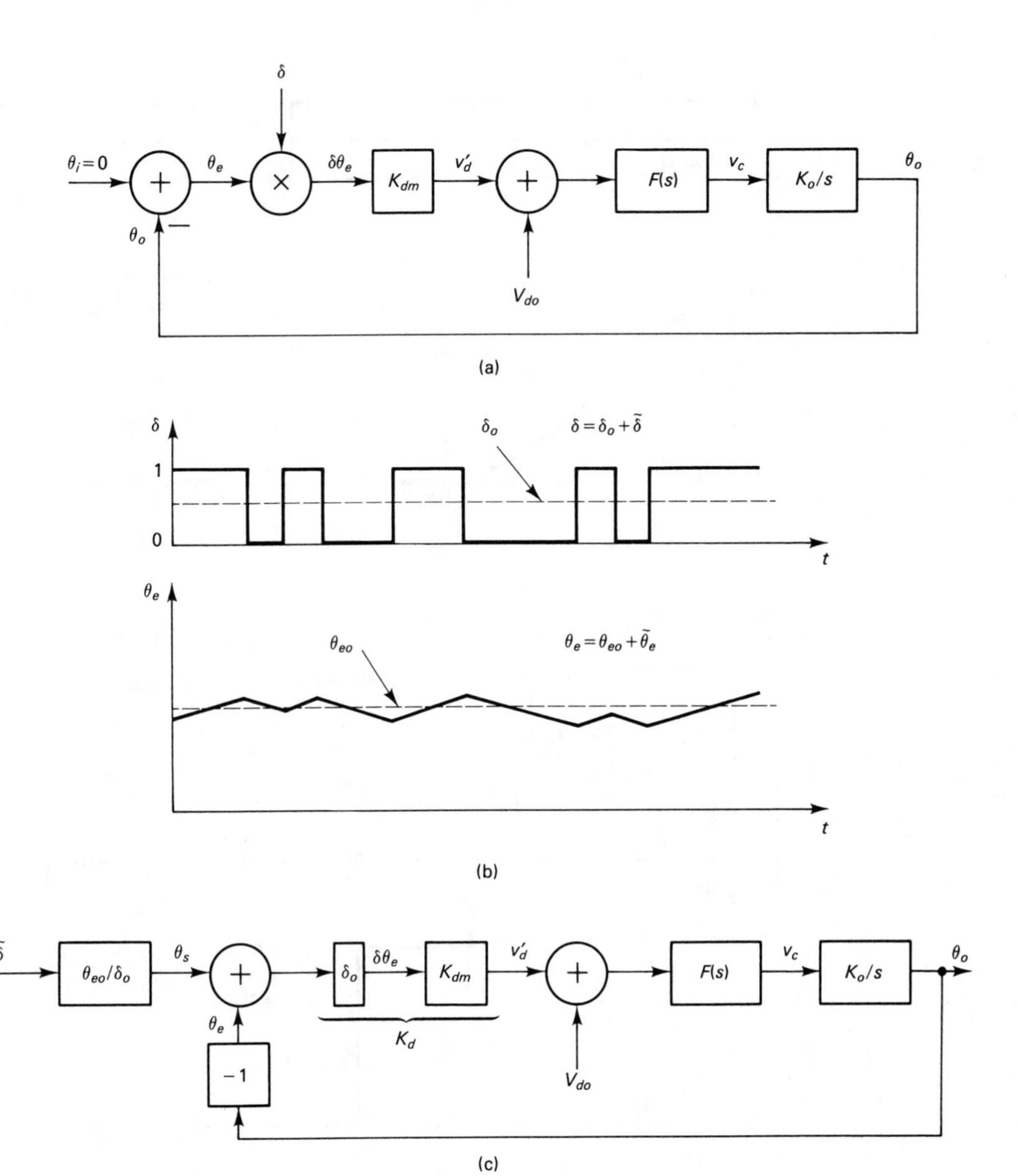

FIGURE 10–12 Offset jitter θ_o due to θ_{eo}

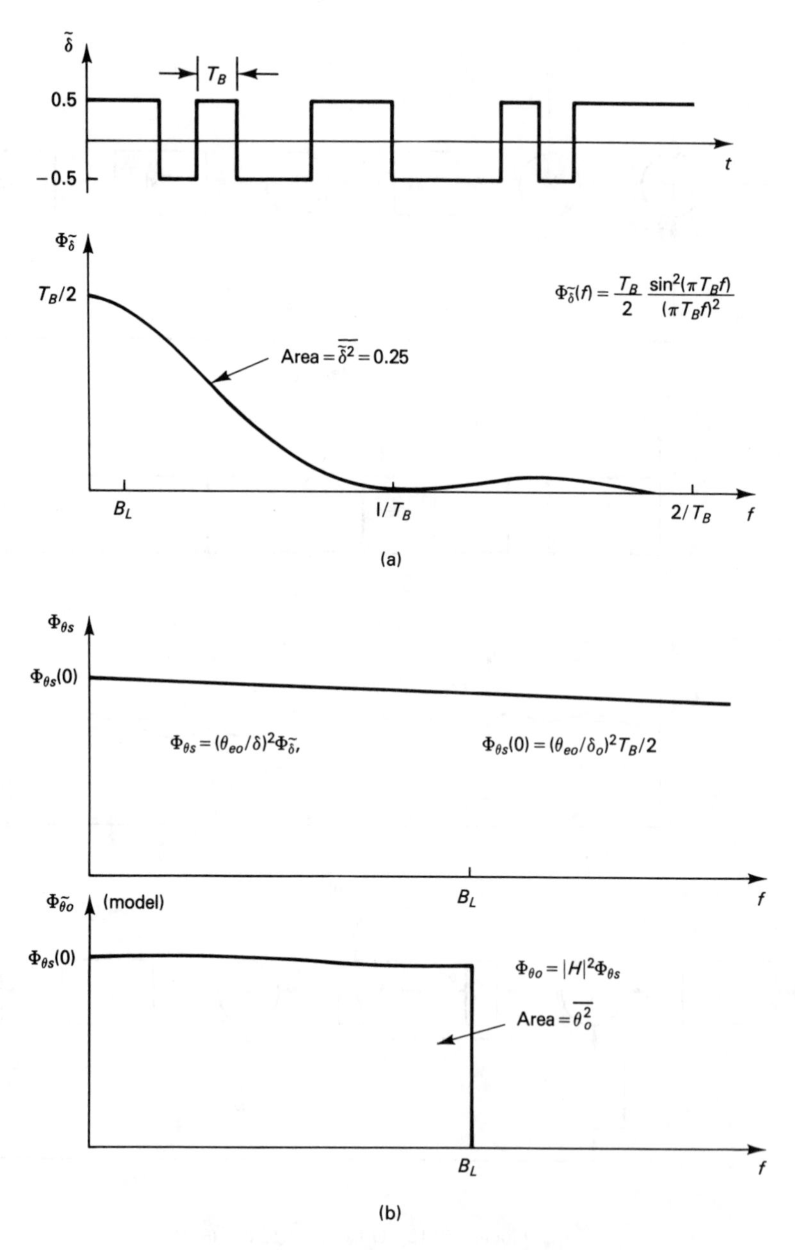

FIGURE 10–13 Spectral densities in the analysis of offset jitter

and with an input phase of $\tilde{\delta}\theta_{eo}/\delta_o$. Although this is not a phase that can actually be measured anywhere, let us define this effective input phase as

$$\theta_s \equiv (\theta_{eo}/\delta_o)\tilde{\delta} \tag{10-8}$$

One of these sources of this effective phase is the pattern of gaps in the data reflected by $\tilde{\delta}$. The other source is the static phase error θ_{eo} caused by dc offset voltage:

$$\theta_{eo} = -V_{do}/K_d \tag{10-9}$$

[See Eq. (3-7).] Let $\tilde{\theta}_o$ be the ac component of the output phase θ_o. Using the PLL phase transfer function $H(s)$ developed in section 3–5, we can find $\tilde{\theta}_o$ in terms of θ_s:

$$\tilde{\theta}_o(s) = \theta_s(s)\mathrm{H}(s) \tag{10-10}$$

This unwanted phase is called *offset jitter*. If we can find the spectral density of $\tilde{\delta}$, then from Eqs. (10-8) and (10-10) we can get the spectral density of the jitter $\tilde{\theta}_o$ and its rms value.

A typical waveform for $\tilde{\delta}$ is shown in Fig. 10–13a. It has zero mean and varies between 0.5 and -0.5 with the data pattern. Its mean square value is 0.25, and it can be shown (see Lathi [3] for example) that its spectral density is

$$\Phi_{\tilde{\delta}}(f) = \frac{T_B}{2} \frac{\sin^2(\pi T_B f)}{(\pi T_B f)^2} \tag{10-11}$$

as illustrated in Fig. 10–13a. Since the bandwidth B_L of the PLL is usually much less than $1/T_B$, we will be able to approximate

$$\Phi_{\tilde{\delta}}(f) \approx \Phi_{\tilde{\delta}}(0) = T_B/2 \tag{10-12}$$

But from Eq. (10-8), the spectral density of θ_s is

$$\Phi_{\theta s}(0) = (\theta_{eo}/\delta_o)^2\Phi_{\tilde{\delta}}(0) = (\theta_{eo}/\delta_o)^2 T_B/2 \tag{10-13}$$

As determined in section 6–2, the noise bandwidth of the PLL is

$$B_L = K/4 \tag{10-14}$$

Then the spectral density $\Phi_{\tilde{\theta}o}$ of $\tilde{\theta}_o$ can be modeled by cutting off $\Phi_{\theta s}$ abruptly at B_L, as shown in Fig. 10–13b. The area under the curve gives the mean-square jitter:

$$\overline{\tilde{\theta}_o^{\,2}} = \Phi_{\theta s}(0)B_L = (\theta_{eo}/\delta_o)^2 T_B K/8 \tag{10-15}$$

For random data, both the 1's density and the transition density are $\delta_o = 0.5$, and Eq. (10-15) becomes

$$\overline{\tilde{\theta}_o^2} = \Phi_{\theta s}(0)B_L = 0.5\theta_{eo}^2 K\, T_B \tag{10-16}$$

Since $\theta_i = 0$ in this analysis and $\theta_e = \theta_i - \theta_o$, therefore $\tilde{\theta}_e = -\tilde{\theta}_o$, and we also have

$$\overline{\tilde{\theta}_e^2} = \Phi_{\theta s}(0)B_L = 0.5\ \theta_{eo}^2 K\, T_B \tag{10-16'}$$

A typical PLL bandwidth might be $K = 0.02/T_B$. Then $\overline{\tilde{\theta}_e^2} = 0.01\ \theta_{eo}^2$, or $\tilde{\theta}_{e\ \mathrm{rms}} = 0.1\ \theta_{eo}$. This justifies our assumption that $\tilde{\theta}_e$ is negligible in Eq. (10-6).

While Eqs. (10-16) and (10-16′) give a measure of the offset jitter, they tell very little about the probability distribution of the offset jitter. If the distribution ware Gaussian, then the mean (θ_{eo}) and the standard deviation ($\tilde{\theta}_{e\ \mathrm{rms}}$) would completely characterize the distribution of θ_e. But the distribution is *not* Gaussian. Figure 10–14 shows probability distribution for θ_e for pseudorandom data. This was obtained by a computer simulation of a first-order PLL ($\omega_2 = 0$) with a pseudorandom data pattern of length $2^{20} - 1$. (See Golomb [4] for a description of pseudorandom pattern generators.) The case simulated here is for $KT_B = 0.02$. From Eq. (10-16′), this results in a normalized standard deviation $\tilde{\theta}_{e\ \mathrm{rms}}/\theta_{eo} = 0.1$. Figure 10–14 also shows a Gaussian distribution with the same mean and

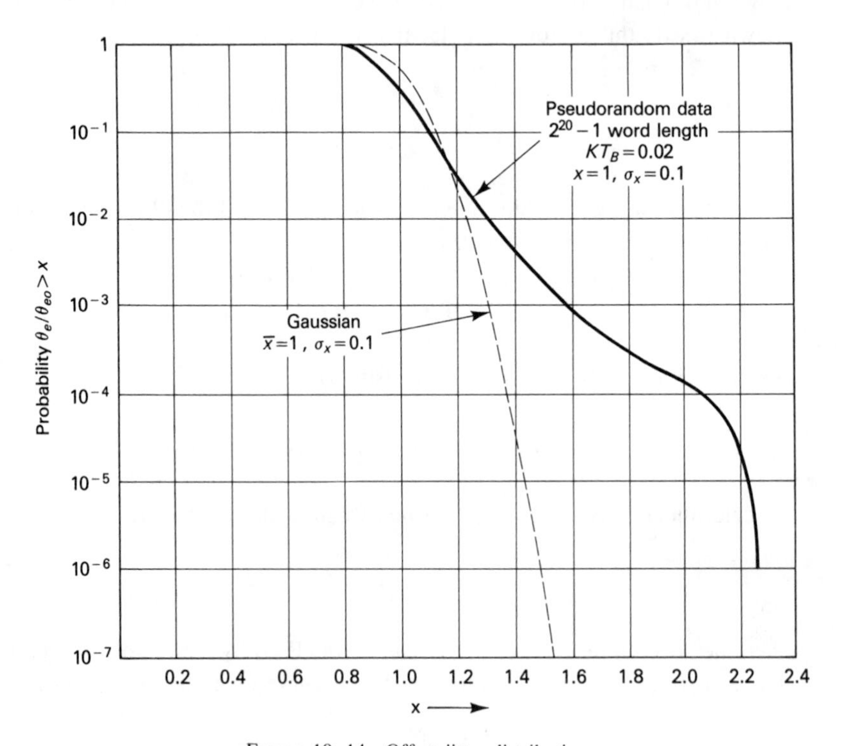

FIGURE 10–14 Offset jitter distribution

standard deviation for comparison. While the Gaussian distribution exceeds five standard deviations ($x = 1.5$ on the abscissa) with a probability of only 7.2×10^{-7}, the actual θ_e/θ_{eo} exceeds 1.5 with a probability of 0.002. While the Gaussian distribution exceeds $x = 2.27$ with a probability of only 3×10^{-37} (this is off the plot), the actual θ_e/θ_{eo} exceeds 2.27 with a probability of 10^{-6}.

EXAMPLE 10–2

Clock is recovered from a 1.544 Mb/s RZ data stream by a PLL with a bandwidth $K = 2\pi \times 5$ kHz $= 31.4$ krad/s. The phase detector in the PLL is a CMOS exclusive-OR gate with logic levels $V_H = 5$ V and $V_L = 0$ V. The PD offset voltage can be as great as $V_{do} = -0.15$ V. The error rate after sampling the data with the recovered clock is to be 10^{-6} or less. How much has V_{do} reduced the timing margin for alignment of the clock?

The optimum clock alignment is with the rising clock edge centered on the RZ data pulse, as in Fig. 10–1. Then the timing margin is $\pi/2$ radians before the clock edge reaches the edge of the data pulse and causes errors. The PD offset voltage $V_{do} = -0.15$ V causes a static phase offset $\theta_{eo} = -V_{do}/K_d$ [see Eq. (10-9)]. The exclusive-OR phase detector has a gain $K_d = (V_H - V_L)/\pi = 5$ V/3.14 rad $= 1.59$ V/rad [see Eq. (10-1)]. Then $\theta_{eo} = 0.0942$ rad, which reduces the timing margin by this much. But offset jitter further reduces the timing margin. The data spacing is $T_B = 1/1.544$ Mb/s $= 648$ ns. Then Eq. (10-16′) gives the mean square jitter: $\tilde{\theta}_e^2 = 0.5\ \theta_{eo}^2 K\ T_B = 0.01\ \theta_{eo}^2$, and taking the square root gives $\tilde{\theta}_{e\ \text{rms}} = 0.1\ \theta_{eo} = 0.00942$ rad. This is the case covered by Fig. 10–14, which shows that θ_e reaches about 12.7 $\tilde{\theta}_{e\ \text{rms}}$ away from the mean θ_{eo} with a probability 10^{-6}. Then the total reduction of timing margin is $\theta_{eo} + 12.7\ \tilde{\theta}_{e\ \text{rms}} = 0.214$ radians. The margin with optimum alignment is $\pi/2 = 1.57$ radians, so the reduction is $0.214/1.57 = 13.6\%$

A few comments need to be made to allow generalizations from Example 10-2. It was assumed that the 10^{-6} error rate specification was for a pseudorandom data pattern with length $2^{20} - 1$, for which Fig. 10–14 applies. The bandwidth $K = 0.02/T_B$ was also chosen in Example 10–2 so that Fig. 10–14 applied. For other bandwidths, the 10^{-6} point is roughly 13 $\tilde{\theta}_{e\ \text{rms}}$ away from θ_{eo}. For example, if K were increased to $2\pi \times 40$ kHz, then $KT_B = 0.16$, and from Eq. (10-16′), $\tilde{\theta}_{e\ \text{rms}} = 0.0267$ radians. Then the timing margin reduction would be $\theta_{eo} + 13\ \tilde{\theta}_{e\ \text{rms}} = 0.441$ radians, or 28.1%. This is a rough estimate; the reader should write his own program for other pseudorandom patterns and other values of KT_B. The discrete-time difference equation is

$$\chi_{n+1} = \chi_n(1 - 2K\ T_B D_n) + K\ T_B$$

where $\chi = \theta_e/\theta_{eo}$, and D_n is a pseudorandom sequence of 1's and 0's.

Note that if the recovered clock samples NRZ data rather than RZ data, the timing margin for optimal alignment is π rather than $\pi/2$.

10–7 JITTER ACCUMULATION

While offset jitter is seldom a problem in one PLL, it can accumulate to a significant level in a series of tandem PLLs. This is the case in a chain of data repeaters, shown in Fig. 10–15a. Data v_i transmitted at the head end has the clock recovered by a PLL, and the data is regenerated (cleaned up) as v_i. After some distance of transmission, distortion and noise require that the signal again be regenerated. Repeated cock recovery may occur as many as 1000 times in long-distance transmission. Each clock recovery circuit in the transmission path adds its own offset jitter to the total jitter.

The accumulation of jitter is modeled in Fig. 10–15b. Each PLL has an effective phase θ_s due to offset and the data pattern. This is added to the phase jitter of the data from the previous repeater. The PLL then filters the combined phase by its transfer function *H(s)*. The data pattern is the same for each PLL, and if we assume in the worst case that the static phase error θ_{eo} is the same for each PLL, then they all have the same θ_s. As can be seen from Fig. 10–15b, the transfer function from θ_s to the last output phase θ_N is

$$\frac{\theta_N(s)}{\theta_s(s)} = H(s) + H^2(s) + \ldots + H^N(s)$$

$$= \frac{H(s)}{1 - H(s)} [1 - H^N(s)] \tag{10-17}$$

For $\omega_2 = 0$, $H(s) + 1/(s/K + 1)$, and Eq. (10-17) becomes

$$\theta_N/\theta_s = (K/s)[1 - 1/(s/K + 1)^N] \tag{10-18}$$

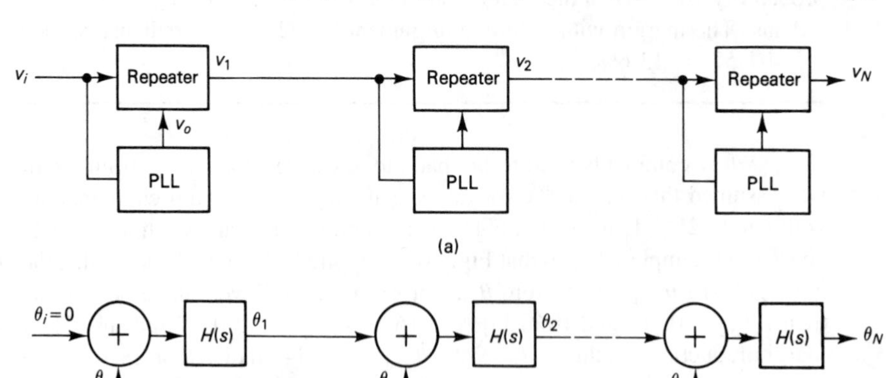

$$\frac{\theta_N(s)}{\theta_s(s)} = H(s) + H^2(s) + \ldots + H^N(s)$$

$$= \frac{H(s)}{1 - H(s)} [1 - H^N(s)]$$

FIGURE 10–15 Jitter accumulation in a chain of *N* regenerators

Then, as in Chapter 6, the spectral densities of θ_N and θ_s are related by

$$\Phi_{\theta N}/\Phi_{\theta s} = |\theta_N/\theta_s|^2 \tag{10-19}$$

For the case $\omega_2 = 0$, Eqs. (10-18) and (10-19) give

$$\Phi_{\theta N}/\Phi_{\theta s} = (K/2\pi f)^2 \, |1 - 1/(j2\pi f/K + 1)^N|^2 \tag{10-20}$$

This normalized power spectral density is plotted in Fig. 10–16 for various N, where N is the number of repeaters. For $2\pi f/K > 1/N$, $(2\pi f/K + 1)^N >> 1$, and $\Phi_{\theta N}/\pi_{\theta s}$ approaches $(K/2\pi f)^2$, as can be seen from Fig. 10–16. For $2\pi f/K < 1/N$, $\Phi_{\theta N}/\Phi_{\theta s} \approx N^2$. As a result, cumulative jitter tends to be heavy in low-frequency content.

The mean-square cumulative jitter is proportional to the area under the curves in Fig. 10–16. Byrne et al. [5] have evaluated the integral of Eq. (10-20):

$$\int_0^\infty \Phi_{\theta N}/\Phi_{\theta s} \, df = 0.5 \; N \; S(N) \; K \tag{10-21}$$

where $S(N)$ is a function of N that ranges from 0.5 to 1 as N increases (see Fig. 10–17). Over the range of f for which $\Phi_{\theta N}(f)$ is important, $\Phi_{\theta s}(f)$ is usually constant enough that

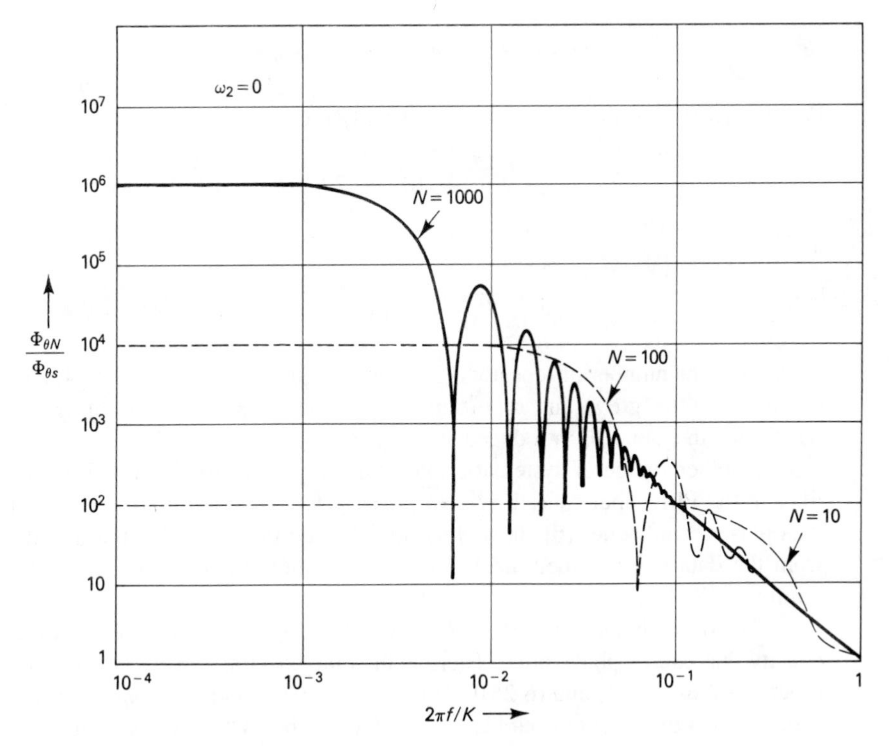

FIGURE 10–16 Accumulated jitter power density ($\omega_2 = 0$)

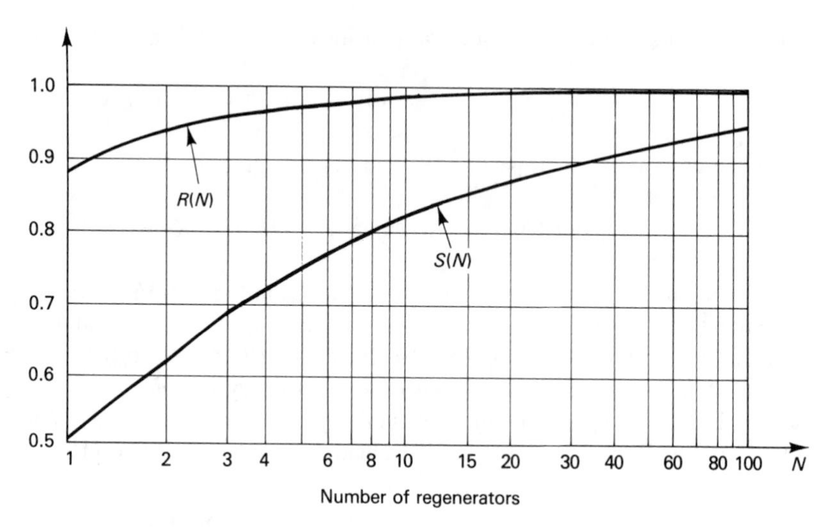

FIGURE 10–17 Factors for calculating accumulated jitter $\overline{\theta_N^2}$

we can approximate $\Phi_{\theta s}(f) \approx \Phi_{\theta s}(0)$. Then the mean-square cumulative jitter for $\omega_2 = 0$ is

$$\overline{\theta_N{}^2} = \int_0^\infty \Phi_{\theta N}\, \mathrm{df} = 0.5N\ S(N)\ K\ \Phi_{\theta s} \tag{10-22}$$

For the usual case of $\delta_o = 0.5$, Eq. (10-13) gives

$$\Phi_{\theta s}(0) = 2\theta_{eo}{}^2/T_B \tag{10-23}$$

and

$$\overline{\theta_N{}^2} = N\ S(N)\ K\ T_B\ \theta_{eo}{}^2 \tag{10-24}$$

where N is the number of repeaters. Taking the square root of Eq. (10-24) shows the rms cumulative jitter grows in proportion to $\sqrt{N}$. This is characteristic of *systematic jitter,* which has the same θ_s introduced in each PLL (see Fig. 10–15).

Another source of systematic jitter is *intersymbol interference* (ISI). This is phase jitter at the PLL input caused by data pulses effectively shifting the phase of following data pulses. Duttweiler [6] shows how to calculate the phase spectral density due to ISI from the data pulse shape, and Rosa [7] shows how to shape the channel response to minimize ISI jitter.

The offset-induced $\Phi_{\theta s}(0)$ given in Eq. (10-13) is only one source of phase spectral density that causes phase jitter. There is also the spectral density Θ_o due to noise at the input [see Eqs. (6-23) and (6-26)]. This is nonsystematic or *random jitter* since the noise is different at the input of each PLL. DeLange [8] shows that random jitter accumulates as

$0.282\sqrt{N}\,R(N)\,K\Theta_o$, where $R(N)$ is a function of N that grows from 0.886 to 1.0 as N increases (see Fig. 10–17). Then with Eq. (10-22), the total accumulated jitter is

$$\overline{\theta_N^2} = 0.5N\,S(N)\,K\,\Phi_{\theta s}(0) + 0.282\sqrt{N}\,R(N)\,K\,\Theta_o \tag{10-25}$$

where $\Phi_{\theta s}$ includes all systematic jitter—the offset jitter in Eq. (10-13), and ISI jitter, and pattern-dependent jitter described in section 10–4. Since the random jitter [the second term in Eq. (10-25)] grows more slowly with N, it is usually negligible for $N > 30$.

EXAMPLE 10–3

A transmission line with 30 repeaters carries RZ data at 1.544 Mb/s. Each repeater has an input filter with bandwidth $B_i = 3$ MHz, a signal-to-noise ratio $\text{SNR}_i = 12$, and a clock-recovery PLL with a bandwidth $K = 31.4$ krad/s and $\omega_2 = 0$. The PD in the PLL has a figure of merit $M = 10.6$. Find the accumulated jitter at the end of the transmission line.

From Eqs. (3-7) and (4-14), $\theta_{eo} = 1/M = 0.0942$ radians. The data spacing is $T_B = 1/1.544\ \text{Mb/s} = 648$ ns. For random data with $\delta_o = 0.5$, Eq. (10-13) gives $\Phi_{\theta s}(0) = 2\theta_{eo}^2 T_B = 1.15 \times 10^{-8}$ rad²/Hz. For $N = 30$, Fig. 10–17 gives $S(N) = 0.9$. Then the first term of Eq. (10-25) is 0.00487 rad².

From Eq. (6-26) the random phase spectral density is $\Theta_o = 1/B_i\text{SNR}_i = 2.08 \times 10^{-8}$ rad²/Hz. From Fig. 10–17, $R(N) = 0.99$. Then the second term in Eq. (10-25) is 0.000706 rad², which is only 14.5% of the first term. The total accumulated jitter is $\overline{\theta_N^2} = 0.00487 + 0.000706 = 0.00558$ rad², or $\theta_{N\ \text{rms}} = \sqrt{0.00558} = 0.0747$ radians.

Accumulated jitter is removed by a clock-recovery PLL with a VCXO (see section 5–6) to achieve a very small bandwidth K (< 1 rad/s). The recovered clock with no jitter samples the data with accumulated jitter θ_N, and the reclocked data therefore has no jitter (practically). If the phase θ_N between the clock and the data exceed $\pi/2$ for RZ data, errors will occur. For $N = 30$, the Central-Limit Theorem [9] makes θ_N about Gaussian, and it exceeds $x\,\theta_{N\ \text{rms}}$ with a probability approximated [10] by

$$P(x) \approx 0.4\,(1/x - 1/x^3)\,e^{(-x^2/2)} \tag{10-26}$$

For the $\theta_{N\ \text{rms}} = 0.0747$ radians in Example 10-3, $\theta_N = \pi/2$ corresponds to $x = 1.57/0.0747 = 21$. Then θ_N exceeds $\pi/2$ with a probability $P(x) = 3 \times 10^{-98}$, and there are essentially no errors due to reclocking of the data. If θ_N were large enough to cause a significant error rate, an *elastic store* [11] would be necessary in the jitter removal.

Equation (10-24) gives the accumulated offset jitter for the case $\omega_2 = 0$. For the more usual case of $\omega_2 \neq 0$, the peak value of $H(s)$ exceeds unity, and the H^N in Eq. (10-17) can be quite large for large N. As shown in Chapter 2, the peak value of H is $H_p \approx 1 + \omega_2/K$ for $\omega_2 << K$. Then

$$H_P^N \approx (1 + \omega_2/K)^N \approx \exp(N\omega_2/K) \tag{10-27}$$

and the phase transfer function θ_N/θ_s grows about exponentially with N. This can lead to much greater jitter accumulation than predicted by Eq. (10-24) if ω_2 is not kept sufficiently small. For $\omega_2 \neq 0$, the PLL transfer function is

$$H(s) = \frac{Ks + K\omega_2}{s^2 + Ks + K\omega_2}$$

Using this expression for $H(s)$ in Eqs. (10-17) and (10-19) leads to the normalized spectral densities in Fig. 10–18. As ω_2/K increases, the peaking and the area under the curve for a given N increase.

The area under a curve in Fig. 10–18 gives the normalized mean-square jitter:

$$\overline{\theta_N^2}/K\Phi_{\theta s}(0) = \int_0^\infty \Phi_{\theta N}/K\Phi_{\theta s}(0) \quad df$$

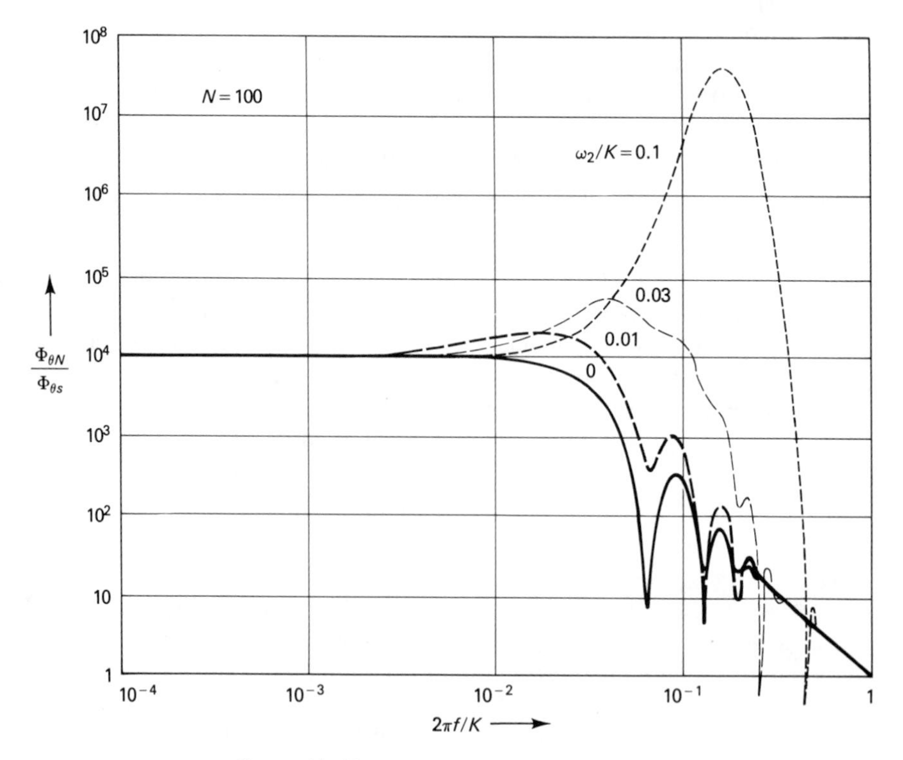

FIGURE 10–18 Accumulated jitter power density

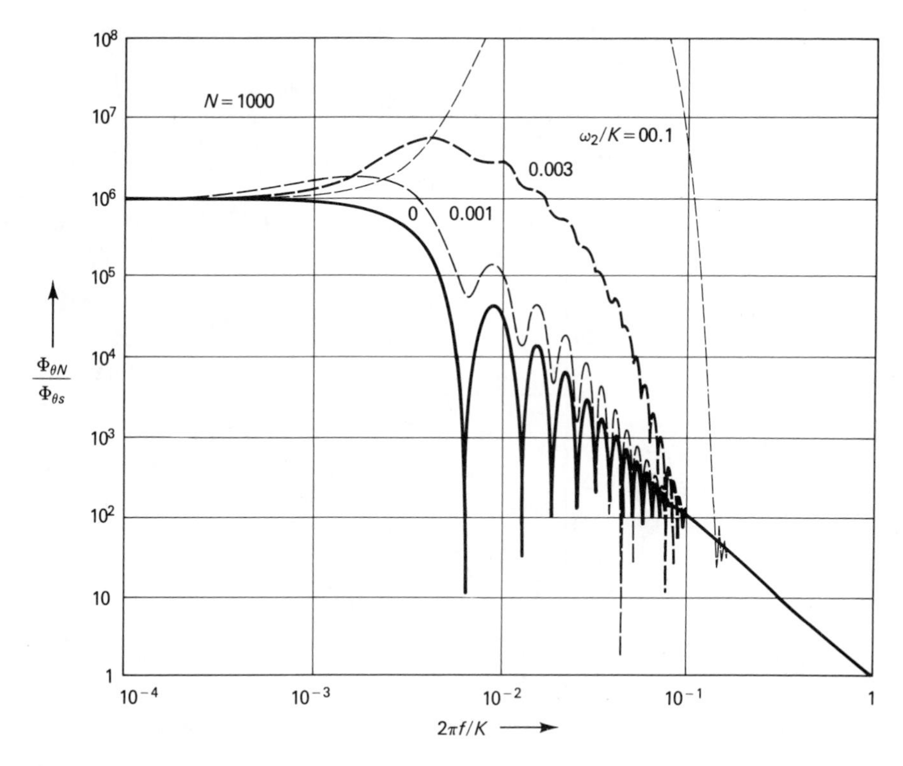

FIGURE 10–18 (*Cont'd*)

A computer was used to evaluate the areas by numerical methods, and the results are plotted in Fig. 10–19 as a function of N for several values of ω_2/K. A good rule of thumb is to make

$$\omega_2/K = 1/N \tag{10-28}$$

so that $H_P^N \approx 2.7$ [see Eq. (10-27)]. It can be seen from Fig. 10–19 that for $\omega_2/K = 1/N$, $\overline{\theta_N^2}$ is increased by a factor of only 2.3 over that given by Eq. (10-24), independent of N. Then for $\omega_2/K = 1/N$, the accumulated offset jitter is

$$\overline{\theta_N^2} \approx 2.3\ K\ N\ S(N)\ \theta_{eo}^2 T_B \tag{10-29}$$

EXAMPLE 10–4

A transmission line with 30 repeaters carries RZ data at 1.544 Mb/s. Each repeater has a clock-recovery PLL with a bandwidth $K = 31.4$ krad/s and a PD with a figure of merit $M = 10.6$. The initial frequency error before acquisition is $\omega_{eo} = 1$ Mrad/s. Find the

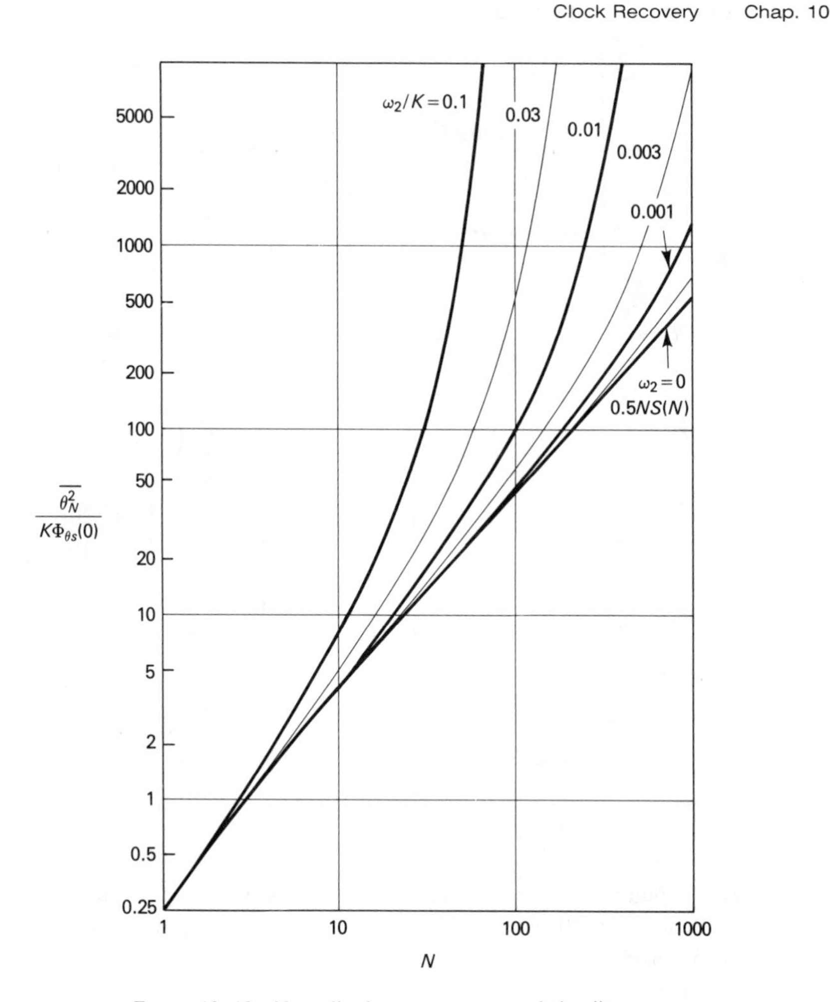

FIGURE 10–19 Normalized mean-square cumulative jitter

accumulated offset jitter at the end of the transmission line and the pull-in time for three cases: $\omega_2 = 0.01\ K$, $\omega_2 = 0.03K$, and $\omega_2 = 0.1K$.

For $\omega_2 = 0.1K$, Fig. 10–19 gives $\overline{\theta_N{}^2}/K\Phi_{\theta s}(0) = 115$. As in Example 10-3, $\Phi_{\theta s}(0) = 1.15 \times 10^{-8}$. Therefore, $\overline{\theta_N{}^2} = 0.0415\ \text{rad}^2$, or $\theta_{N\ \text{rms}} = \underline{0.204\ \text{radians}}$. The input frequency to which the PLL is to lock is $\omega_i = 2\pi f_B = 9.7$ Mrad/s. For an exclusive-OR PD, the lock-in frequency is $\omega_L = (\pi/2)K = 49.3$ krad/s. From Eq. (8-46b), the pull-in time for a rotational frequency detector is

$$T_p \approx (2\pi/\omega_L)\ell\text{n}(\omega_i/20\omega_L) + (40\pi\omega_{eo}/\omega_i - 2\pi)/\omega_L$$

$$= 0.291\ \text{ms} + 0.135\ \text{ms} = \underline{0.426\ \text{ms}}$$

For $\omega_2 = 0.03K$, Fig. 10–19 gives $\overline{\theta_N^2}/K\Phi_{\theta s}(0) = 31$. Therefore, $\overline{\theta_N^2} = 31 \times 31.4$ krad/s $\times 1.15 \times 10^{-8} = 0.0112$ rad^2 or $\theta_{N\ \mathrm{rms}}$ = 0.106 radians. The pull-in time is the same since it doesn't depend on ω_2. However, since $\omega_2 = 1/R_2C$, the value of C must be about triple that for $\omega_2 = 0.1K$, and the frequency detector must provide about triple the current to charge C during acquisition.

For $\omega_2 = 0.01\ K$, Fig. 10–19 gives $\overline{\theta_N^2}/K\Phi_{\theta s}(0) = 18$. Therefore, $\overline{\theta_N^2} = 18 \times$ 31.4 Mrad/s $\times 1.15 \times 10^{-8} = 0.0065$ rad^2, or $\theta_{N\ \mathrm{rms}}$ = 0.081 radians. The pull-in time is the same. However, the current required of the frequency detector is now ten times that for $\omega_2 = 0.1\ K$.

Decreasing ω_2 from $0.1\ K$ to $0.03\ K$ reduced $\theta_{N\ \mathrm{rms}}$ by 50%. Decreasing ω_2 further to $0.01\ K$ only reduced $\theta_{N\ \mathrm{rms}}$ by another 20%. This is probably not worth the extra current required of the frequency detector. Therefore, the best choice for ω_2 is $0.03K$, as recommended by Eq. (10-28).

REFERENCES

[1] P. R. Trischitta and E. L. Varma, *Jitter in Digital Transmission Systems,* Artech House: Norwood, Mass., 1989.

[2] B. P. Lathi, *Modern Digital and Analog Communication Systems,* HRW: Philadelphia, 1989, section 2.9.

[3] Lathi, *Modern Digital and Analog Communication Systems.*

[4] S. W. Golomb, *Shift Register Sequences,* Holden-Day: San Francisco, 1967.

[5] C. J. Byrne, et al., "Systematic Jitter in a Chain of Digital Regenerators," *B.S.T.J.*, vol. 42, no. 6 (November 1963), pp. 2692–2714.

[6] D. L. Duttweiler, "The Jitter Performance of Phase-Locked Loops Extracting Timing from Baseband Data Waveforms," *B.S.T.J.* vol. 55, no. 1 (January 1976), pp. 37–58.

[7] E. Rosa, "Analysis of Phase-Locked Timing Extraction Circuits for Pulse Code Transmission," *IEEE Trans. on Communications,* vol. COM-22, no. 9 (September 1974), pp. 1236–1249.

[8] O. E. DeLange, "The Timing of High-Speed Regenerative Repeaters," *B.S.T.J.* vol. 37, no. 6 (November 1958), pp. 1455–1486.

[9] Lathi, *Modern Digital and Analog Communication Systems,* section 5.4.

[10] National Bureau of Standards, *Handbook of Mathematical Functions,* U.S. Government Printing Office: Washington, D.C., 1964, p. 932.

[11] Z. Kitamura, K. Terada, and K. Asada, "Asynchronous Logical Delay Line for Elastic Stores," *Electronics and Communications in Japan,* vol. 50 (November 1967), pp. 90–99.

CHAPTER

11

FREQUENCY SYNTHESIZERS

A frequency synthesizer generates any of a number of frequencies by locking a VCO to an accurate frequency source such as a crystal oscillator. Most quality FM radios now use a frequency synthesizer to generate the 101 different frequencies necessary to tune to the various stations. For proper tuning, the synthesized frequency should be accurate to within 10 parts per million (ppm)—the accuracy of a crystal oscillator. Since it is impractical to have 101 crystal oscillators, a frequency synthesizer is used to generate any one of the frequencies from just one crystal oscillator.

11–1 SINGLE-LOOP SYNTHESIZER

The simplest form of a frequency synthesizer is shown in Fig. 11–1. It is a PLL with $\div N$ frequency divider in the feedback path. When the PLL is in lock, the fed-back frequency f_o/N equals the input frequency—the *reference frequency* f_r. Therefore, the output frequency is

$$f_o = N f_r \tag{11-1}$$

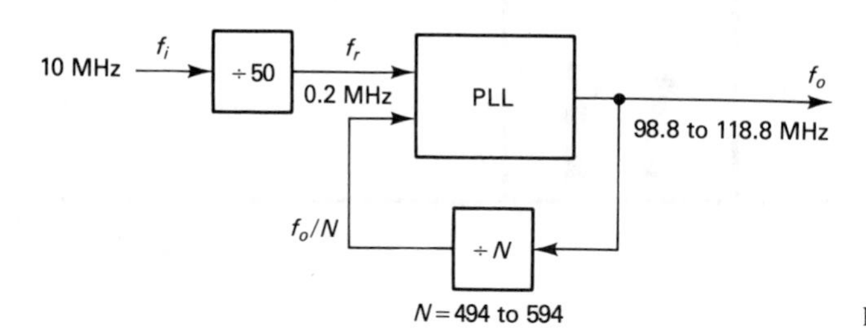

FIGURE 11–1 Simple frequency synthesizer

For a fixed f_r, the desired f_o is generated by selecting the proper integer N. Circuits that divide a frequency by a value N that is controlled by a signal are called *programmable dividers* or *programmable counters*. Motorola [1] manufactures an integrated programmable counter—the MC4018. See Rohdel [2] for a complete discussion of programmable counters.

EXAMPLE 11–1

Design a single-loop synthesizer to generate 98.8 MHz, 99.0 MHz, 99.2 MHz, . . . 118.8 MHz. That is, f_o is to have a *range* from 98.8 MHz to 118.8 MHz with a *resolution* of 0.2 MHz.

All the frequencies are multiples of 0.2 MHz. Then from Eq. (11-1), $f_r = 0.2$ MHz. The lowest frequency is 98.8 MHz $= 494 \times 0.2$ MHz, and the highest frequency is 118.8 MHz $= 594 \times 0.2$ MHz. Then N must range from 494 to 594.

Since it is difficult to make crystals that resonate as low as 0.2 MHz, the reference frequency here would probably be obtained by dividing down a higher crystal frequency such as $f_i = 10$ MHz (see Fig. 11–1). If the ÷50 is made a selectable $\div M$, then $f_r = f_i/M$, and Eq. (11-1) becomes

$$f_o = (N/M)f_i \tag{11-2}$$

Integrated circuits are available commercially that provide the $\div M$ and $\div N$ frequency dividers and the PLL's phase detector in one package. [3]

11–2 CHOOSING THE BANDWIDTH K

The bandwidth K of a PLL in a frequency synthesizer affects four parameters of the performance: the pull-in time after a new N is selected, the suppression of phase noise, the suppression of spurious modulation, and the resistance to injection locking. The pull-in time, given by Eq. (8-28′), is approximately

$$T_p = 8\, f_{eo}/NK^2$$

for the choice $\omega_2 = K/4$. The phase noise in most cases will be governed by Eq. (6-63):

$$\overline{\theta_o^2} = a/K$$

where a is a constant. Spurious modulation, analyzed in section 9–1–2, is also present in frequency synthesizers. We will show in Eq. (11-14′) that the spurious phase modulation is

$$\Delta\theta = \pi\delta NK^2/f_r^2$$

where δ is proportional to the PD offset voltage. Avoidance of injection (section 5–9 and Example 9-2) is usually not a dominant consideration compared with the three above.

The strategy in designing for N and K is to minimize the maximum N to reduce $\Delta\theta$. Then K is chosen as a compromise between a small value to reduce $\Delta\theta$ and a large value to reduce $\overline{\theta_o^2}$. If the T_p for the chosen N_{max} and K is unacceptable, a decrease in T_p will have to be traded off against an increase in $\Delta\theta$.

The $\div N$ frequency divider is still considered part of the phase detector, as it was in Fig. 4–12. For a three-state PD with a $\div N$, the phase detector gain is $K_d = V_{dm}/2\pi N$. Therefore, the bandwidth $K = K_d K_h K_o$ is inversely proportional to N. As N is varied in a frequency synthesizer to change the frequency, K also varies. However, the product NK remains constant. In particular,

$$N_{min}K_{max} = N_{max}K_{min} \tag{11-3}$$

This quick discussion provides a rough overview of the factors influencing the choice of the bandwidth K. The following sections will provide more rigorous analysis and some design examples.

11–3 SYNTHESIZER WITH MIXER

It is sometimes possible to reduce N by introducing a mixer into the PLL, as in Fig. 11–2. This will reduce the spurious phase modulation $\Delta\theta$ while achieving the same range and resolution as the frequency synthesizer in Fig. 11–1.

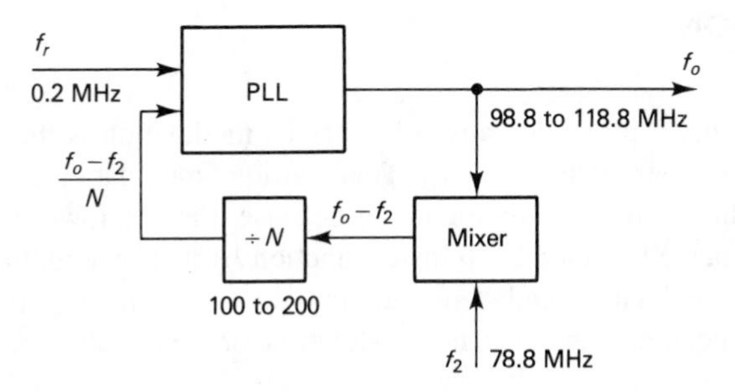

FIGURE 11–2 Frequency synthesizer with mixer

The mixer in Fig. 11–2 produces the sum and difference of the frequency f_o and a second reference frequency f_2. The sum $f_o + f_2$ is removed by a low-pass filter (not shown), and the difference $f_o - f_2$ remains as the desired output of the mixer. This difference is divided by N and compared with the reference f_r. Then in lock, $(f_o - f_2)/N = f_r$, or

$$f_o = N f_r + f_2 \tag{11-4}$$

Here, f_2 provides most of f_o, and N provides the smaller, variable portion of f_o. In Fig. 11–1, f_o has a range from 98.8 MHz to 118.8 MHz and a resolution of 0.2 MHz. This same range is realized by Eq. (11-4) for $f_r = 0.2$ MHz, $N = 100$ to 200, and $f_2 = 78.8$ MHz (see Fig. 11–2). Compared with the synthesizer in Fig. 11–1, the largest value of N is now only 200 rather than 594.

Any of the phase detectors described in Chapter 4 can serve as a mixer. The three-state PD has a property that is especially useful—it produces an output only when f_o is *greater* than f. This prevents the PLL from locking at $(f_2 - f_o)/N = f_r$ rather than $(f_o - f_2)/N = f_r$.

In practice, the mixer produces not only $f_o - f_2$ and $f_o + f_2$ but also some spurious intermodulation products. Manassewitsch [4] gives a thorough discussion of these spurious frequencies and how to minimize them. In particular, the f_o and f_2 components are sometimes strong, and the low-pass filter must separate these from the desired $f_o - f_2$. Therefore, it is necessary to maintain

$$f_o - f_2 < f_2 \tag{11-5}$$

For the case studied here, $f_2 = 78.8$ MHz, and the highest $f_o - f_2$ is 40 MHz. Suppose that, before filtering, the amplitudes of the $f_o - f_2$ component and the f_2 component are equal (as with a three-state PD mixer). Then a seventh-order low-pass filter at 40 MHz will increase the ratio of the amplitudes to $(78.8/40)^7 = 115/1$. As presented in Chapter 6, the additive f_2 "noise" component causes phase modulation with an amplitude of 1/115 radian and a frequency of $78.8 - 40 = 38.8$ MHz. The PLL transfer function $H(s)$ further reduces this spurious phase modulation before it reaches the output.

11–4 SPURIOUS MODULATION

We saw in section 9–1–2 that a principle cause of spurious modulation is the spurious frequencies produced by the phase detector at $\tilde{v}_d$. This spurious modulation $\Delta\theta$ of θ_o is usually more serious than that introduced by the mixer because it is of a lower frequency and therefore more readily passed by the PLL transfer function $H(s)$. The amplitude of $\Delta\theta$ for applications involving modulation and demodulation was given by Eq. (9-14). In frequency synthesizer applications, there is no modulation or demodulation, and the spurious modulation will be less. In this section, we will derive the amplitude of $\Delta\theta$ when the input is a fixed reference frequency.

A typical circuit for a single-loop synthesizer (with or without a mixer) is shown in Fig. 11–3a. The phase detector is a three-state PD (see Fig. 4–8b). The loop filter is active with essentially infinite gain at dc. Therefore, the average of its input $\tilde{v}_d$ must be zero, where

$$\tilde{v}_d \equiv v_U - v_D$$

If there is no offset voltage V_{do}, then $\tilde{v}_d$ consists of pulses of zero width. In practice there is always some V_{do} (shown negative in Fig. 11–3b), and $\tilde{v}_d$ must have pulses wide enough to make the average be zero. Then the duty cycle δ must be

$$\delta = V_{do}/V_{dm} \tag{11-6}$$

where V_{dm} is the maximum value of $\tilde{v}_d$. For a three-state PD with a $\div N$ frequency divider, V_{dm} is related to the PD gain K_d by

$$V_{dm} = 2\pi N K_d \tag{11-7}$$

The period of the waveform is

$$T_d = 1/f_r$$

where f_r is the reference frequency at the input to the PLL.

It is the pulses in the $\tilde{v}_d$ waveform that cause phase jitter at θ_o. The suppression of these pulses is called *reference suppression* because their frequency is the reference frequency f_r. The $\tilde{v}_d$ waveform experiences a (high-frequency) gain of K_h in the loop filter and a gain of K_o in the VCO to produce a pulse in $\Delta\omega_o$ of height

$$\Delta\omega = V_{dm}K_hK_o = 2\pi N K_d K_h K_o = 2\pi N K \tag{11-8}$$

and width δT_d (see Fig. 11–3b). Since θ_o is the integral of $\Delta\omega_o$, the change in θ_o is the area under this pulse:

$$\Delta\theta = \Delta\omega\delta T_d = 2\pi N K \delta/f_r \tag{11-9}$$

Then for a fixed offset V_{do} and a fixed bandwidth K, the jitter increases in proportion to N. (The mixer is transparent to phase; it doesn't enter into the analysis.)

It is possible to reduce the spurious modulation $\Delta\theta$ still further by adding a low-pass function to the loop filter. One circuit design to do this is shown in Fig. 11–4a. The capacitor C_3 together with R_1 form a low-pass filter with a cutoff at

$$\omega_3 = 4/R_1C_3 \tag{11-10}$$

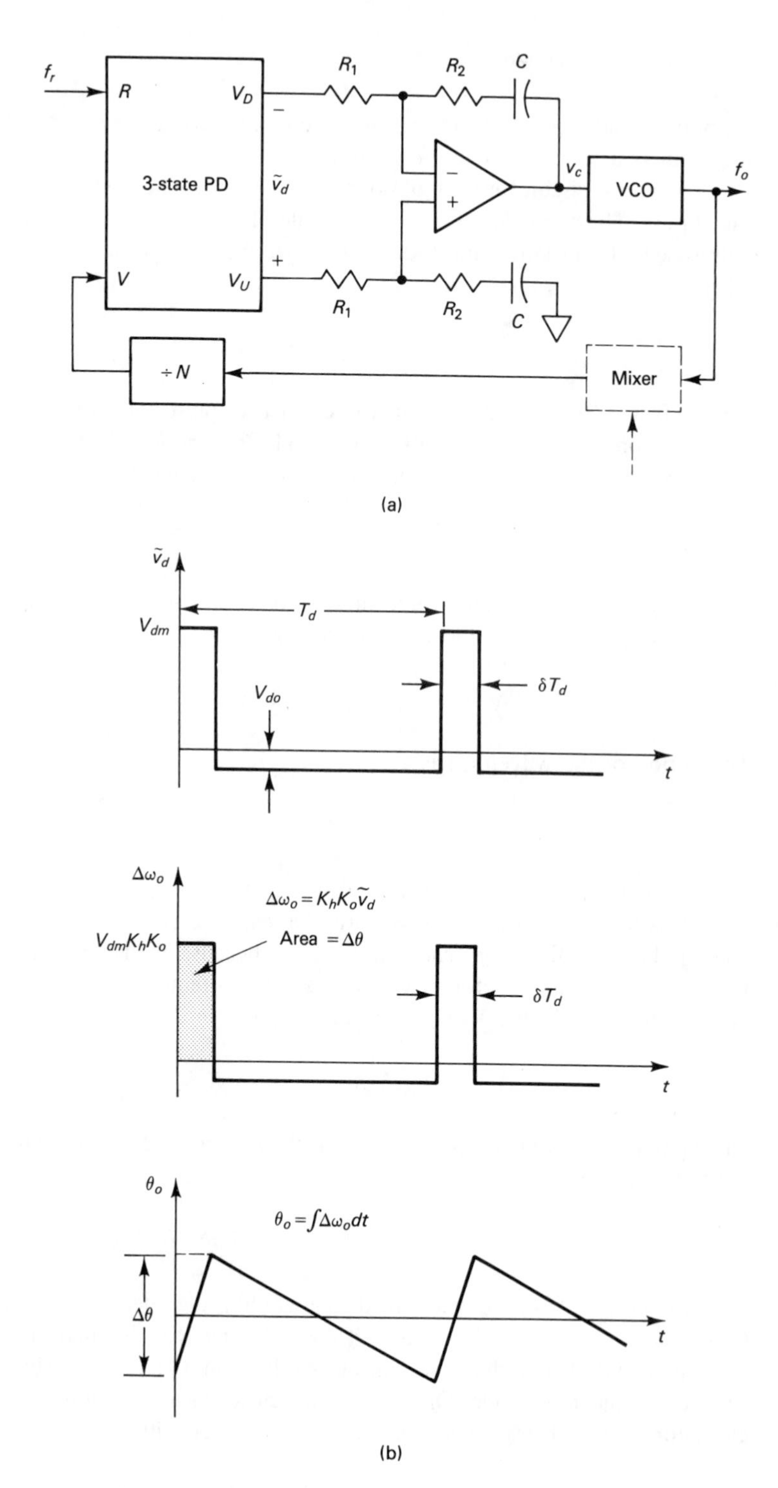

FIGURE 11–3 Phase jitter due to spurious modulation

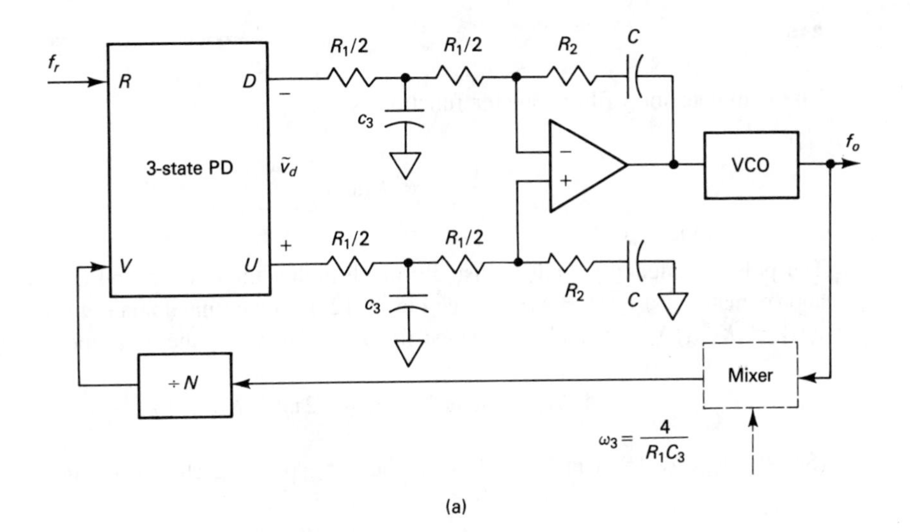

(a)

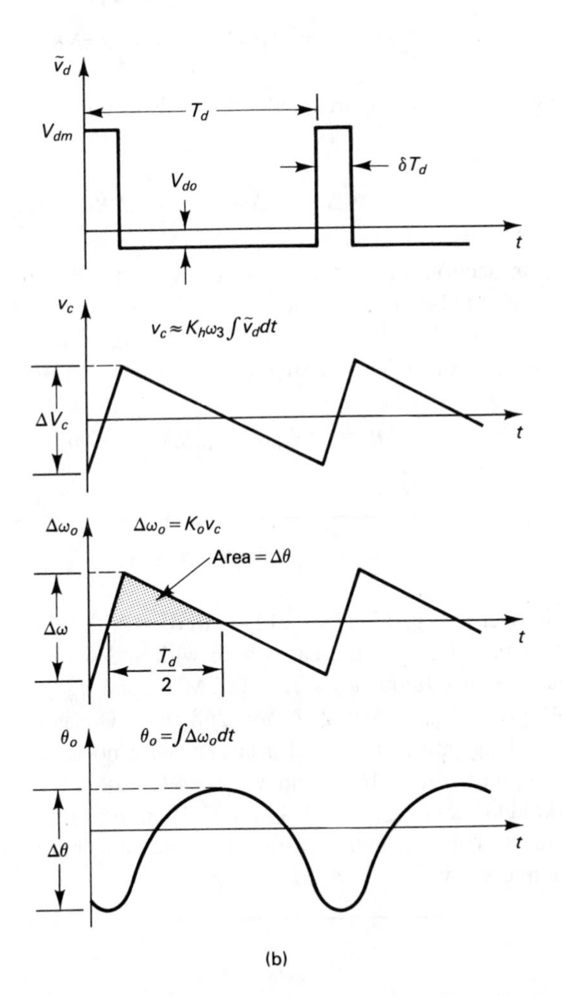

(b)

FIGURE 11–4 Phase jitter with reference supressed by pole at ω_3

The complete loop filter transfer function is then

$$\frac{v_c(s)}{\tilde{v}_d(s)} = K_h\omega_3 \frac{s + \omega_2}{s(s + \omega_3)}$$

The pulse frequency f_r of $\tilde{v}_d$ is usually much higher than $\omega_3/2\pi$ and $\omega_2/2\pi$, so we can approximate $v_c(s)/\tilde{v}_d(s) \approx K_h\omega_3/s$ (see Fig. 3–12a). In the time domain this corresponds to $v_c(t) = K_h\omega_3\int\tilde{v}_d(t)dt$, and the change in v_c is $K_h\omega_3$ times the area under a $\tilde{v}_d$ pulse:

$$\Delta v_c = K_h\omega_3 V_{dm}\delta T_d = 2\pi N K_d K_h \omega_3 \delta T_d \tag{11-11}$$

(See the v_c waveform in Fig. 11–4b.) The corresponding change in $\Delta\omega_o$ is

$$\begin{aligned}\Delta\omega &= K_o\Delta v_c = 2\pi N K_d K_h K_o \omega_3 \delta T_d \\ &= 2\pi N K \omega_3 \delta T_d = 2\pi N K \omega_3 \delta / f_r \end{aligned} \tag{11-12}$$

The change $\Delta\theta$ in θ_o is the area under the positive portion of the $\Delta\omega_o$ waveform (see Fig. 11–4b):

$$\Delta\theta = \Delta\omega T_d/8 = \pi N K \omega_3 \delta / 4 f_r^2 \tag{11-13}$$

As noted in section 11–2, the product NK is not dependent on the programmed N. Therefore, Eq. (11-13) may be evaluated for any consistent pair—$N_{\min}$ and $K_{\max}$, for example. The smallest jitter is realized by making ω_3 as small as possible. For stability reasons (see section 3–7), the smallest practical ω_3 is $4K_{\max}$. For this choice of ω_3,

$$\Delta\theta = \pi N_{\min}\delta(K_{\max}/f_r)^2, \qquad \omega_3 = 4K_{\max} \tag{11-14}$$

EXAMPLE 11–2

The synthesizer in Fig. 11–1 has a PD with $\delta = 0.01$. Find the necessary K so that $\Delta\theta$ is just 0.05 radian. Look at the cases both with and without a pole at ω_3.

The reference frequency is $f_r = 0.2$ MHz, and $N_{\max} = 594$. Then for no pole at ω_3, Eq. (11-9) gives $K_{\min} = \Delta\theta f_r/2\pi N\,\delta = \underline{268 \text{ rad/s}}$. This is a very small bandwidth that will make for a long pull-in time and a larger phase noise.

For a pole at $\omega_3 = 4K_{\max}$ and $N_{\min} = 494$, Eq. (11-14) gives $K_{\max}{}^2 = \Delta\theta f_r{}^2/\pi N_{\min}\delta = 129\ (\text{krad/s})^2$, and $K_{\max} = 11.4$ krad/s. Then from Eq. (11-3), $K_{\min} = (494/594)K_{\max} = \underline{9.5 \text{ krad/s}}$. This is a more reasonable bandwidth, but it can still be improved, as the next example shows.

EXAMPLE 11–3

The synthesizer in Fig. 11–2 has a PD with $\delta = 0.01$. Find the necessary K with a pole at ω_3 so that $\Delta\theta$ is 0.05 radian.

For a pole at $\omega_3 = 4K_{max}$ and $N_{min} = 100$, Eq. (11-14) gives $K_{max}{}^2 = \Delta\theta f_r{}^2/\pi N_{min}\delta = 636\ (\text{krad/s})^2$, and $K_{max} = 25.2$ krad/s. Then $K_{min} = (100/200)K_{max} = \underline{12.6 \text{ krad/s}}$. This is some improvement over the 9.5 krad/s in Example 11–2.

EXAMPLE 11–4

The synthesizer in Fig. 11–5 achieves the same range and resolution of f_o as that in Fig. 11–2, but N changes by a factor of 101 rather than a factor of 2 over the range. For a PD with $\delta = 0.01$, find the necessary K with a pole at ω_3 so that $\Delta\theta$ is 0.05 radian.

For a pole at $\omega_3 = 4K_{max}$ and $N_{min} = 1$, Eq. (11-14) gives $K_{max}{}^2 = \Delta\theta f_r{}^2/\pi N_{min}\delta = 6.36 \times 10^{10}\ (\text{rad/s})^2$, and $K_{max} = 252$ krad/s. Then $K_{min} = (1/101)K_{max} = \underline{2.5 \text{ krad/s}}$. This is not even as large as the K_{min} in Example 11-2.

The three examples above illustrate a couple of points. Both the mixer and the divider reduce the output frequency before it is applied to the PD. What share of the job should each have? For a given reference suppression, K is larger if N is kept small. But if N is so small that N_{max}/N_{min} is large, then K also varies over a large range. This forces K_{min} to be much smaller than K_{max} and ω_3. It can be shown from Eqs. (11-3) and (11-14) that for a given $\Delta\theta$, K_{min} is maximized for

$$N_{max}/N_{min} = 2$$

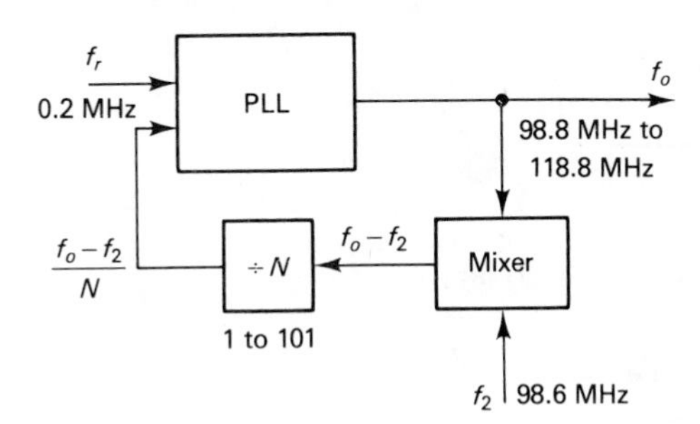

FIGURE 11–5 Frequency synthesizer for Example 11–4

11–5 DIVIDED OUTPUT

In this section, we will look at a way of reducing the pull-in time by synthesizing a higher frequency than necessary and then dividing down to the desired frequency. We will look first at a straightforward synthesis with no division at the output. Suppose we want to generate a spectrum of audio tones in the octave from $f_o = 2092$ to 4184 Hz. (This corresponds to the top octave of a piano.) The resolution is to be 4 Hz—about the limit of a human's ability to tell the difference between two tones in that range. The simplest solution in terms of circuitry is the synthesizer in Fig. 11–6. Equation (11-1) applies, so $f_o = N(4\text{ Hz}) = 2092$ to 4184 Hz for $N = 523$ to 1046. However, $f_r = 4$ Hz is a very low reference frequency. We will see that this requires a very small K, which results in a very long pull-in time.

The whole design can be scaled up by a factor of 1000 by increasing f_r from 4 Hz to 4000 Hz, as in Fig. 11–7. Then $f_o = N(4000\text{ Hz}) = 2.092$ MHz to 4.184 MHz as N goes from 523 to 1046. A $\div 1000$ frequency divider at the output produces

$$f_o' = f_o/1000 = N(4\text{ Hz}) \tag{11-15}$$

which varies from 2092 Hz to 4184 Hz over the range of N. The higher f_r will allow a higher PLL bandwidth K, and the pull-in time will not be so long.

11–6 PULL-IN TIME

The pull-in time T_p for a PLL with a three-state PD and $\div N$ frequency dividers is given by Eq. (8-28′):

$$T_p = \frac{\omega_{eo}/K - 2\pi N}{\pi N \omega_2} = \frac{f_{eo}/K - N}{0.5 N \omega_2} \tag{11-16}$$

where $f_{eo} \equiv \omega_{eo}/2\pi$ is the initial frequency error in Hz. If the N in the numerator is neglected and we choose $\omega_2 = K/4$, then T_p is approximated by $T_p \approx 8f_{eo}/NK^2$, as indicated in section 11–2.

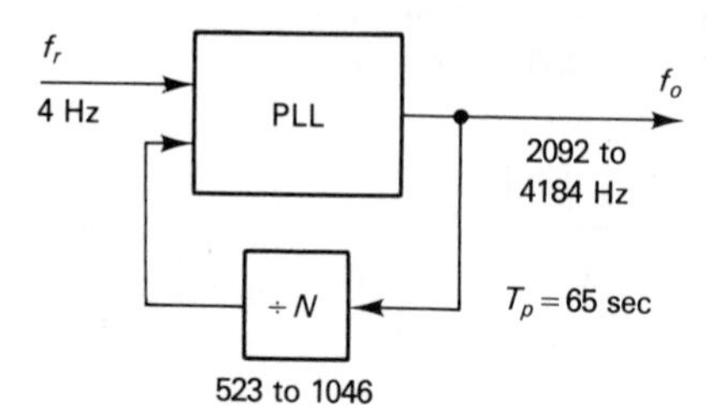

FIGURE 11–6 Audio frequency synthesizer

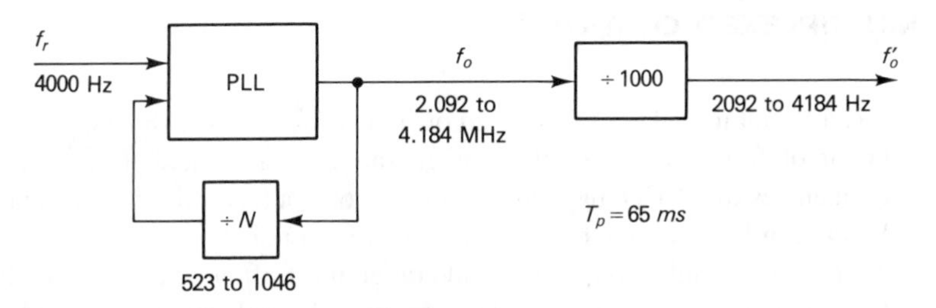

FIGURE 11–7 Synthesizer with divided output

EXAMPLE 11–5

The audio synthesizer in Fig. 11–6 has a PD with $\delta = 0.01$. Choose K and ω_2 so the frequency jitter $\Delta\omega$ is only $2\pi(4\text{ Hz})$. Find the pull-in time when N is changed from 1046 to 523.

For $\omega_3 = 4K_{\max}$, Eq. (11-12) gives $\Delta\omega = 8\pi N_{\min} K_{\max}{}^2 \delta / f_r$. For $N_{\min} = 523$ and $f_r = 4$ Hz, $\Delta\omega = 8\pi$ rad/sec requires $K_{\max} = 0.875$ rad/s and $K_{\min} = (523/1046)K_{\max} = 0.437$ rad/sec. Choose $\omega_2 = K_{\min}/4 = 0.109$ rad/s—as high as possible for fast pull-in.

The initial frequency is $f_o = 4184$ Hz, and the final frequency is 2092 Hz. Then $f_{eo} = 4184 - 2092$ Hz $= 2092$ Hz. The final divider value is $N = N_{\min} = 523$, corresponding to $K = K_{\max} = 0.875$ rad/sec. Then from Eq. (11-16), T_p = 65.5 seconds, which is certainly excessive for most applications.

EXAMPLE 11–6

The audio synthesizer in Fig. 11–7 has a PD with $\delta = 0.01$. Choose K and ω_2 so the frequency jitter $\Delta\omega'$ is only $2\pi(4\text{ Hz})$. Find the pull-in time when N is changed from 1046 to 523.

For the synthesizer in Fig. 11–7, all frequencies to the PLL are scaled up by a factor of 1000 from those in Fig. 11–6. Then if we choose $K_{\max} = 875$ rad/s, and $\omega_2 = 109$ rad/s (scaled up by a factor of 1000 from those in Example 11–5), the spurious modulation will be $\Delta\omega = 2\pi(4\text{ KHz})$. After the ÷ 1000 frequency divider, the spurious modulation is $\Delta\omega' = 2\pi(4\text{ Hz})$, as desired.

For $N = 1046$, the initial frequency is $f_o = 4.184$ MHz, and the final frequency is 2.092 MHz. Then $f_{eo} = 4.184 - 2.092$ MHz $= 2.092$ MHz. The final divider value is $N = 523$. Then from Eq. (11-16), the pull-in time is T_p = 65.5 ms, which is barely noticeable in a human time frame.

11–7 MULTIPLEXED OUTPUT

It is difficult to make a VCO based on an L-C oscillator with a range of much more than a factor of four because of the limited range of a varactor (see section 5–3). At high frequencies, the VCO may have a range of only a factor of two (an octave). Multivibrator VCOs can have a wide range, but they are too noisy for many synthesizer applications. Therefore, a synthesizer with a wide range must often be realized by dividing the output frequency by various amounts. For example, the audio synthesizer in Fig. 11–7 generates the tones in the top octave of a piano. The next lower octave can be obtained by dividing by two, and the next octave below that is obtained by dividing by four, etc. The circuit in Fig. 11–8 shows a binary counter used to divide f_o by 2^M, where M is 0 to 6 depending on the stage of the counter. Thus, the six stages generate the next six octaves below the top octave f_o'. A multiplexer (controlled by the digital number M) selects the desired octave, so from Eq. (11-15) the final output frequency is

$$f_o'/2^M = N(4\text{ Hz})/2^M \tag{11-17}$$

For $N = 523$ to 1046 and $M = 0$ to 6, this gives a range from 32.7 Hz to 4184 Hz (about the range of a piano). In the top octave, the resolution is 4 kHz/1000 = 4 Hz. In the bottom octave, the resolution is 4 Hz/64 = 0.06 Hz (this is the same percentage resolution as in the top octave).

11–8 MULTIPLE-LOOP SYNTHESIZERS

If a synthesizer with a frequency resolution as fine as 0.01% is needed, more than one PLL must be used in the design to keep the spurious modulation and pull-in time reasonable. Suppose a frequency synthesizer is to have a 1-kHz resolution for a frequency range from 10 MHz to 20 MHz. An attempt at realizing this with a single-loop synthesizer

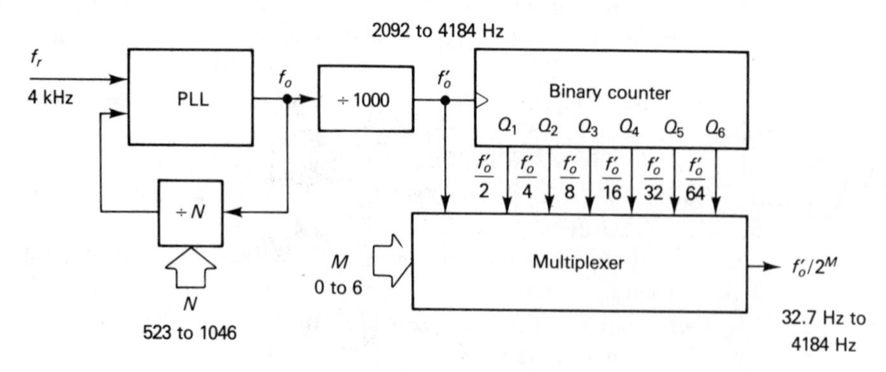

FIGURE 11–8 Synthesizer with multiplexed output

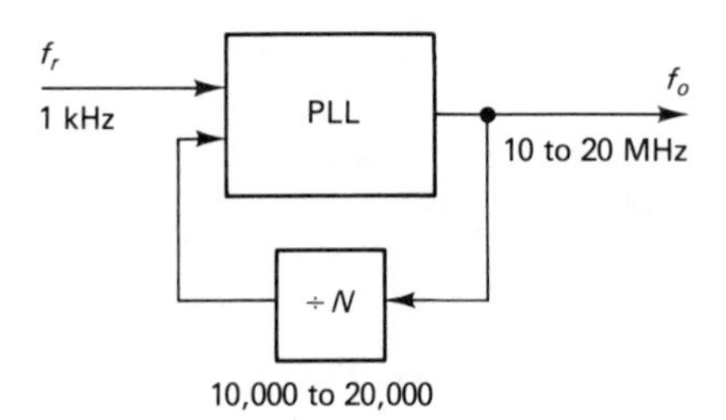

FIGURE 11–9 High-resolution synthesizer (impractical realization)

is shown in Fig. 11–9. The output frequency is $f_o = N(1\text{ kHz})$, where N goes from 10,000 to 20,000. However, with this large an N and with an f_r as small as 1 kHz, an extremely small K is needed to suppress spurious modulation [see Eq. (11-14)]. This small K would result in a long pull-in time [see Eq. (11-16)] and in larger phase noise.

One solution is to split up the resolution between two PLLs, as in Fig. 11–10. Here N_1 provides the fine resolution, generating $f_i' = N_1(1\text{ kHz})$, and N_2 provides the coarse resolution, generating $f_2 = N_2\,(100\text{ kHz})$. The third PLL sums these two frequencies to generate

$$f_o = f_2 + f_1' = (100\,N_2 + N_1)(1\text{ kHz}) \tag{11-18}$$

For example, if $f_o = 15.573\text{MHz}$ is desired, the proper settings are $N_2 = 154$ and $N_1 = 173$.

Note that the maximum value of N in any loop is only 200, and f_r is not less than 100 kHz for any loop. The price is that there are now three PLLs rather than just one. Integrated circuits such as the Motorola MC145157 (see reference [5]) help reduce the

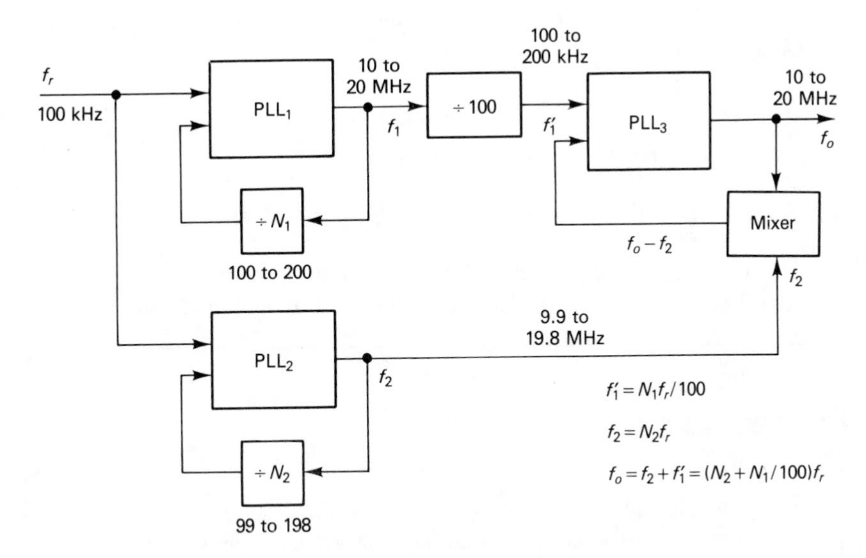

FIGURE 11–10 Multiple-loop synthesizer

cost and board space. These ICs include a three-state PD and all necessary frequency dividers.

Which of the three PLLs in Fig. 11–10 is critical in determining the phase jitter? The following example lends some insight.

EXAMPLE 11–7

The bandwidth of each PLL in Fig. 11–10 is $K_{max} = 2\pi(5.0 \text{ kHz}) = 31.4$ krad/s corresponding to $N = 100$. This is a compromise between a smaller K, which is better for reference suppression, and a larger K, which is better for phase noise suppression. Also, for each PLL, $\omega_3 = 4K_{max} = 126$ krad/s, and $\delta = 0.01$. Find the peak-to-peak phase jitter $\Delta\theta$ at the output due to spurious modulation.

Let the peak-to-peak jitter at f_1, f_1', and f_2 be represented by $\Delta\theta_1$, $\Delta\theta_1'$, and $\Delta\theta_2$. From Eq. (11-14), $N_1 = 100$ gives $\Delta\theta_1 = 0.31$ radian, and $N_2 = 99$ gives $\Delta\theta_2 = 0.30$ radian. But $f_1' = f_1/100$, so $\Delta\theta_1' = \Delta\theta_1/100 = 0.0031$ radian.

Let the peak-to-peak jitter at f_o be $\Delta\theta_3$ when f_1' and f_2 are jitter-free. Then Eq. (11-14) applies to PLL_3 with $N = 1$ (no divider). The minimum reference for PLL_3 is $f_1' = 100$ kHz, and Eq. (11-14) gives $\Delta\theta_3 = 0.0031$ radians. When f_1' and f_2 are *not* jitter-free but have jitter $\Delta\theta_1'$ and $\Delta\theta_2$, then the total phase jitter at the output is

$$\Delta\theta = \Delta\theta_3 + (\Delta\theta_1' + \Delta\theta_2)\,|H(j\omega_r)| \tag{11-19}$$

where ω_r is the frequency of the phase jitter from PLL_1 and PLL_2; that is, $\omega_r \equiv 2\pi f_r = 2\pi(100 \text{ kHz})$. This makes the approximation that the phase jitter waveform is sinusoidal, which it is not (see θ_o in Fig. 11–4b). Therefore, Eq. (11-19) is conservatively large in its estimate. For $\omega_r >> \omega_3 = 4K$, as in our case, then $|H(j\omega_r)| \approx K\omega_3/\omega_r^2 = 4(K/\omega_r)^2 = (K/\pi f_r)^2$. Then Eq. (11-19) becomes

$$\begin{aligned}\Delta\theta &= \Delta\theta_3 + (\Delta\theta_1' + \Delta\theta_2)(K/\pi f_r)^2 \qquad (11\text{-}20)\\ &= 0.0031 + (0.0031 + 0.30)\,0.01\\ &= 0.0031 + 0.000031 + 0.0030 = 0.006131 \text{ radian}\end{aligned}$$

The middle term is contributed by PLL_1, and this is negligible. The first and last terms are contributed by PLL_3 and PLL_2, and these are about equal.

11–9 PHASE NOISE

The principle sources of phase noise in a synthesizer are the VCOs and to a lesser extent the reference frequencies. To find the contribution from each of these sources, we need the phase transfer functions for the PLLs making up the synthesizer.

A PLL with a mixer has two reference frequencies—f_r and f_2, as shown in Fig. 11–11a. The corresponding signal flow graph in Fig. 11–11b diagrams the equations relating the phases θ_o, θ_r, and θ_2. Note that the gain $V_{dm}/2\pi$ of the three-state PD is considered only part of the total PD gain $K_d = V_{dm}/2\pi N$. Solving the flow graph gives

$$\theta_o = \frac{(V_{dm}/2\pi)\ F(s)\ K_o/s}{1 + (V_{dm}/2\pi N)\ F(s)\ K_o/s}\ \theta_r + \frac{(V_{dm}/2\pi N)\ F(s)\ K_o/s}{1 + (V_{dm}/2\pi N)\ F(s)\ K_o/s}\ \theta_2$$

$$= \frac{N\ K_d\ F(s)\ K_o/s}{1 + K_d\ F(s)\ K_o/s}\ \theta_r + \frac{K_d\ F(s)\ K_o/s}{1 + K_d\ F(s)\ K_o/s}\ \theta_2$$

$$= N\ H(s)\ \theta_r + H(s)\ \theta_2$$

so the transfer functions are

$$\theta_o/\theta_r = N\ H(s) \tag{11-21}$$

$$\theta_o/\theta_2 = H(s) \tag{11-22}$$

The corresponding frequency responses are shown in Fig. 11–11c. We already have from section 6–6 the transfer function from the VCO noise θ_n to the PLL output:

$$\theta_o/\theta_n = H_e(s) \tag{11-23}$$

The frequency response plotted in Fig. 3–16 is repeated here in Fig. 11–11c. The total spectral density of θ_o (see section 6–1) is therefore

$$\Phi_{\theta o} = |H_e|^2\Phi_{\theta n} + N^2|H|^2\Phi_{\theta r} + |H|^2\Phi_{\theta 2} \tag{11-24}$$

where $\Phi_{\theta r}$, $\Phi_{\theta 2}$, and $\Phi_{\theta n}$ are the special densities of θ_r, θ_2, and θ_n. The area under $\Phi_{\theta o}$ gives the mean square phase noise $\overline{\theta_o^{\,2}}$.

EXAMPLE 11–8

The synthesizer in Fig. 11–10 has $N_2 = 198$ and $N_1 = 200$ so $f_o = 20$ MHz. PLL_1 and PLL_2 both have $\omega_2 = 2\pi(500\text{ Hz})$ and bandwidths $K_1 = 2\pi(2.5\text{ kHz})$. PLL_3 has $\omega_2 = 2\pi(500\text{ Hz})$ and bandwidth $K_3 = 2\pi(5.0\text{ kHz})$. The spectral density $\Phi_{\theta n}$ of the VCO phase noise in each PLL has $\Theta_o = 10^{-12}\text{ rad}^2/\text{Hz}$, $f_a = 500$ Hz, $f_b = 200$ kHz, and $f_c = 1.0$ MHz (see Fig. 11–12). The 100 kHz reference frequency has phase noise with a flat spectral density $\Phi_{\theta r} = 10^{-16}\text{ rad}^2/\text{Hz}$. Find the rms phase noise at the synthesizer output.

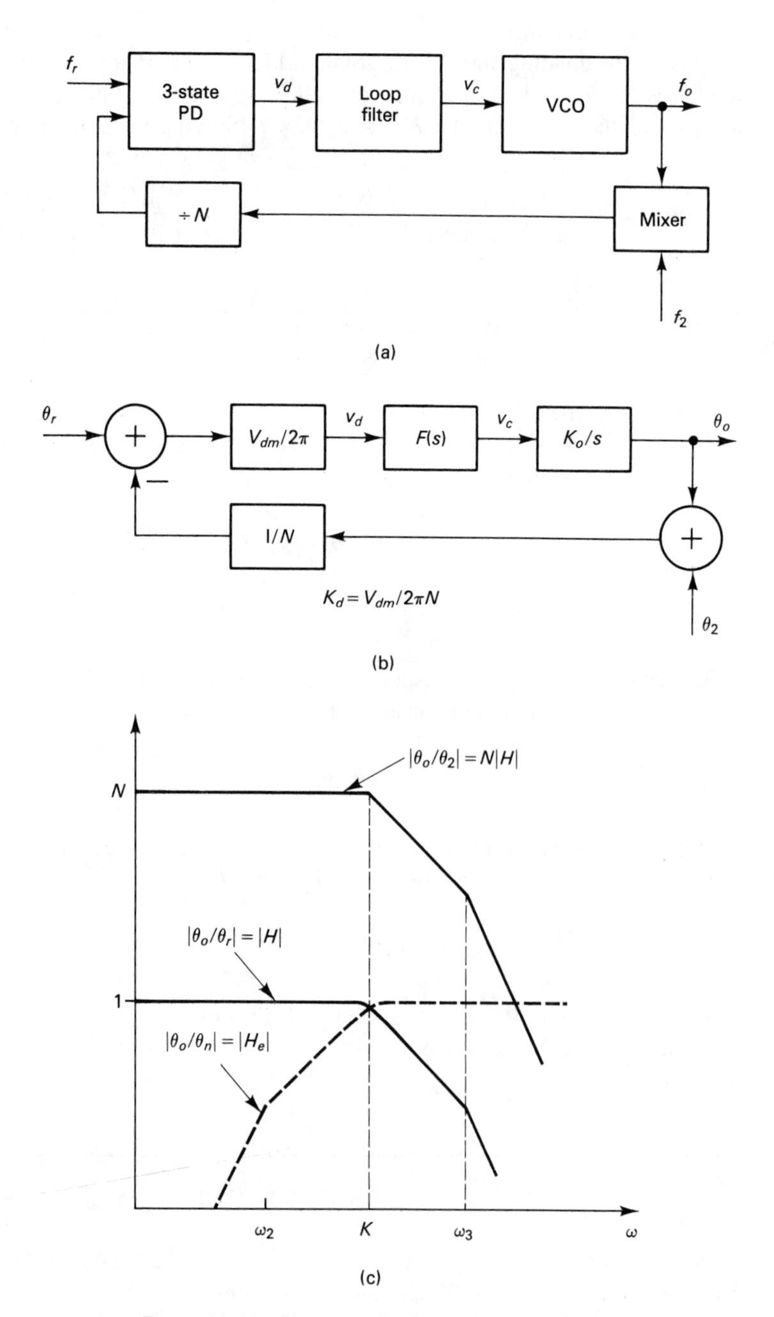

FIGURE 11–11 Phase transfer functions θ_o/θ_r and θ_o/θ_2

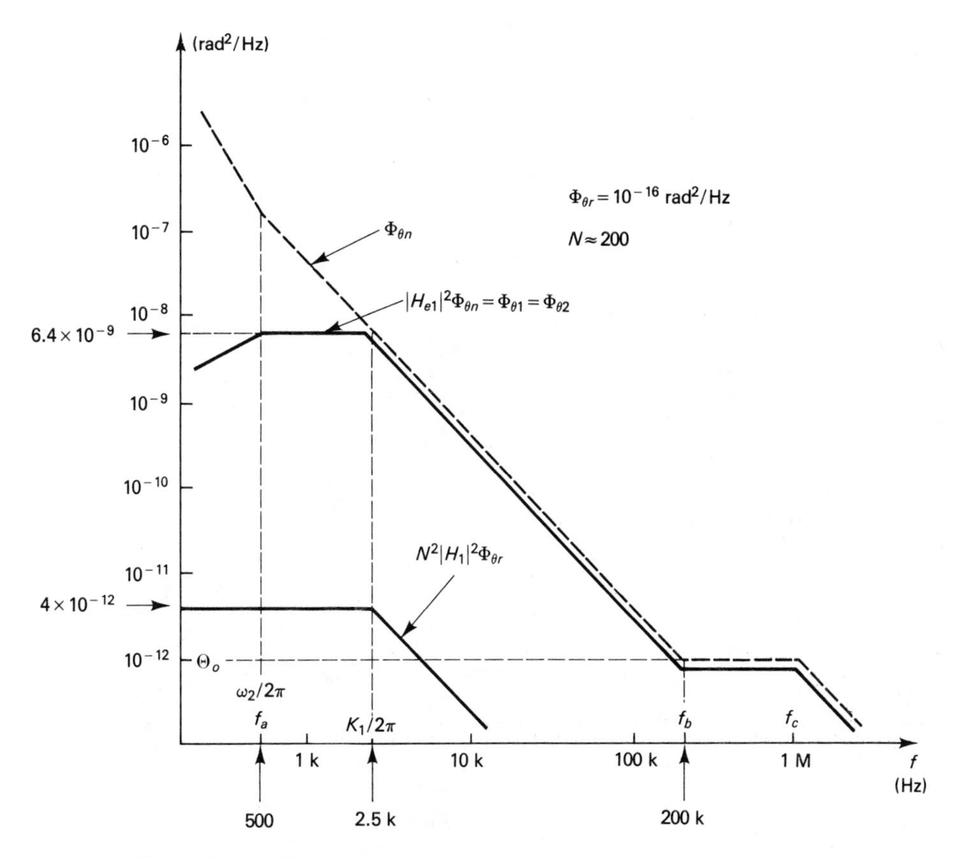

FIGURE 11–12 Phase spectral densities for PLL_1 and PLL_2 in Example 11–8

If we approximate $N \approx 200$ for both PLL_1 and PLL_2, then the analysis is the same for both loops:

$$\Phi_{\theta 1} = \Phi_{\theta 2} = |H_{e1}|^2 \Phi_{\theta n} + N^2 \, |H_1|^2 \Phi_{\theta r}$$

where H_{e1} and H_1 are H_e and H corresponding to bandwidth K_1. The third term from Eq. (11-24) is not present because there is no mixer. The two remaining terms are plotted in Fig. 11–12; it is clear that the first term dominates. Therefore, $\Phi_{\theta 1} = \Phi_{\theta 2} \approx |H_{e1}|^2 \Phi_{\theta n}$.

One reference frequency for PLL_3 is $f_1{}'$ with $\theta_1{}' = \theta_1/100$ and spectral density $\Phi_{\theta 1}{}' = \Phi_{\theta 1}/(100)^2$. Therefore, Eq. (11-24) for PLL_3 becomes

$$\Phi_{\theta o} = |H_{e3}|^2 \Phi_{\theta n} + |H_3|^2 \Phi_{\theta 1}{}' + |H_3|^2 \Phi_{\theta 2}$$

where H_{e3} and H_3 are H_e and H corresponding to bandwidth K_3. These three terms are plotted in Fig. 11–13; it is clear that the second term is negligible. Then

$$\overline{\theta_o^{\,2}} = \int_0^\infty \Phi_{\theta o} \, df = \int_0^\infty |H_{e3}|^2 \Phi_{\theta n} \, df + \int_0^\infty |H_3|^2 \Phi_{\theta 2} df$$

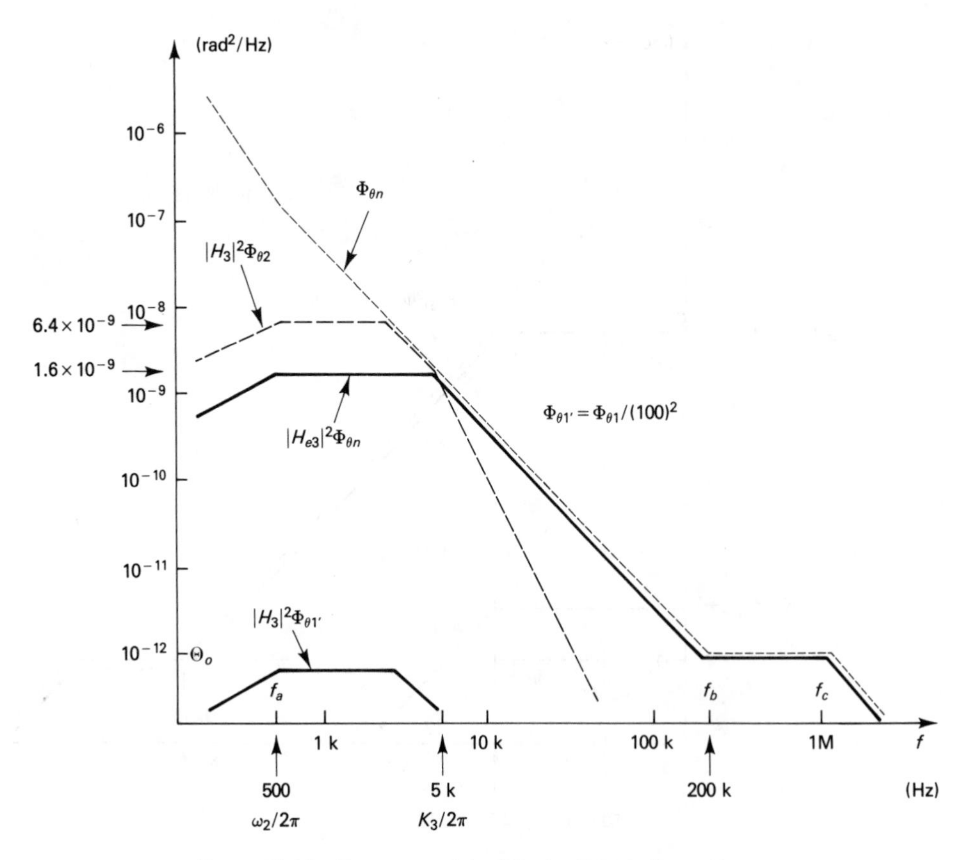

FIGURE 11–13 Phase spectral densities for PLL_3 in Example 11–8

The evaluation of the first integral is given by Eq. (6-55):

$$\int_0^\infty |H_{e3}|^2 \Phi_{\theta n}\, df = \Phi_{\theta n}(K_3/2\pi) \left[\frac{K_3}{4} + \frac{f_a}{y}\, \ell n \frac{1+y}{1-y} \right] + \frac{\pi}{2} f_c\, \Theta_o$$

$$= 16.3 \times 10^{-6} \text{ rad}^2$$

where $\Phi_{\theta n}(K_3/2\pi) = (2\pi f_b/K_3)^2 \Theta_o = 1.6 \times 10^{-9}$, and $y = \sqrt{1 - 4\omega_2/K_3} = 0.775$. The evaluation of the second integral is given by a modified form of Eq. (6-55):

$$\int_0^\infty |H_3|^2 \Phi_{\theta 2}\, df = \Phi_{\theta n}(K_1/2\pi) \left[\frac{K_1}{4} \cdot \frac{K_3}{K_3 + K_1} + \frac{f_a}{y}\, \ell n \frac{1+y}{1-y} \right]$$

$$= 23.6 \times 10^{-6} \text{ rad}^2$$

where $\Phi_{\theta n}(K_1/2\pi) = (2\pi f_b/K_1)^2\Theta_o = 6.4 \times 10^{-9}$, and $y = \sqrt{1 - 4\omega_2/K_1} = 0.447$. The term involving Θ_o is negligible, and the factor of $K_3/(K_3 + K_1)$ is necessary because of the additional break at $K_3/2\pi$ that $|H_3|^2$ causes. Then the total $\overline{\theta_o^2} = 16.3 \times 10^{-6} + 23.6 \times 10^{-6} = 39.9 \times 10^{-6}$ rad^2, and $\theta_{o\ \mathrm{rms}} = \underline{0.00632 \text{ rad}}$.

11–10 PRESCALING

One practical consideration in designing frequency synthesizers is the speed limitation of the components. The programmable $\div N$ frequency dividers, in particular, are typically limited to a maximum frequency of 25 MHz. (See, for example, the specifications for the MC4018 with an "early decode" feature for reprogramming. [5] This restriction has led to some tricks in realizing the $\div N$ in high-frequency synthesizers.

Consider the single-loop synthesizer in Fig. 11–14. It generates $f_o = N(1 \text{ MHz}) =$ 100 MHz to 200 MHz for $N = 100$ to 200. The $\div N$ is not too large, and f_r is not too small a fraction of f_o, so spurious modulation and pull-in time should not be a problem. But the 200 MHz output frequency appears directly at the input of the $\div N$, which can handle only 25 MHz maximum.

One possible solution would be to use a fixed $\div 10$ prescaler before the $\div N$ so it sees only 20 MHz, as in Fig. 11–15. If f_r is reduced to 100 kHz, then we again have $f_o = 10N_1(100 \text{ kHz}) = N_1(1 \text{ MHz}) = 100$ MHz to 200 MHz for $N_1 = 100$ to 200. But the overall $N = 10N_1 = 1000$ to 2000, and f_r is only $f_o/1000$. This will lead to problems with spurious modulation and pull-in time.

It is not such a problem to make a fixed divider, such as the $\div$ 10 in Fig. 11–15, that can handle high frequencies. It is the *programmable* dividers that are restricted to about 25 MHz. A compromise is to use a divider that can be programmed for only two values. Such *two-modulas prescalers* can be made to work at frequencies as high as 600 MHz. [6]

The synthesizer in Fig. 11–16 uses a two-modulas prescaler to realize a $\div N$ that handles high frequencies. The price is that it takes four devices to realize the division: a $\div N_1$ frequency divider, an A counter, a $\div 10/\div 11$ prescaler, and a control unit such as the MC12014. Under command of the control unit, the prescaler divides by either 10 or

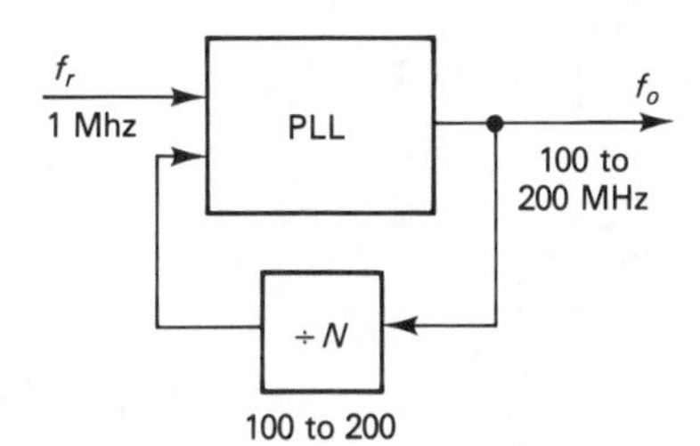

FIGURE 11–14 High-frequency synthesizer (impractical realization)

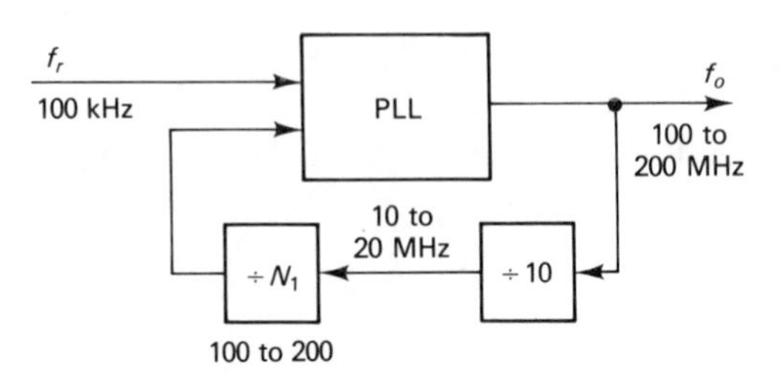

FIGURE 11–15 Synthesizer with fixed prescaler

11. The control unit monitors the outputs of the $\div N_1$ and the A counter to know when to select the $\div 10$ or the $\div 11$.

Suppose we want to divide f_o by 128 in order to synthesize 128 MHz. Then N_1 is set to 12, and A is set to 8. The sequence begins with the control loading the A counter with 8 and selecting $\div 11$ for the prescaler. Then after 11 cycles of f_o, the A counter decrements to 7. This continues until the A counter reaches zero, signaling the control to select $\div 10$ for the prescaler. So far the $\div N_1$ (a $\div 12$ here) has received 8 pulses from the prescaler, so it needs 4 more before it puts out one pulse itself. Each of these next 4 correspond to 10 cycles of f_o. Therefore, the total cycles of f_o required to produce one pulse from the $\div N_1$ is

$$11 + 11 + 11 + 11 + 11 + 11 + 11 + 11 + 10 + 10 + 10 + 10 = 128$$

When the control senses this pulse, it loads the A counter with 8, selects $\div 11$ for the prescaler, and the sequence starts over. From this example, it should be clear that the circuit realizes a frequency division of $10N_1 + A$. Therefore

$$f_o = (10\,N_1 + A)f_r \tag{11-20}$$

Note that the operation requires that

$$A \leq N_1 \tag{11-21}$$

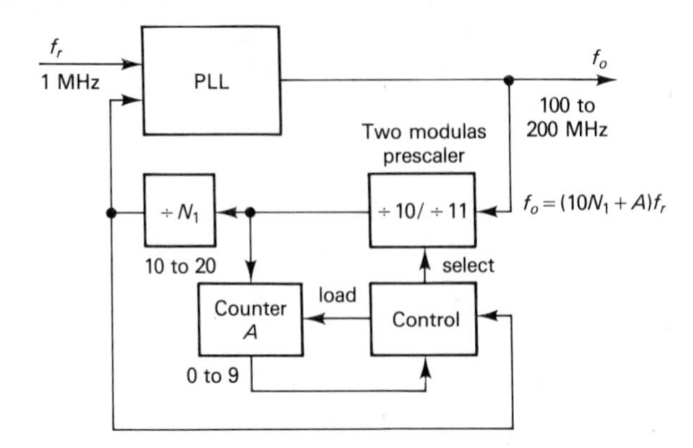

FIGURE 11–16 Synthesizer with two modulas prescaler

Therefore, the circuit can't divide by 79, which would require $N_1 = 7$ and $A = 9$. However, prescalers are available with $\div M / \div (M+1)$ capabilities, where M is not necessarily 10. Then the generalization of Eq. (11-20) is

$$f_o = (MN_1 + A)f_r \tag{11-22}$$

For example, for $M = 6$, the prescaler can divide by either 6 or 7. Then $f_o = (6N_1 + A)$ 1 MHz, and $f_o = 79$ MHz requires $N_1 = 13$ and $A = 1$, which satisfies Eq. (11-21). Note that with $M = 6$, f_o can now go no higher than 150 MHz if the $\div N_1$ is to see no more than 25 MHz. Therefore, there is a tradeoff in selecting M.

REFERENCES

[1] *MECL Device Data,* Motorola Semiconductor Products Inc., Phoenix, 1988, section 6.

[2] U. L. Rohde, *Digital PLL Frequency Synthesizers,* Prentice-Hall: Englewood Cliffs, NJ, 1983, section 4–6.

[3] *CMOS/NMOS Special Functions Data,* Motorola Semiconductor Products Inc., Phoenix, 1988, section 6.

[4] V. Manassewitsch, *Frequency Synthesizers,* Wiley: New York, 1997, sections 2–1 and 6–2.

[5] *CMOS/NMOS Special Functions Data.*

[6] *MECL Device Data.*

List of Symbols

B_i	Noise bandwidth of input bandpass filter
B_L	Noise bandwidth of PLL tansfer function H
B_m	Message bandwidth
C_T	Varactor capacitance
C_x	Capacitance in crystal model
C_x	External capacitor for multivibrator VCO
$F(s)$	Loop filter transfer function
f_a	Offset frequency below which flicker noise dominates
f_b	Half-bandwidth of resonant oscillator's tank
f_c	Half-bandwidth of filter following oscillator
f_m	Offset frequency from "carrier"
f_o	PLL output frequency in Hz
f_r	Reference frequency for synthesizer
$G(s)$	Forward gain of PLL control loop
$H(s)$	PLL phase transfer function
$H_e(s)$	Transfer function from θ_i to θ_e
H_p	Peak value of transfer function $\|H(j\omega)\|$
K	PLL 3-dB bandwidth
K_d	Phase detector gain
K_h	High-frequency gain of loop filter
K_m	Multiplier gain
K_o	VCO gain
M	Phase detector figure of merit
$n(t)$	Noise
N	Frequency divider ratio
N_o	Power spectral density of white noise
T_p	Pull-in time
T_s	Mean time between cycle slips
v_c	VCO control voltage
v_d	Phase detector output voltage (average)
$\tilde{v}_d$	Phase detector output voltage (instantaneous)
V_{dm}	Maximum value of v_d
V_{do}	Phase detector free-running voltage, or offset voltage
V_H	Logic "high" voltage
v_i	PLL input voltage
V_i	Peak value of v_i
v_I	Injection voltage into VCO
V_I	Peak value of v_I
V_L	Logic "low" voltage
v_o	VCO output voltage
V_o	Peak value of v_o
v_p	Pull-in voltage from phase detector during acquisition
v_1	Oscillation voltage in VCO
V_1	Peak value of v_1
v_2	Voltage across R_2 in loop filter
v_3	Voltage across C in loop filter
δ_o	Density of ones in an RZ data signal
$\Delta\theta$	Spurious phase modulation (peak-to-peak)
$\Delta\omega$	Spurious frequency modulation (peak-to-peak)
$\Delta\omega_i$	Input frequency deviation
$\Delta\omega_o$	Output frequency deviation
$\Phi_x(f)$	Power spectral density of $x(t)$
θ_d	Phase difference between input and output signals
θ_e	Phase error between input and output signals
θ_{em}	Value of θ_e for maximum phase detector voltage
θ_i	Phase of PLL input signal
θ_n	Oscillator phase noise when not in a closed loop
θ_o	Phase of PLL output signal
Θ_o	Power spectral density of white phase noise
ω_c	Average frequency error during acquisition
ω_d	Detector frequency—fundamental frequency of $\tilde{v}_d$
ω_e	Frequency error during acquisition
ω_{eo}	Initial frequency error during acquisition
ω_i	Average (carrier) frequency of input signal
ω_L	Lock-in frequency—ω_c for which acquisition is complete
ω_m	Modulation frequency of input signal
ω_n	Oscillator frequency noise when not in a closed loop
ω_o	Output frequency of VCO
ω_p	Pull-in frequency—maximum ω_{eo} for acquisition
ω_p	Peaking frequency of $\|H(j\omega_m)\|$
ω_1	Loop filter pole frequency less than K
ω_2	Loop filter zero frequency
ω_3	Loop filter pole frequency greater than K

Index